The Faunas of Hayonim Cave, Israel

A 200,000-Year Record of Paleolithic Diet, Demography, and Society

American School of Prehistoric Research
Bulletin 48

Peabody Museum of Archaeology and Ethnology
Harvard University

The Faunas of Hayonim Cave, Israel

A 200,000-Year Record of Paleolithic Diet, Demography, and Society

■

Mary C. Stiner
University of Arizona, Tucson

With contributions by

Ofer Bar-Yosef
Anna Belfer-Cohen
Paul Goldberg
Steven L. Kuhn
Amy V. Margaris
Liliane Meignen
Natalie D. Munro
Todd A. Surovell
Bernard Vandermeersch
Stephen Weiner

Peabody Museum of Archaeology and Ethnology
Harvard University
Cambridge, Massachusetts 2005

Original design: Janis Owens

Cover design: Mary Sweitzer

Composition: Donna Dickerson

Editing: Jane Kepp

Proofreading: Janice Herndon

Production management: Donna Dickerson and
Joan K. O'Donnell

Printed by IBT Global

Manufactured in the United States of America

Second printing 2009

Library of Congress Cataloging-in-Publication Data

Stiner, Mary C., 1955-

The faunas of Hayonim Cave (Israel) : a 200,000-
year record of Paleolithic diet, demography, and
society / Mary C. Stiner ; with contributions by Ofer
Bar-Yosef ... [et al.].
p. cm. -- (Bulletin ; 48)
Includes bibliographical references.
ISBN 0-87365-552-4 (pbk. : alk. paper)
1. Hayonim Cave (Israel) 2. Meged Rockshelter
(Israel) 3. Mousterian culture--Israel--Galilee
Region. 4. Paleolithic period--Israel--Galilee Region.
5. Prehistoric peoples--Food--Israel--Galilee Region.
6. Excavations (Archaeology)--Israel--Galilee Region.
7. Animal remains (Archaeology)--Israel--Galilee
Region. 8. Galilee Region (Israel)--Antiquities.
I. Bar-Yosef, Ofer. II. Title. III. Bulletin (American
School of Prehistoric Research) ; no. 48.

GN772.2.M6S85 2005

933--dc22

2005006703

For Eitan, dear friend and colleague

Eitan Tchernov at Hayonim Cave, 1997.

Contents

■

CHAPTER THREE

Experiments in Fragmentation and Diagenesis of Bone and Shell   39

*Mary C. Stiner, Ofer Bar-Yosef, Steven L. Kuhn,
and Stephen Weiner*

CHAPTER FOUR

Bone, Ash, and Shell Preservation in Hayonim Cave   59

*Mary C. Stiner, Steven L. Kuhn, Todd A. Surovell,
Paul Goldberg, Amy V. Margaris, Liliane Meignen,
Stephen Weiner, and Ofer Bar-Yosef*

CHAPTER FIVE

Vertebrate Taphonomy and Evidence of Human Modification   81

CHAPTER NINE

Prey Diversity and Changes in Human
Dietary Breadth 165

CHAPTER TEN

Food Utility, Attrition, and Transport of
Ungulate Body Parts 177

CHAPTER ELEVEN

Evidence of Large Game Hunting from
Ungulate Mortality Patterns 197

CHAPTER TWELVE

The Changing Shape of Paleolithic Society

APPENDIXES

List of Figures

■

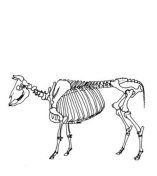

List of Tables

■

Acknowledgments

■

I NEVER EXPECTED THIS ZOOARCHAEOLOGICAL PROJECT TO expand the way it did, to last as long as it did, or to teach me as much as it did. Much of the stimulus and a good share of the credit for the research are attributable to the many colleagues and students with whom I have collaborated since 1992. Perhaps nowhere is this more true than for the complex issues of vertebrate taphonomy and site formation processes. Steve Weiner has been particularly influential on this subject, having taught me the possibilities of infrared spectroscopy and served as a constant source of inspiration and criticism in experiments and archaeological analyses. Without his guidance and help on topics in biochemistry and geochemistry, I would have gotten nowhere at all. Likewise, my interactions with Steve Kuhn, Natalie Munro, Todd Surovell, and Joe Beaver on problems in predator-prey simulation modeling and human dietary breadth contributed tremendously to the positive outcomes of this study. Other major contributions to the research program came from Eitan Tchernov and his remarkable work in Pleistocene animal community turnover, from Paul Goldberg on sediment micromorphology and issues relating to site formation processes, from Liliane Meignen and Steve Kuhn on Paleolithic technology and spatial analysis, and from Helene Valladas, Norbert Mercier, Jack Rink, and Henry P. Schwarcz on issues of chronology. I am indebted to Ofer Bar-Yosef in more ways that I can describe. In the interest of brevity, I thank him here for including me in the most stimulating research context of my still short career, for his generous advice and compelling ideas, and for his ingenious way of making complex projects peopled by strong-minded folk function so fruitfully.

Two manuscript reviewers, John D. Speth and Manuel Domínguez-Rodrigo, along with Richard Meadow of the Peabody Museum Press, provided valuable advice, comments, and criticisms on the penultimate draft of this book. I owe a special debt to John for his many years of encouragement, assistance, and data sharing. He had much to do with my becoming involved in the Hayonim project, and I am grateful for his open and generous collaboration on all intellectual levels. Four other colleagues also contributed substantially to conceptual aspects of the study, each in a distinct way: Steve Kuhn, Natalie Munro, Lee Lyman, and Larry Keeley. Natalie provided a remarkable data set for the Natufian period of the Wadi Meged, and I am deeply grateful for her many years of valuable assistance in the laboratory and in the field.

Thanks also to Aaron Stutz for his generous help in the Hayonim project and for access to unpublished

information from his study of dental eruption and wear stages in mountain gazelles. I am also grateful to my research assistants and other University of Arizona graduate students who were involved in the Hayonim project over the years: Joe Beaver, Laura Berstresser, P. Jeffrey Brantingham, Rebecca Dean, Xing Gao, Kris W. Kerry, Amy Margaris, Mentor Mustafa, Todd Surovell, and Patrick Wrinn. I thank my colleagues Anna Belfer-Cohen, Erella Hovers, Na'ama Goren-Inbar, Nigel Goring-Morris, Lior Grosman, Idit Saragusti, Liora Horwitz, Theodora Bar-El, and Rivka Rabinovich (Hebrew University of Jerusalem), Ruty Shahack-Gross (Weizmann Institute), Tamar Dayan, Yoel Rak, and Baruch Arensburg (Tel Aviv University), Guy Bar-Oz, Mina Weinstein-Evron, Danny Kaufman, and Daniella Bar-Yosef (University of Haifa), and Bernard Vandermeersch, Anne Marie Tillier, and the late Henri Laville (University of Bordeaux I) for their assistance and stimulating discussions on the subjects of paleo-anthropology, zooarchaeology, and ecology. The comparative collections of the Zoology Laboratory in the Department of Evolution, Systematics, and Ecology of the Hebrew University of Jerusalem were an invaluable resource for this project. Thanks also to the Israel Antiquities Authority for permission to export faunal samples temporarily from Hayonim Cave to the University of Arizona (USA) for the FTIR and fragmentation analyses.

The zooarchaeological research was supported from 1995 through 2004 by a CAREER and Creativity Extension Grant made to me by the National Science Foundation (SBR-9511894) and by a 1997–2000 grant to Eitan Tchernov and me from the Israel–United States Binational Science Foundation.

The larger Hayonim project, co-organized by Ofer Bar-Yosef, Liliane Meignen, and Bernard Vandermeersch, was supported by the French Ministry of Foreign Affairs and the French CNRS, the National Science Foundation (grants DBS-9208163 and SBR-9409281), and the American School of Prehistoric Research (Peabody Museum) of Harvard University. The 1994–1997 excavations of Meged Rockshelter were supported by grants to Steven Kuhn and Anna Belfer-Cohen from the Irene Levi-Sala CARE Foundation, the L. S. B. Leakey Foundation, and the Wenner-Gren Foundation for Anthropological Research.

All members of the Hayonim and Meged Rockshelter projects are deeply indebted to Mario Chech (CNRS), who managed all of the technical aspects of the excavations with energy, ingenuity, and humor.

CHAPTER ONE

Goals and Approaches

MY GOAL IN WRITING THIS BOOK and compiling the contributions of close colleagues was to present new findings on early Middle Paleolithic exploitation of ungulates and small animals in the upper Wadi Meged, or Meged Valley (Galilee), at the sites of Hayonim Cave and Meged Rockshelter. More than a faunal report, this study was designed to address changes in the shape of Paleolithic human foraging adaptations and society in the eastern Mediterranean Basin over roughly 200,000 years, a few temporal gaps notwithstanding.

The Hayonim project lasted nearly a decade (1992–2001) and in its first formulation was to have been about Neandertal ecology in particular. We found no Neandertals in Hayonim Cave, at least not in pieces large enough to be distinguished from other forms of humanity. What we did find was a rich faunal record (table 1.1) associated with early Mousterian lithic assemblages and hearths, making possible a detailed study of hominid economics in the late Middle Pleistocene without the distraction of knowing what those hominids looked like. Very little is known about the Levantine zooarchaeological record of the early Middle Paleolithic in the Near East. The Galilee preserves many later Paleolithic faunas, however, and I use these to set the early Middle Paleolithic of Hayonim Cave in evolutionary perspective.

What one can hope to accomplish in a book is a synthesis that raises general understanding of the data to a higher level, something that is almost never possible in scattered short articles. It is difficult to imagine the closure of any large research program without this final step's being taken. The chapters that follow deal principally with two families of problems, the taphonomic history of the faunal accumulations (chapters 3–5) and the foraging ecology of early humans (chapters 6–12). Diverse information is brought to bear on these questions, including data on molecular diagenesis and dissolution of skeletal materials in sediments, density-mediated bone attrition, and bone macrodamage such as normally results from burning, contact with stone tools, gnawing by carnivores or rodents, weathering, and fracturing.

With regard to human foraging ecology, species abundance and diversity reflect mainly the nature of local animal communities, but these data may also reveal important changes in human dietary breadth. Also of interest are patterns of prey body part representation and the intensity of marrow processing, mortality patterns and sex ratios in the most common prey species, and the spatial distributions of skeletal materials in the deposits relative to stone tools and

Table 1.1
Faunal sample sizes (number of identified specimens, or NISP) for the Paleolithic assemblages from Hayonim Cave and Meged Rockshelter

Period	Site	Layer	Culture	Depth Range	NISP
Epipaleolithic	Hayonim Cave	B	Natufian[a]	Variable	16,814
	Hayonim Cave	C	Kebaran	Variable	3,901
	Meged Shelter	—	Kebaran	0–199 cm bd	2,019
Upper Paleolithic	Meged Shelter	—	Pre-Kebaran (UP)	>200 cm–bedrock	608
	Hayonim Cave	D	Aurignacian[b]	Variable	≥ 10,834
Middle Paleolithic	Hayonim Cave	E, unit 1	Mousterian	250–299 cm bd	161
	Hayonim Cave	E, unit 2	Mousterian	300–349 cm bd	352
	Hayonim Cave	E, unit 3	Mousterian	350–419 cm bd	1,870
	Hayonim Cave	E, unit 4a	Mousterian	420–444 cm bd	4,854
	Hayonim Cave	E, unit 4b	Mousterian	445–464 cm bd	4,054
	Hayonim Cave	E, unit 5	Mousterian	465–494 cm bd	3,176
	Hayonim Cave	E, unit 6	Mousterian	495–529 cm bd	2,351
	Hayonim Cave	F, unit 7	Mousterian	>530 cm bd	270

NOTE: The Middle Paleolithic (Mousterian) sample is from the central trench only. It was subdivided into vertical units, or cuts, in centimeters below datum (cm bd) according to variations in lithic distributions and sediment micromorphology.

[a] Data for the Hayonim Natufian are from Munro 2001.

[b] Data on the Hayonim Aurignacian are based on Stiner's NISP counts of small game taxa in the old collection and R. Rabinovich's (1998) counts of carnivores and ungulates.

wood ash residues. Some components of this research have already appeared in journal articles (Stiner 2002b; Stiner et al. 1995, 1999, 2000, 2001a, 2001b), but the bulk of the information is new and the data sets are now complete.

In collaboration with other investigators, I also pursue some unexpected observations that emerged early in the research, mainly through controlled experiments in bone diagenesis and through predator-prey simulation modeling. In the first of these trajectories we explore how bone mineral diagenesis affects the quantification of archaeofaunal records (chapter 3). This work, which employs infrared spectroscopy, is allied closely with investigations by Stephen Weiner (Weizmann Institute of Science, Rehovot, Israel), Paul Goldberg (Boston University), and Ofer Bar-Yosef (Harvard University) of mineral microenvironments and microstratigraphy in the sediments of Hayonim Cave (see chapter 2; also Goldberg and Bar-Yosef 1998; Weiner et al. 1995, 2002; on other sites, Karkanas et al. 1999, 2000; Weiner and Bar-Yosef 1990; Weiner and Goldberg 1990; Weiner et al. 1993). The second area of exploration is that of Paleolithic predator-prey dynamics involving small game animals and its implications for changes in human population densities during the Middle and Late Pleistocene (chapter 8). The computer

simulation modeling was done at the University of Arizona in Tucson (USA) in collaboration with Todd Surovell and Natalie Munro (Stiner et al. 1999, 2000).

BACKGROUND TO THE ISSUES

Middle Paleolithic adaptations and the processes that gave rise to anatomically modern humans in Eurasia are widely debated in human origins research (compare Bar-Yosef 1992; Chase and Dibble 1987; Clark and Lindly 1989a; Davidson and Noble 1989; Gamble 1986; Klein 1989; Mellars 1989; Simmons and Smith 1991; Soffer 1989a; Stringer and Andrews 1988; Tchernov 1992b; Trinkaus 1986; Vandermeersch and Bar-Yosef 1988; Whallon 1989; White 1982; Wolpoff 1989). So-called modern human behavior, most clearly associated with Upper Paleolithic and Late Stone Age cultures, emerged and spread rather suddenly throughout the Old World—ironically, some 50,000 years or more after the onset of skeletal gracilization in anatomically modern populations. Although the genus *Homo* certainly evolved in Africa, the geographical origins of our species/subspecies are less clear. We can be sure that anatomically modern humans did not first evolve in Europe, and most recent genetic evidence points to a principally African origin for these populations (e.g., Hewitt 2000; but see Eswaran 2002). Some cases for the early beginnings of modern human behavior—as evidenced, for example, by rare decorative objects and early bone implements—are reported for southern and east-central Africa (reviewed by McBrearty and Brooks 2000; see also Ambrose 1998; d'Errico et al. 2001). Some of these cases may indeed pre-date anything of the kind elsewhere in the Old World, although early examples have also surfaced in adjacent areas of Asia and in Eastern Europe (Kuhn et al. 2001; Kozlowski 1982, 2000). More to the point, most researchers seem to agree that early glimmerings of modern behavioral traits on the African continent generally appeared later than modern skeletal features, as they also did in western Asia.

Between Africa and Europe lies the Levant, a region that figures prominently in discussions of modern human origins by virtue of its position at the intersection of three continents, its rich fossil and archaeological records, and the contradictions it currently poses to the standard chronicle of change in hominid morphology and lifeways (Akazawa 1987; Bar-Yosef 1989a, 1989b, 1992; Bar-Yosef et al. 1986, 1992; Garrod 1954, 1957; Garrod and Bate 1937; Howell 1952, 1959; Jelinek 1982; Jelinek et al. 1973; Schick and Stekelis 1977; Suzuki and Takai 1970; Tchernov 1984b; Turville-Petre 1932; Watanabe 1968). Two closely related hominid variants of the late Pleistocene—*Homo (sapiens) neanderthalensis* and *H. s. sapiens*—were once thought to be chronospecies. However, more recent evidence indicates that these hominid populations were semicontemporaneous subspecies (or species), separated geographically roughly where the Eurasian and Afro-Arabian landmasses meet in the Levant (Bar-Yosef et al. 1992; Bar-Yosef and Vandermeersch 1981; Jelinek 1982; Tchernov 1988, 1989, 1992a, 1992b, 1992c; Valladas et al. 1988; Vandermeersch 1982). Nowhere else in the world have paleoanthropologists encountered such striking, complex temporal and geographic relationships between these hominid populations.

The earliest known skeletal representatives of anatomically modern humans in the Levant come from the caves of Qafzeh and Skhūl (Howell 1952; Howells 1976; Trinkaus 1984; Vandermeersch 1982) and have been dated by the thermoluminescense technique to between roughly 90,000 and 110,000 years ago (Meignen et al. 2001; Valladas et al. 1988). Dated Neandertal skeletal remains in the Levant, in contrast, cluster between 70,000 and 36,000 years ago (Bar-Yosef 1989a, 1989b, 1992; Valladas et al. 1988, 1998), apparently correlating with somewhat cooler, wetter climatic

conditions in the region (cf. Bate 1942, 1943; Garrard 1980; Tchernov 1988, 1989). Assuming that the dates for these Levantine cases are correct, we are faced with the possibility that whereas anatomically modern humans appeared quite early and sporadically in southwestern Asia—that is, around 100,000 years ago—they did not flood into Europe for another 50,000 years(cf. Grun and Stringer 1991; Grun et al. 1991; Mellars et al. 1987; Stringer 1988; Stringer et al. 1984; Valladas et al. 1988). What is more, hominids of relatively modern appearance continued to use Middle Paleolithic technology for quite some time.

These observations imply that the main barrier to modern human demographic expansion prior to about 45,000 years ago was behavioral, perhaps lying in the ways resources were used and the mechanisms by which human groups insulated themselves from unpredictability in the supply of those resources. Neandertaloid humans already present in Europe were doing well for themselves until perhaps 50,000 to 35,000 years ago. This robust human variant may have been adapted most specifically to the climatic and foraging conditions of periglacial environments. Following this logic, periodic southward expansion of Neandertal populations into the Levant would have been limited partly by climate regimes and prevailing habitat structures (Bar-Yosef 1992; Bar-Yosef and Vandermeersch 1981; Howell 1959; Tchernov 1989, 1992a, 1992b, 1992c). Paleontologists familiar with the Levant are perhaps most comfortable with such an idea, because it is consistent with some broader patterns in animal community dynamics in the eastern Mediterranean Basin. The Palaearctic, Irano-Turanian, and Saharo-Arabian biogeographic provinces intersect there, creating a unique biological "suture zone" (chapter 6). Of course human cultural geography later came to defy the constraints typical of mammalian biogeography, but this may have been a uniquely Upper Paleolithic development (e.g., Tchernov 1992d).

The story just outlined is no doubt an oversimplification of the demographic and biogeographic processes involved, but it is a useful point of departure for refining the questions being asked of the Levantine data. Modern animal subspecies are distinguished by subtle differences in physical morphology, by details of their foraging adaptations, or by both, particularly if the species is distributed over a large geographic area (*sensu* O'Brien and Mayr 1991; MacArthur and Levins 1967). Whether the two varieties of Paleolithic humans were ecologically confined to certain biomes at any time in the past or were equally successful throughout the total geographic range of humans is an important distinction. There is also the possibility that these populations differed somewhat in their potentials for demographic growth, due to differences in resource use strategies or systems of cooperation. In either case, it currently is difficult to comprehend from existing information how the ecology of archaic humans differed from that of anatomically modern humans such that one population might periodically flood an area at the expense of the other and even become the sole living representative of the species worldwide. Obviously, archaeological interpretations of these evolutionary processes hinge upon knowledge of preexisting variations in hominid ecology during the late Middle Pleistocene—a topic that generally leaves us fumbling in the dark.

Whether Neandertals and other robust hominids should be considered separate species or merely sub-species is an academic question for which there seems to be no satisfactory answer. Questions about how early humans interacted with the world around them and how Paleolithic populations compare ecologically and behaviorally do seem answerable. If archaic and anatomically modern human populations were very distinct from one another in an ecological sense, then the differences should be evident from the ways in which they exploited food resources over an appreciable environmental gradient (latitudinal, coastal-continental, altitudinal). By exploring multiple dimensions of the human econiche, one can learn whether some of the major transitions in hominid lifeways prefaced, co-occurred with, or postdated shifts in skeletal morphology or technology and whether these trends were linked to changes in global climate.

We can be sure that Middle Paleolithic foragers lived by hunting and gathering, but they appear to have been a different sort of forager from the humans of any recent hunter-gatherer society (Kuhn and Stiner 1998b, 2001; Stiner 1994). There is little reason to think that "hunter-gatherers" sprang into the world as a single entity. Many potential dimensions of the human niche are likely to have developed independently, and some rather essential changes that are thought to have made us human—such as big game hunting—took place well before the Upper Paleolithic and the Late Stone Age. Upper Paleolithic and later foragers are distinctive from their predecessors, however, for the very existence of cultural geography (including stylistic expression) among groups, for their greater technological investment in the face of seasonal or unpredictable food supplies, and, probably, for their greater manipulation of social ties as insurance against the uncertainties of a foraging way of life. Also

remarkable was the ability of some Upper Paleolithic and Epipaleolithic cultures to pack landscapes with people as never before. The low demographic potential of Middle Paleolithic populations and the repetitiveness of their stone industries across vast areas of the Old World are stunning by comparison (chapters 7–9).

Nearly all publications on Paleolithic foraging have been about large game hunting, and many discussions of zooarchaeological evidence seem permanently lodged on the hunting-scavenging debate. Clearly, large mammal exploitation was important in the Paleolithic, and zooarchaeologists will continue to squeeze their data for differences between the Middle and Upper Paleolithic adaptations, counting tool marks and ungulate body parts. New insights from these approaches are few, however, or well short of clear. I hope to convince readers that other aspects of the faunal record are also worthy of emphasis. Small game exploitation is evidenced in many Mediterranean Paleolithic sites, including Middle Paleolithic horizons—more often than most readers are aware. The taphonomic noise is as much an issue with these remains as with those of large mammals, but the truth is that most evidence of small animal exploitation has simply been overlooked. I do not ignore large game hunting, especially since there was plenty of it during the Middle Paleolithic. Rather, the point is that this is only one of several important dimensions of the hominid foraging niche during this long, remote culture period.

THE SITES AND EXCAVATIONS

Chapter 2 presents a much fuller account of the archaeological projects from which the study samples come, but a quick introduction is in order here. Hayonim Cave and Meged Rockshelter lie within 1 km of each other in the Wadi Meged, a drainage that empties on the modern Mediterranean coast roughly 20 km to the west. Both shelters face generally south-southeast. Hayonim Cave ("cave of the pigeons") lies in the north bank of the Wadi Meged in the western Galilee (fig. 1.1). It is one of several large, bell-shaped caves with open chimneys in the immediate area, and it contains early Middle Paleolithic, Levantine Aurignacian, Kebaran, Natufian, and Byzantine cultural layers. Earlier excavations in the cave, between 1965 and 1979, established the antiquity and cultural associations of the deposits (Bar-Yosef 1991b; Bar-Yosef and Belfer-Cohen 1988; Belfer-Cohen 1988a, 1988b; Belfer-Cohen and Bar-Yosef 1981; Goldberg 1979; Goldberg and Laville 1988; Henry

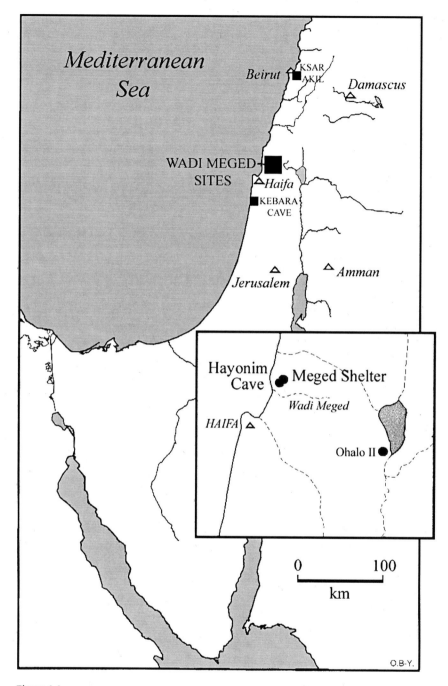

Figure 1.1
Location of the Wadi Meged sites in the Galilee region of the Levant, along with Kebara Cave (Mount Carmel) and Ksar 'Akil (near Beirut).

et al. 1981; Tchernov 1968, 1981; Valla et al. 1989, 1991). Middle Paleolithic layers constitute the thickest portion of the stratigraphic sequence—minimally 8 m in the rear-central portion of the cave, thinning to 3–4 m in its center. Part of the original vault collapsed long ago, but geological studies by Paul Goldberg and the late Henri Laville indicated that most of the sediments in the Middle Paleolithic layer accumulated under the protection of the cave roof (Goldberg 1979; Goldberg and Bar-Yosef 1998; Goldberg and Laville 1988).

Since the first excavation campaign at Hayonim Cave, new questions have been raised about the Paleolithic sequence in the Levant, calling for closer examination of the relations between chronostratigraphy, technology, and subsistence at this site. Tchernov's (1988, 1989) work on the microfaunas suggested that most of the Mousterian deposit of Hayonim Cave formed before 100,000 years ago, and much of it extends backward into the Middle Pleistocene. Apparently coeval Mousterian layers in Tabun Cave on Mount Carmel (Mercier et al. 1995a; Valladas et al. 1998) yielded little faunal information, owing to a combination of poor preservation and Dorothy Garrod's excavation methods of the time (Bar-Yosef 1989a; Jelinek et al. 1973; Margaris 2000). For now at least, Hayonim Cave is the only major source of faunal correlates for early Mousterian technology and culture in the region.

I participated in the renewed excavations at Hayonim Cave and Meged Rockshelter from 1992 to 2000. The Hayonim excavations were directed by Ofer Bar-Yosef, Liliane Meignen, and Bernard Vandermeersch, and those at Meged Rockshelter, by Steven Kuhn and Anna Belfer-Cohen. My role in these projects was to run a complete study of the macrofaunal remains, seeing them out of the ground and through a suite of laboratory analyses. Much of the identification work was done in the Zoology Laboratory of the Department of Evolution, Systematics, and Ecology of the Hebrew University of Jerusalem (Israel), where a large comparative collection is housed. Some of the more specialized taphonomic studies (mainly infrared spectroscopy and fragmentation and burning analyses) were performed in my laboratory at the University of Arizona (Tucson, USA).

The Mousterian Layer E spanned a great period of time, and its youngest portions may date to 70–90 thousand years ago (KYA). A uranium-thorium (U-Th) date on a fallen stalagmite in the northeast corner of the central excavation area indicates that the sediments of Middle Paleolithic (MP) unit 2 must be younger than 135 KYA. The rest of the sediment column dates to between 135 and 170–200 KYA. Layer F, which also contained early Mousterian material, dates to greater than 200 KYA. The internal radiation dose in the Mousterian sediments was exceptionally low, making estimates of the external dose critical to determining thermoluminescence (TL) and electron spin resonance (ESR) radiometric ages. The ESR results do not supply information about temporal variation within Layer E, owing to poor resolution overall. Specifically, age determinations throughout Layer E by the ESR technique

are essentially the same, but this does not necessarily mean that the thick column of sediment accumulated rapidly, because the calculated error is high and the ambiguities of the external dose are considerable. The TL dates offer somewhat better temporal resolution within Layer E (see chapter 2). The early Mousterian industries from Layer E generally resemble the "Tabun C" type of Tabun Cave on Mount Carmel, and those of Layer F are not unlike the "Tabun D" type. Layer G contained Acheulo-Yabrudian assemblages of the sort described by Jelinek at Tabun Cave (Jelinek 1981, 1982; Jelinek et al. 1973; Meignen 1998; Meignen et al. 2001).

A coherent chronostratigraphic series existed for the early Mousterian in the central excavation area of Hayonim Cave. There were few, if any, clear-cut stratigraphic breaks in the sediment column, however, which required us to make semiarbitrary divisions on the basis of variations in sedimentation rates, sediment microstructure, and artifact content and densities. Faunal data were not used in dividing up the sediment column, because the preservation conditions for bone, mollusk shell, and wood ash varied in vertical and horizontal space (chapters 3 and 4). Seven of the Mousterian vertical units in Hayonim Cave—units 1 through 7—contained bones in association with lithic artifacts, and the very rich unit 4 was subdivided into sections 4a and 4b for some of the faunal analyses (table 1.1). Layers F and G lacked bone almost entirely, and that which existed tended to be altered chemically (Weiner et al. 2002; chapter 2). Layer F represents a sediment regime categorically different from that of Layer E with respect to the source of sediments and the mode of sedimentation, suggesting that Hayonim was a very different sort of cave at that time. There may have been a long hiatus between the formation of Layers E and F. Layer G, although rich in stone artifacts, contained no bones in the areas we excavated.

The macrofaunas from Hayonim Cave and Meged Rockshelter are clearly of anthropogenic origin, to the exclusion of other agencies (chapters 4 and 5). So, too, are major constituents of the sediments, which are dominated by wood ash products (chapter 2; see also Albert et al. 1999; Schiegl et al. 1994, 1996; Weiner et al. 1995, 2002). Most of the faunal remains in Hayonim Layers B through E and in the Paleolithic layers of Meged Rockshelter are in a good state of preservation, and large samples are available for analysis. The patterns of bone damage invariably point to hominid activities (chapter 5). Remains of large carnivores and traces of gnawing on herbivore bones are remarkably rare, an unusual situation for Mediterranean cave sites (compare Gamble 1983, 1986; Lindly 1988; Stiner 1991b, 1994; Stiner et al. 1996; Straus 1982; Tozzi 1970) but one that is not unknown (e.g., Stiner 1994). Owls appear to have been the most important collectors of micromammals (mainly small rodents) in the cave. They frequently roosted in small solution cavities of the vault while the Paleolithic deposits were forming, and barn (*Tyto alba*) and other owls continue to inhabit the cave system today. Owls sample habitats while hunting for prey and then regurgitate indigestible products at their roosts, thereby providing invaluable records of Pleistocene animal community structures and local environments (e.g., Andrews 1990; Brain 1981; Stuart 1982, 1991; Tchernov 1989, 1996). The microfaunas from the Paleolithic layers of Hayonim Cave are rich and chronologically extensive because the cave provided so many attractive roosts for owls.

Hayonim Cave and the nearby rockshelter called Meged also contained Upper Paleolithic and Epipaleolithic archaeofaunas (table 1.1). In Hayonim Cave, the Mousterian was overlain by a Levantine Aurignacian layer (D) inside the cave, a Kebaran layer (C) oriented toward the cave entrance, and finally a widespread Natufian layer (B) overlain by historic material (see chapter 2). Meged Rockshelter contained early Kebaran and pre-Kebaran horizons along with a late Upper Paleolithic layer dating to about 20 KYA (chapter 2; also Kuhn et al. 2004). Three consecutive assemblages were defined in Meged Rockshelter on the basis of depth ranges and changes in lithic artifact associations: 0–199 cm below datum (bd), 200–214 cm bd, and 215 cm bd to bedrock. Oddly, the frequency of faunal remains in the sediments of this shelter declined before bedrock was reached in the 1997 excavation units, whereas many lithic artifacts were found throughout the deposits and in contact with bedrock. This discrepancy does not appear to be the product of differential preservation, because all of the sediments are rich in unaltered calcite, which is indicative of a stable chemical environment, and the bone mineral is in uniformly good condition on the basis of the criteria presented in chapter 4. The Upper Paleolithic and Epipaleolithic faunas generally are well-preserved and were collected by humans alone; no raptor- or owl-collected assemblages were mixed with the cultural remains.

THE WADI MEGED FAUNAL SAMPLES

The total Wadi Meged faunal sequence begins at roughly 200 KYA and ends at around 11 KYA (table 1.2). As explained earlier, Hayonim Cave preserves multiple Paleolithic phases, whereas Meged Rockshelter is relatively small and preserves a few Upper and Epipaleolithic phases. The "Wadi Meged series" is a concatenation of faunal assemblages from the two sites, arranged in chronological order (table 1.1).

The chronology of the faunal series is presented in a deliberately generalized way in table 1.2 and in the faunal analyses to come. More specific information on available dating results, such as they are, is provided in chapter 2. Accurate radiometric age estimates are highly desirable in principle, but the picture one gains from these results can be misleading if only one or a few dates are available for a cultural layer, as is the case for the Natufian, early Kebaran, and Aurignacian layers in the Wadi Meged series. The radiocarbon technique was used exclusively for the Upper Paleolithic and Epipaleolithic layers, and one or more different techniques—ESR (see appendix 3), U-Th, and TL—were applied to the Mousterian layers. The numbers of dates obtained and their levels of precision and replicability vary considerably between periods, and the comparability of results is limited. Hence, I avoid any literal use of dating results in this study of long-term change in hominid subsistence and instead consider variation in terms of well-ordered time ranges and general oxygen isotope stages.

The strengths of the Wadi Meged sample include its tight geographic placement and the high quality of archaeological recovery and documentation at the two sites. Fine screening was employed systematically during the excavations, and sediment chemistry relevant to bone preservation conditions was tested throughout (chapter 4). Faunal samples of unclear origin or representing mixed cultural entities were removed from consideration, lending further clarity to the temporal assignments claimed in this study. Because all of the assemblages are from shelters, one may assume that any components attributable to humans were carried to the sites from other locations and potentially hold information about foragers' transport decisions. Shelter faunas are particularly rich sources of economic and paleoecological information in Mediterranean environments because they were formed in limestone holes or fissures, in which terra rosa sediments often favor skeletal preservation. Although shelter sites represent

Table 1.2

Cultural, radiometric, and probable oxygen isotope chronologies for the faunal assemblages from Hayonim Cave and Meged Rockshelter

Site and Layer	Time Range (Youngest to Oldest, Years Ago)	Oxygen Isotope Stage
Hayonim, Natufian (Layer B)	11,000–3,000	1
Hayonim, Kebaran (Layer C)	14,000–17,000	2[a]
Meged, early Kebaran (<200 cm)	18,000–19,000	2[a]
Meged, pre-Kebaran (>199 cm)	19,000–22,000	2[a]
Hayonim Aurignacian (Layer D)	26,000–28,000	2[a]
Hayonim, Mousterian Units 1–2 (Layer E)	70,000–100,000	5–6
Hayonim, Mousterian Unit 3 (Layer E)	~150,000	6[a]
Hayonim, Mousterian Unit 4 (Layer E)	~170,000	6–7
Hayonim, Mousterian Units 5–6 (Layer E)	~200,000	7?
Hayonim, Mousterian Unit 7 (Layer F)	>200,000	7

SOURCES: Oxygen isotope stages are from Shackleton and Opdyke 1973 and Martinson et al. 1987.

NOTE: The Upper Paleolithic chronology uses generous age ranges based on uncalibrated radiocarbon dates. The Middle Paleolithic chronology is based on a combination of thermoluminescence, electron spin resonance, and uranium/thorium techniques.

[a] Generally colder and/or drier climatic conditions.

only one part of the total site use spectrum of Paleolithic peoples, this limitation of the sample presents certain advantages in that some conditions of site use (e.g., travel end-points) are knowable.

The Mousterian sample in the Wadi Meged series comes from the central (main trench) area in Hayonim Cave (squares G18–K24) (fig. 1.2). We know a great deal about the sediment formation and bone preservation history of this part of the site, and it is the area least altered overall by chemical or mechanical processes. Moreover, the layers there are essentially horizontal. The only exception may be the G-row squares in the excavation grid, where the boundary between Layers E and F often is lower than 530 cm bd. Eliminating G-row faunal data from the sample would barely affect the sample sizes for Mousterian units in Layer E.

The Mousterian faunal sample from Layers E and F in the central area exceeds 17,000 identified specimens (table 1.1) and was studied in its entirety. Unidentified fragments were many more in number and therefore were sampled for damage and related characteristics (chapters 4 and 5). A separate, smaller Mousterian assemblage was obtained from the deep sounding at the front of Hayonim Cave (see chapter 2), but I do not consider these data because of lingering ambigui-

ties in the sediment formation history, including down-filling and bone dissolution in this area of the deposits. Nearly all the Mousterian and Kebaran samples to be discussed came from only the most recent excavations. A few specialized analyses, such as the osteometric study of the tortoises, were expanded to include material from the Mousterian, Aurignacian, and Natufian excavations conducted between 1965 and 1979. In addition, I use Rabinovich's (1998) counts of numbers of identified specimens (NISP; see table 1.3, appendixes 1 and 2) of Aurignacian ungulates and carnivores, in combination with my counts of the small game animals. The data presented in this book complement rather than repeat prior work by Davis (1978, 1980a), Rabinovich, Sillen (1981), Tchernov, and others. The Natufian data come from recent research by Natalie Munro (1999, 2001) on a massive sample that integrates newly excavated and older collections from Hayonim Cave.

Potential time-averaging effects on the faunal series were addressed mainly through stratigraphic observations, analyses of sediment formation processes (mechanical and diagenetic), and removal of ambiguous units from the study sample. In Hayonim Cave, the major chronostratigraphic divisions (Layers B, C, D, E, and F) were relatively clear. Layer E, however, was

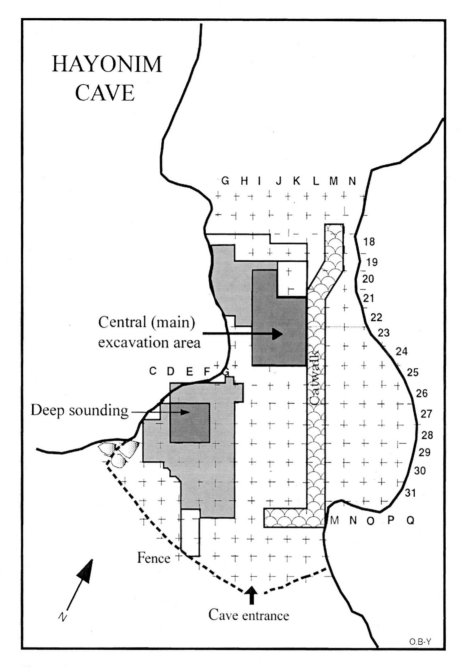

Figure 1.2
Hayonim Cave excavation plan.

exceptionally thick and, as explained earlier, was further subdivided on the basis of distributions of stone
artifacts and other information. Limited mixing of
material significantly older or younger than the unit in
question was estimated from the amount of downward migration of time-diagnostic Upper Paleolithic
and Epipaleolithic artifacts. These issues are addressed
in chapters 2 and 4.

A gap of about 70,000 years separates the Middle
and Upper Paleolithic components in the Wadi Meged
faunal series. This gap corresponds to the late Middle
Paleolithic and Ahmarian (early Upper Paleolithic)
periods. Data on these phases are available from
Kebara Cave on Mount Carmel (Bar-Yosef et al. 1992;
Speth and Tchernov 1998, 2001, 2002), filling the
breach in the Wadi Meged series for some of the com-

Table 1.3
Short definitions of the main zooarchaeological counting units

Variable	Definition	Counting Unit Type
NSP	Number of specimens	Primary
NUSP	Number of unidentified specimens	Primary (subset of NSP)
NISP	Number of identified specimens	Primary (subset of NSP)
PE	Portion of a skeletal element	Primary (subset of NISP)
MNE	Minimum number of skeletal elements	Derived
MNI	Minimum number of individual animals	Derived

NOTE: See appendix 1 for complete definitions. An element is a discrete (potentially separable) skeletal member. An individual is a living, breathing organism from which skeletal materials derive. A specimen is any skeletal item, broken or whole, damaged or undamaged.

parisons to follow. Nearly the same team of scientists excavated Hayonim Cave and Kebara Cave. Together the two caves present a nearly consecutive Middle Paleolithic sequence for the Galilee region (see Bar-Yosef and Meignen n.d.).

COMPARISONS WITH KEBARA CAVE

A number of caveats must be borne in mind when the Wadi Meged data are compared with those from Kebara Cave. First, the faunal analyses of Kebara Cave by John Speth and Eitan Tchernov involved a very different sampling strategy (Speth and Tchernov n.d.) and concentrated to a large extent on certain taxonomic groups and body parts with high identifiability, such that relative variation in damage frequencies among taxa and excavation units is informative in its own right but difficult to compare directly with that for Hayonim Cave, where all of the material was studied. Second, the quantity of material in Kebara was vastly greater and the chronological span much shorter than at Hayonim. The Mousterian layer of Kebara Cave appears to preserve "piles" of material that were pushed or thrown into particular areas of the cave (Bar-Yosef and Meignen n.d.; Bar-Yosef et al. 1986, 1992; Speth and Tchernov n.d.). No such concentrations of material or arrangements of living space were discernible in Hayonim Cave, with the possible exception of apparent hearth areas—although even this is questionable, since hearth feature visibility is very much dependent on preservation conditions.

Because of major differences in bone abundance and research history between the two sites, the sampling strategies used for analyzing the faunas were not identical. A complete study of samples from large excavated areas was feasible for Hayonim Cave but not for Kebara Cave. Thus, at least for now, only the Hayonim analysis is a truly comprehensive characterization of the macrofaunas, including identifiable and unidentifiable fractions and small and large species. The research at Kebara Cave at first focused on questions specific to gazelle and fallow deer exploitation (Speth and Tchernov 1998), later turning to the relations between large and small game exploitation and paleoclimate (Speth and Tchernov 2001, 2002, n.d.) in response to issues raised by the Hayonim study of small game use.

BONE RECOVERY

The excavation techniques employed at Hayonim Cave and Meged Rockshelter followed in part the tradition developed by French prehistorians (e.g., Leroi-Gourhan 1973) and now widely used by Paleolithic archaeologists. This strategy places much emphasis on the horizontal distributions of artifacts and bones, recorded via "piece-plotting" or "point-proveniencing" in three dimensions—northing, easting, and depth in centimeters below an arbitrary datum. Because many Paleolithic sites are too rich to permit plotting of every find, excavators tend to prioritize in favor of objects that retain formal, identifiable qualities or that exceed a certain size. There are many advantages to piece-plotting, not least of which are the possibilities for refitting fragments from different spatial units (e.g., Villa and Mahieu 1991), back-plotting assemblages to reveal subtle layering or the existence of lenses in sediments already excavated (chapter 2), and tracing variation in the broader distributions of faunal and other materials (chapter 4).

Identifiable bones, teeth, and gastropod shells, as well as bone fragments larger than 2.5 cm, were

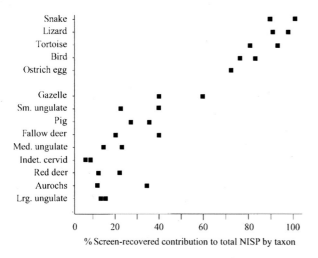

Figure 1.3
The relation between the original body sizes of prey animals (ordinally ranked from smallest to largest) and the animals' representation in the piece-plotted and screen-recovered fractions of the faunal assemblages (percentage of total NISP) from Mousterian units 3–4 in Hayonim Cave.

mapped in three dimensions during the excavations of Hayonim Cave. All other skeletal remains, including microfauna, were recovered in fine screen mesh (3.0 mm and 1.5 mm) from 50 × 50 × 5 cm subsquare units. Sediment samples were wet-sieved and floated in search of charcoal, plant parts, microartifacts, and microfauna, the final extraction of which took place in the Hebrew University Zoology Laboratory each winter. Bone specimens from larger animals were gently washed in water, dried, packaged, and labeled on-site; they are now stored at the Hebrew University of Jerusalem. A computerized inventory of the piece-plotted material was updated daily in the field by the author in preparation for detailed study in the laboratory. The piece-plotted faunal material from Mousterian Layer E, however, made up less than half the total identifiable material recovered and only a minor fraction of all bone. The balance of the collection, obtained via screening, was separated and studied at the University of Arizona, with taxonomic classifications subsequently cross-checked at the Hebrew University of Jerusalem.

Apart from the time that the piece-plotting of bones consumes, this practice in faunal and technological research has at least one other potential disadvantage. The problem begins with the fact that specimens not plotted in three dimensions at the time of excavation must be retrieved from the screens (sieves). For practical reasons these samples go into different bags, and

their provenience designations are volumetric in nature. The absolute sizes of specimens affect their visibility in the sediments, independently of whether or not they retain diagnostic morphological features. This fact, compounded by excavators' practical size thresholds for point-proveniencing, leads to the preferential plotting of large animal remains over those of most small animals, even though any or all animals could be related to hominid activities. Figure 1.3 reveals how this bias plays out in large samples, exemplified for Mousterian units 3 and 4 of Hayonim Cave: fragments of larger-bodied taxa are much more likely to be recognized and recorded in situ by excavators; the screen-recovered fractions contain larger proportions of small animals such as reptiles, birds, and small mammals.

Piece-plotting potentially, then, restructures an archaeological collection in a way that must be undone deliberately for zooarchaeological research on Paleolithic diets. This recovery and curation problem is not unique to the Hayonim excavation project, nor does it pose a major obstacle to research there or elsewhere if compensated for by recombining the piece-plotted and screen-recovered samples in later analyses. Unfortunately, in many earlier studies in the Levant the two components of Paleolithic faunal samples were not rejoined. The widespread oversight of reptile remains as economic refuse in Middle and Upper Paleolithic sites is one result; tortoises were widely exploited during the Paleolithic of the Near East and very clearly so in the case of Hayonim Cave (Stiner and Tchernov 1998).

Thus, in the end, the analyses of the Wadi Meged faunal series reverted to volumetric spatial units, 50 × 50 × 5 cm, rather than point-provenienced units. There is no reason to think that true living floors were preserved in either of the two sites, and this was the only way to integrate identifiable bones recovered from the screens with piece-plotted finds. Sedimentary units of consistent volume are best suited to evolution-oriented research on deeply stratified faunal series.

TAPHONOMIC APPROACHES

A great variety of animals may concentrate bones in shelters, especially in Mediterranean caves, where sediments and preservation chemistry also tend to be heterogeneous. Hayonim Cave lives up to some aspects of this reputation, providing interesting opportunities for taphonomic research due mainly to variable mineral environments but surprisingly little in the way of carnivore interference. In reporting the taphonomic results for the Wadi Meged series—one of the two families of research problems addressed in this book, together with the foraging ecology of early humans—I begin by assessing the agents responsible for the skeletal accumulation and the extent of bone disturbance and loss following primary deposition. These assessments rely on cross-referenced data on bone damage, species representation, and spatial associations among materials. The question of in situ bone attrition—what material may have been lost to decomposition (chapters 3 and 4)—is a pivotal aspect of the study. A later part of the taphonomic investigation focuses on the economic behaviors of the human occupants, on the basis of patterns in food and other cultural refuse.

The ideal situation for this research would be that hominids and owls were the sole occupants and collectors of macrofaunal and microfaunal remains, respectively, in Hayonim Cave and Meged Rockshelter. Although this scenario proves to have been close to true in the Wadi Meged sites, it is a risky assumption for Pleistocene cave sequences in principle (Bar-Yosef et al. 1992; Brain 1981; Gamble 1983; Lindly 1988; Speth and Tchernov 1998; Stiner 1994; Stiner et al. 1996; Straus 1982), and so other potential explanations must be excluded systematically. Data on species representation provide a reasonable starting point for identifying bone collectors and modifiers, because the remains of bone-gathering species themselves are often found among the bones of prey taken to or disgorged in shelters. Spatial distributions of skeletal and artifactual materials, surface damage on bones, and patterns of fragmentation are then used to infer agent dominance, sequences of effects of agents, and agent behavior. Of foremost importance are traces of burning, tool marks, gnawing marks from carnivores and rodents, scars from root etching, indications of weathering, and fracture forms associated with compression, shearing, and other mechanical forces.

The question of in situ bone attrition requires a conceptual slant distinct from that concerned with identifying agents of bone accumulation or disturbance. Rather than seeking positive indications of bone modification activity, the problem is to understand what might have disappeared from the assemblages after they accumulated in a shelter. Such losses can occur through mechanical or chemical means.

Because ungulates do not normally inhabit caves (though exceptions are known), their remains must be transported to such places by other agents in order to be present at all. Humans, for example, may partition ungulate carcasses at acquisition sites in prepara-

tion for the transport of meat and other carcass products to processing camps. Parts of skeletons that never reached a shelter therefore have behavioral implications very different from those of parts that were destroyed subsequent to their arrival on-site. In this study, a combination of techniques was used to distinguish the results of in situ attrition from the results of differential transport: infrared analysis of bone and bone decomposition by-products in sediments (in collaboration with Stephen Weiner), thin-section analysis of sediments (in collaboration with Paul Goldberg), and skeletal portion analysis in relation to bone density and skeletal macrostructure classes (compact bone, cancellous bone, dentin, and tooth enamel).

Bone mineral condition was examined at the molecular level using Fourier-transform infrared spectroscopy, or FTIR (Smith 1996). This tool is particularly helpful for understanding the potential effects of chemical dissolution, heat alteration, and fossilization of bone and other skeletal materials such as mollusk shell. In combination with infrared data on sediment composition, the FTIR technique addresses questions about in situ bone attrition that cannot be answered by patterns of vertebrate body part representation alone (chapter 4). Factors that might explain the patchy bone distributions in Hayonim Cave potentially include local variation in diagenetic and dissolution activity in sediments, erosion, and space-delimited human behavior. The first of these factors proved especially relevant: as bone decomposes through chemical or mechanical alteration, its identifiable qualities are lost to the naked eye. Moist sediments with high organic content may experience changes in pH, which may cause the dissolution of bone mineral and collagen (DeNiro 1985; DeNiro and Weiner 1988a, 1988b, 1988c; Masters 1987; Weiner and Bar-Yosef 1990). As these soluble compounds migrate through the sediments, they leave behind chemical signatures that testify to the original presence of bone. The chemical transformations, including subtle alterations in the crystalline structure and composition of bone and sedimentary minerals, can be traced effectively by infrared spectroscopy (Weiner and Goldberg 1990; Weiner et al. 1993, 2002). Likewise, the chemical and mechanical decomposition products of bone, though not readily visible to the excavator, can be seen in soil thin-sections at low magnification.

A closely related theme is the formation and nature of Mousterian hearths. Some Paleolithic caves in Israel, most notably Kebara Cave, are famous for the hearths preserved in them (Bar-Yosef et al. 1992; Goldberg and Bar-Yosef 1998; Meignen et al. 1989; on

Tabun Cave, see Albert et al. 1999; Jelinek 1981). The early Middle Paleolithic levels of Hayonim Cave were rich in ash and ash by-products, charcoal, burned artifacts, and some reddened earth. Most of these features represented "hearth areas" rather than discrete fireplaces; their vertical and horizontal dimensions indicated multiple burning events, and their margins were often ambiguous (chapters 2, 4). Some of the hearth areas contained recognizable bone fragments—roasted tortoise remains in some, calcined bone powder in others. Other hearth areas contained only silica compounds, the most resistant components of wood ash (Schiegl et al. 1994, 1996). Indeed, the great spatial heterogeneity of the faunal remains in Hayonim Cave was matched only by the patchy distributions of preserved wood ash; lithic debris was more widely and evenly distributed throughout the excavated portions of the site.

It was in response to these observations that we began a program of experiments in the diagenetic alteration of bone (chapter 3; Stiner et al. 1995) and wood ash (Schiegl et al. 1994, 1996). The bone burning experiments simulated the mechanical (trampling) and thermal conditions that might normally occur around small campfires on the ground surface, as opposed to conditions within the underlying sediments. In still other experiments we explored the diagenetic signatures of chemical dissolution, atmospheric weathering, and fossilization from macroscopic and molecular perspectives. Many of the results were consistent with those of prior studies, such as that by Shipman et al. (1984). Other results were more surprising and have put us in a better position to evaluate the circumstances and temporal scope of bone assemblage and hearth formation in Hayonim and other cave sites.

ECOLOGICAL AND ECONOMIC APPROACHES

Significant changes in human behavior and physiology certainly took place during the Middle and Upper Pleistocene. The majority of these changes were expressed as frequency shifts in extant behaviors (Stiner 1994), whether we are talking about hunting, scavenging, or the collecting of game, technological behaviors, or those governed by human biomechanics (Bar-Yosef et al. 1992; Boëda et al. 1998; Burke 2000; Chase 1989; Clark and Lindly 1989b; Dibble 1987; Farizy 1990; Gamble 1986; Geneste 1985; Klein 1989; Kuhn 1995; Meignen 1988; Moncel et al. 1998; Shea 1989; Simmons and Smith 1991; Smith 1984; Stiner

and Kuhn 1992; Stringer 1989; Trinkaus 1986; Wolpoff 1989). Showing that a predator hunts for a living says little about how that predator occupies a given ecosystem. Researchers interested in the evolutionary differences among hominid predatory adaptations should therefore consider the geographic and temporal contexts in which certain behaviors were emphasized, as well as the limits of variation of those behaviors.

Differences among hominid populations could be evident from several dimensions of the predatory niche they occupied, such as techniques for obtaining food and the substrates commonly searched, the prey age groups procured and the body parts of large game prey transported to shelter sites for additional processing, patterns of small animal exploitation, food processing techniques, and seasonal responses to food supply. Cross-species comparisons of predators provide the framework for this search for ecological differences among hominids (Stiner 1990, 1991a, 1992, 1994). Comparisons of ethnographically documented modern human cases help define the breadth of the human niche occupied by contemporary mobile peoples across geographic clines (Kuhn and Stiner 2001), drawing on works by Binford (1978, 2001), Brain (1981), Bunn et al. (1988), Gifford (1977), Hawkes (1987; Hawkes et al. 1997), Hudson (1991), Keeley (1988), Kelly (1995), Lupo (1995), O'Connell et al. (1988a, 1988b), Oswalt (1976), Speth and Spielmann (1983), Yellen (1977, 1991a, 1991b), and others. Although the objective is to learn about cultural systems that no longer exist, the modern reference data provide expectations about the magnitude of differences that might reasonably distinguish hominid behavioral adaptations in evolutionary time.

A predator's adaptations limit its selection of prey according to body size, defense characteristics, and habitat preferences. However, the array of prey species that a predator consumes and their relative abundances in its diet are conditioned foremost by local availability. The composition of animal communities is the ecological backdrop of human predatory economics: if the observed variation in species representation in archaeofaunas cannot be related to climate change or animal community turnover, then it may potentially signal changes in human predatory adaptations such as permanent increases in dietary breadth. The analysis of diet breadth in this study began with comparisons of the relative frequencies, trends, and associations between species over time (chapters 6 and 7), followed by classic diversity analysis and controlled manipulation of the criteria for categorizing prey (chapter 9).

Predator-prey simulation modeling was another important feature of the investigation, specifically for the questions of hunting pressure and changes in human population densities (chapter 8). Human demography is very relevant to the origins of modern human behavior, not to mention the forager-farmer transition in later prehistory (Bar-Yosef 1995, 1996; L. Binford 1968, 1999; Cohen 1977; Davis et al. 1988, 1994; Flannery 1969; Henry 1985, 1989; Keeley 1988; Munro 2001; Stiner 2001; Stiner et al. 1999, 2000). Demographic arguments have slipped in and out of vogue in anthropology. Dissatisfactions with these arguments are due at least partly to the circular explanations they occasionally foster, although they are no more subject to frailties of reasoning than any other genre of argument in anthropological publications. Refinements in our basic understanding of how population phenomena (such as size and density) interact with specific physical and social environments bring demographic concepts to the fore once again, without trappings of environmental determinism (see, for example, Harpending and Bertram 1975; Winterhalder and Goland 1993). Built from an independent empirical base, our predator-prey simulation models project the outcome of density-dependent harvesting by humans in terms of the relative resilience of certain prey populations (following Stephens and Krebs 1986). The approach is most relevant to small-bodied prey species, because they differ so much in their costs of exploitation and their reproductive potential. Here, the application is confined to the Mediterranean Paleolithic, but the approach and the results are relevant to global questions about the relation between human demography and cultural evolution during the Pleistocene.

Ungulate hunting, a topic widely discussed by paleoanthropologists, arguably emerged in hominids long ago. Large game hunting, for better or for worse, is also one of the more frequently used definitions of modern human behavior. The evolutionary questions raised in this study are more specific, focusing on possible changes in hunting efficiency, the galvanization of a certain niche, and the possibility of greater specialization over time. The early Mousterian faunas of Hayonim Cave are important for evaluating the antiquity of certain large game hunting adaptations that characterized all later humans.

Prior work on Paleolithic faunas in Italy (Stiner 1990, 1994, 1999) revealed significant variations and some evolutionary shifts in the human predatory niche during the Late Pleistocene. A trend toward greater foraging specialization in this region was suggested by hominids' increasing emphasis on prime adult ungulate prey (mainly deer and wild cattle), a focus that apparently evolved from a more generic form of ambush hunting (chapter 11). Habitual prime-age-focused ungulate hunting is ecologically unique to humans among large predators. The behavior certainly was present in human systems in Italy by the late Middle Paleolithic. The tendency became "fixed" in human behavioral repertoires worldwide by Upper Paleolithic times, and it persists in a wide variety of Holocene cultures. This raises the question of when (and where) this predator-prey relationship might first have appeared. The full temporal and geographic scope of prime-age-focused hunting begs for further exploration, beginning at the least with conjoining Mediterranean ecosystems and earlier time ranges. An obvious question for this study in the Levant is about the nature of ungulate procurement in the early Mousterian, allowing of course that this case is but one piece of the larger puzzle. Modern wildlife data, principles of mammalian demography, and the life history characteristics of prey species provide much of the framework for interpreting the prehistoric cases.

Taphonomic questions notwithstanding, body part representation for large prey potentially informs us about the transport strategies of early humans and the contexts of food consumption at shelters (chapter 10). There has been much discussion of how Mousterian humans obtained large game—mainly, whether they hunted it in the sense that later foragers did or whether they relied on scavenging. This may seem a simplistic debate, and to some extent it has been, but more than one kind of predator-prey relationship could have evolved among divergent hominid populations, just as they have among other large predators. Bone transport is perhaps most interesting for what it might say about how prehistoric people staged carcass dismemberment and processing across landscapes. Many potential economic solutions exist, depending on the sizes of social groups, patterns of food dispersion, aridity, fat availability, and so on. Ungulate body part representation is examined in this study according to body size category (small, medium, and large), the total quantity of parts transported per carcass source, and biases in skeletal composition that are largely independent of bone density (chapter 10).

A related point concerns food processing tactics, here examined from element damage patterns, portion-of-element representation, and the extent to which potentially rich marrow reserves were extracted (chapters 5 and 10). The faunal remains from Hayonim Cave and Meged Rockshelter are sufficiently well preserved to permit this kind of study. Although bone surface visibility is sometimes obscured by calcite concretions, old green-break edges and diagnostic fracture forms are readily visible.

In this study, osteometric data were applied to several questions about the characteristics of Pleistocene prey populations and the possibility of body size diminution in relation to climate change (chapter 7). In tortoises, which grow for much of their lives, body size serves as a proxy for individual age (e.g., Klein and Cruz-Uribe 1983; Stiner et al. 1999) and potentially reveals age structure distortions attributable to harvesting pressure from humans. Seasonal mortality is another potentially important dimension of faunal records (see Speth and Tchernov 2001, n.d.), but it was of limited utility for this particular study because of the coarse resolution of the deposits and small number of individual animals involved. In addition, the results of dental cementum (annular ring) analysis are highly controversial when applied to ungulates of lower-latitude environments (cf. Lieberman 1991, 1993; Stutz 2002). Deciduous dentitions and antler development data are useful in principle, but juvenile animals and deer antlers are comparatively few in the Mousterian samples from Hayonim Cave.

Finally, relations between Mousterian technology, game use, and the intensity of site occupations are invaluable for answering questions of land use. General patterns of stone raw material transport, manufacture, and resharpening must be linked to foraging agendas and territoriality (Kuhn 1992b, 1993, 1995; Geneste 1985; Henry 1992; Jelinek 1991; Marks and Volkman 1986; Stiner 1994; Stiner and Kuhn 1992). Lithic analyses by Liliane Meignen (1988, 1994, 1995, 1998) emphasize technological variables relating to economic constraints, including raw material concerns, and to local manufacturing traditions passed from one generation to the next. The main issues taken on here center on the simpler questions of occupation intensity and the extent to which blanks and tools were used up (resharpened or reduced) on-site (chapter 12).

With two related but semidiscrete research trajectories and thousands of faunal specimens for a nearly equal number of elapsed years, this book has a lot to cover. Not for the first time, the Mousterian and Upper Paleolithic prove to be full of surprises, despite what must have been a simpler social life back then.

■

CHAPTER TWO

Archaeological Background to Hayonim Cave and Meged Rockshelter

■

Ofer Bar-Yosef,

Anna Belfer-Cohen,

Paul Goldberg,

Steven L. Kuhn,

Liliane Meignen,

Bernard Vandermeersch,

and Stephen Weiner

■

THIS ZOOARCHAEOLOGICAL STUDY WAS CENTERED ON THE assemblages from Hayonim Cave and Meged Rockshelter, two prehistoric sites in the Wadi Meged in western Galilee. The excavation project at Hayonim Cave was directed in its first years, 1965–1979, by Ofer Bar-Yosef (at the time with the Hebrew University of Jerusalem), Baruch Arensburg (Tel-Aviv University), and Eitan Tchernov (Hebrew University of Jerusalem). The layers immediately below the historical ash deposits were found to contain Natufian remains, with Kebaran, Aurignacian, and Mousterian industries in successive layers below them (Bar-Yosef 1991; Bar-Yosef and Goren 1973). The excavations at Hayonim Cave stopped in 1973 in the Mousterian deposits. When work resumed in 1977–1979, excavations were concentrated on the Natufian layers in response to massive damage caused by local shepherds, which had also blocked the deeper parts of the "central area" (Bar-Yosef 1991; Belfer-Cohen 1988a, 1988b).

New excavations began at Hayonim Cave in 1992 and lasted until 2000, conducted jointly by Ofer Bar-Yosef (Harvard University), Liliane Meignen (Centre de Recherches Archéologiques [CNRS], Valbonne, France), and Bernard Vandermeersch (then at the University of Bordeaux I, France). The target of the renewed excavations was the Mousterian sequence, exposed in the central area, while an Acheulo-Yabrudian layer was reached in the deep sounding near the cave entrance. Only a few additional squares of the Natufian deposits were excavated during these years.

Smaller but important faunal and artifact collections dating to the late Upper Paleolithic and early Kebaran were obtained from Meged Rockshelter, 0.5 km east of Hayonim Cave in the Wadi Meged (Kuhn et al. 2004). This rockshelter was first tested during the early 1970s by the team then working at Hayonim Cave. The main excavation of Meged Rockshelter was conducted from 1994 to 1997 by Steven L. Kuhn (University of Arizona) and Anna Belfer-Cohen (Hebrew University of Jerusalem).

In this chapter we provide a brief review of the region and summarize in some detail the excavations of the sites, the field techniques used, site stratigraphies and dating, and the archaeological assemblages. Research on the artifacts and other materials from Hayonim Cave is ongoing, so the summaries offered here are preliminary, but they are largely representative of what has been learned to date. Additional information, including the results of specialized analyses, can be found in other publications (Albert et al. 2003; Bar-Yosef 1991, 1998; Belfer-Cohen 1988a, 1988b, 1991; Hopf and Bar-Yosef 1987; Kuhn et al. 2004;

Figure 2.1
Entrance of Hayonim Cave.

Meignen 1998, 2000; Meignen et al. 2001; Mercier et al. 1995a; Schiegl et al. 1994, 1996; Stiner et al. 1995; Stiner and Tchernov 1998; Tchernov 1998a; Valladas et al. 1998; Weiner et al. 2002).

THE REGION

Hayonim Cave is located in the western Galilee, at the geographic boundary between what is traditionally called the lower and the upper Galilee. The geomorphological structure of the western Galilee includes, first, a coastal plain composed of two major features. Its northern portion, situated between the ridge of Rosh Ha'Niqra and the town of Akko, lies at elevations of 50–100 m above sea level and is only 3–4 km wide. Farther south is the valley of Akko, which stretches for 7–8 km south of Hayonim Cave to the foot of Mount Carmel. Inland, the foothills, which are in part an elevated Neogene abraded plain, tilt westward into the coastal plain. The hills beyond attain heights of up to about 800 m and were formed by transverse faulting into a series of parallel ridges. The major geological formations (the Judea Group) are mostly of Late Cretaceous age and lithologically are mainly limestone, dolomite, and chalk (Freund 1978; Issar and Kafri

1972). Hayonim Cave itself is located in a limestone formation. A survey by Delage (2001) indicated that various formations within 10 km of Hayonim Cave contain flint nodules, which were used extensively by the prehistoric inhabitants.

The main wadi courses in this region are Nahal Bezet, Nahal Keziv, Nahal Ga'aton, Nahal Beit Ha'emeq, and Nahal Yassaf. The Wadi (or Nahal) Meged is a tributary of the latter.

Soil types range from distinctly sandy soils and grumosols on the coastal plain to terra rosa and brown rendzina types (Dan and Yaalon 1971) in the hilly area. The local climate is one of mild winters and high humidity. Average January temperatures are 8–10 degrees C in the higher altitudes and 12–14 degrees C in the coastal areas. August average temperatures are 24–28 degrees C. Annual precipitation reaches 500–800 mm during the winter months, whereas the summers (May through mid-September) are hot and humid with almost no rain.

The vegetation of the western Galilee is typical of the Mediterranean phytogeographic zone, with stands of *Quercus calliprinos* and *Pistacia atlantica* grading into stands of *Acer obtusifolium, Laurus nobilis, Quercus boissieri,* and *Rhamnus palaestina*. Bushes diminish in number with elevation, whereas the number of ferns

Figure 2.2
View of Hayonim Cave and the Wadi Meged, facing west, with the Mediterranean Sea in the background.

increases, because of the rise in humidity. *Quercus boissieri,* together with *Q. calliprinos,* grows in karstic areas on the northern and western slopes and along wadi courses. The chalky exposures are today covered with pine and *Arbutus andrache,* but under natural conditions *Q. calliprinos* and *Rhamnus* species tend to take over (Aloni 1984). Historical records from the nineteenth century indicate that forest cover was greater in the area than it is today. The composition of the Pleistocene vegetation, of course, might have been quite different, judging from major changes in precipitation recorded in the speleothems from Peqi'in Cave (Bar-Matthews et al. 2003).

The Valley of Akko is a small rift limited by faults along the northern edge of the Mount Carmel block and the Nahal Hilazon fault line. The valley's contact with the eastern hills is an undulating line, due to the series of ridges, each limited by transverse faults. These ridges descend westward, and the widening wadi passes between them to create a landscape of flat alluvial land. The Akko Valley was filled by Pliocene and Pleistocene sediments, and similar sediments continued to accumulate into the late Holocene. Dune movements along the Mediterranean shoreline periodically blocked the wadi outlets to the sea, so that marshes formed. Even without a detailed study of the area, it

seems likely that these flat valleys were covered by open forests and a few small ponds during the Pleistocene, providing habitats suitable for gazelles, deer, and many other animals.

The hilly area east and south of Hayonim Cave is defined by the western edge of Biq'at Beit Hakerem, the Ridge of Har Gilon, and the gorge of Nahal Hilazon, the last of which captured the original Nahal Meged drainage basin and caused the Meged Valley to be open on both its western and eastern sides. The absence of water in the valley today is probably the result of modern drilling into the Cenomanian-Turonian aquifer at the Galilee foothills, in combination with the relatively dry late Holocene climate and the limited catchment area of the drainage.

HAYONIM CAVE AND ITS EXCAVATION

Hayonim Cave (figs. 2.1 and 2.2), situated on the right bank of the Wadi Meged, was formed by karstic activity in hard limestone beds of an eroded Upper Cenomanian shelf in the Yanuch Formation (Freund 1978). The cave originally included at least four solution chambers, only two of which, chambers I and IV, remain intact (fig. 2.3). The extant chamber adjacent to the one in which the excavations were carried out has been used recently as a goat pen. The remnants of the two other chambers (II and III) contain eroded brecciated deposits in which some Mousterian artifacts are visible. A fifth chamber may have existed on the eastern side of the complex, as evidenced by rockfall scars and isolated patches of breccia. The entire area of the complex is not yet known.

The size of the excavated chamber is about 150 square meters inside the drip line (see fig. 1.2). A series of man-made terraces lie in front of the cave, planted since 1968 with olive groves that stretch down to the valley. The top two terraces contain the remains of a Geometric Kebaran station, a large Natufian site known as Hayonim Terrace, and material dating to the Pottery Neolithic (Henry et al. 1981; Khalaily et al. 1993; Valla et al. 1991).

The goals of the 1965–1979 excavations at Hayonim Cave were in line with the traditional questions of regional archaeology (Bar-Yosef and Goren 1973; Bar-Yosef and Tchernov 1967). We sought mainly to achieve the following:

1. Define a well-excavated cave sequence through the systematic use of modern fieldwork techniques, including piece-plotting and careful dry and wet sieving;
2. Provide a cave sequence in the western Galilee that would facilitate cultural correlations between the caves of Mount Carmel and those along the Lebanese coast (e.g., Ksar 'Akil);
3. Retrieve microfaunal assemblages for identifying a biochronological sequence alongside the history of Paleolithic human activities. These data were to be used, in the absence of radiometric dates (and complementary to them as dates became available), as a means of establishing a relative chronology for the cultural layers.

We uncovered Natufian, Kebaran, Aurignacian, and Mousterian layers in Hayonim Cave (fig. 2.4) while also recognizing major gaps in the cultural sequence. The microfaunal assemblages from a depth of circa 2.5–5.5 m below datum (bd) allowed Eitan Tchernov to explore the relative chronology of the Mousterian layers by identifying discrete "biozones" and to present his overview at the first "Préhistoire du Levant" meeting in Lyon, France, in June 1980 (Tchernov 1981).

The 1980s saw new interest in the Middle Paleolithic, mainly in terms of the double-edged debate of modern human origins and the demise of the Nean-

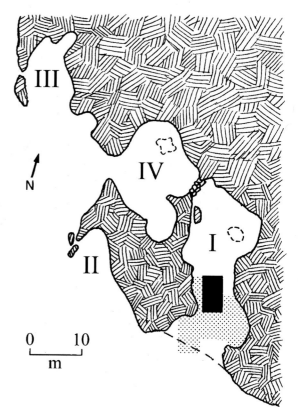

Figure 2.3
Simplified plan of the Hayonim cave chambers.

dertals. Following the large field project at Kebara Cave on Mount Carmel (Bar-Yosef et al. 1988, 1992) and the dating of hominid-bearing layers at Qafzeh Cave (Valladas et al. 1988), returning to excavate in Hayonim Cave became essential for the completion of chronological controls for the regional cultural sequence. Thus, a new project was formulated.

The objectives of the 1992–2000 excavations at Hayonim Cave were to uncover and study the early Middle Paleolithic sequence. Previous observations had made it clear that the uppermost Mousterian deposits (so-called upper Layer E) contained an industry resembling the one from Qafzeh Cave, and the material retrieved from a deep sounding next to the western wall contained numerous elongated blanks similar to those of the Tabun D assemblage of Tabun Cave, which we describe later. Hence, the goal was to continue systematically the excavation of the Mousterian layers and to date them by thermoluminescence (TL), electron spin resonance (ESR), and other techniques. On the whole, radiocarbon dates, TL, ESR, and uranium/ thorium (U-Th) readings indicated that the exposed sequence at Hayonim Cave ranged from possibly 250

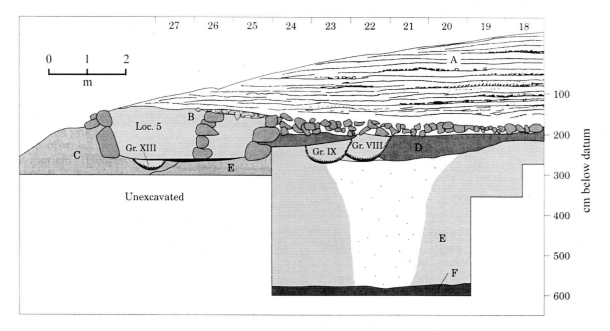

Figure 2.4
North-south cross section of the central trench in Hayonim Cave, showing the positions of the Levantine Aurignacian (D) and younger cultural deposits dating to the Kebaran (C) and Natufian (B) above Mousterian Layers E and F. Graves (Gr.) are of Natufian age. Mousterian Layer F contained little bone of any sort, but many proveniences in Layer E, apart from the white V-shaped area, were rich in bone. The Aurignacian layer, excavated many years ago, was uniformly rich in bone, in contrast to the Mousterian layer below it. There is evidence of an erosional unconformity separating the youngest Mousterian layer and the Aurignacian, a hiatus in deposition that lasted roughly 50,000–70,000 years.

KYA through historic times, with several major temporal gaps therein.

Because the early Mousterian faunas preserved in Hayonim Cave are the only large samples available for the region, we also sought to increase the size of the excavated, stratified faunal assemblages and to document fully the well-preserved hearth areas in the cave's interior. Hayonim Cave also presented us with a unique laboratory in which to pursue issues of diagenesis and site formation processes in detail—stimulated initially by our work at Kebara Cave and employing the same techniques for recording mineralogy and micromorphology. The perspectives provided by these scientific approaches proved especially important for understanding potential distortions in the radiometric dating results and the differential preservation of bones, mollusk shells, and wood ash in the sediments.

The excavation trenches in the Mousterian deposits were situated near the cave entrance (the deep sounding) and in the central area inside the chamber (see fig. 1.2). Three-dimensional mineralogical mapping of the deposits demonstrated the uneven effects of diagenesis in the central area, mainly in a patch running diagonally across it (chapter 4). There, diagenesis had reduced the preservation of bones and mollusk shells but had not affected the lithic distributions. The best preservation conditions for bones were apparent in Mousterian (MP) units 3–5, where most of the well-defined hearth areas were also found (fig. 2.5). Although generally thinner, the hearth features resembled those identified in Kebara Cave (Meignen et al. 1989, 2001; Schiegl et al. 1994, 1996; Weiner et al. 2002) in that they frequently were stacked (fig. 2.6). Micromorphological and geochemical analyses in Hayonim Cave indicate that the majority of the sediments were anthropogenic in origin. Thus, conclusions about past paleoclimatic conditions are difficult to draw from these data, with the exceptions of the unconformity between Layers E and F, a possible unconformity between Layers F and G, and the precipitation of the speleothems and travertines in the uppermost part of Layer E.

EXCAVATION TECHNIQUES

The entire surface of Hayonim Cave was superimposed with a 1 × 1 m grid (see fig. 1.2), and each square meter was subdivided into four quadrants (a, b, c, d). If no structural remains or changes in color were encountered, then arbitrary excavation units were 5 cm thick. All lithic pieces larger than 2 cm were plotted in three dimensions (northing, easting, and depth below datum), as were most of the bones larger than 3–4 cm. The datum was fixed as an arbitrary zero-point on the cave's wall. All sediments were dry- and wet-sieved progressively through screens with 3 mm and 1 mm mesh, and small bone fragments, microfauna, fish and legless lizard scales, shells, and flint chips were retrieved by hand. Flotation techniques were used in an effort to recover floral remains, but few or none were found in most instances. The same recovery techniques were employed at Hayonim Cave and Meged Rockshelter, although few of the sediments from the latter site were wet-screened. We are confident that all the faunal and artifact collections were recovered completely and therefore can be compared with one another without reservation.

STRATIGRAPHY

The excavations in Hayonim Cave (1965–2000) documented the following stratigraphy:

Layer A
Radiocarbon dates, one coin from the second century A.D., a small collection of Roman-Byzantine sherds, and a glass furnace suggest that these ashy deposits, about 3 m thick, span the years from approximately A.D. 200 to the present. The cave was occupied most recently by shepherds and their flocks, mainly in the winter (late October to late April). The ashy deposits were the result of herders burning the uppermost dung layers to clear the shelters of organic detritus every 10 to 20 years.

Layer B
The Natufian complex included a series of built-up rooms, graves, and domestic rubbish stretching across the lighted part of the cave chamber near the entrance. This layer varied in thickness from 0.1 to 1.2 m and was subdivided stratigraphically into five phases of deposition. On the basis of lithic typological comparisons, we assume that the main occupation took place during the Early Natufian, with a more ephemeral occupation in the Late Natufian. The overall age of the Natufian culture in the Levant ranges from 14,800 to 11,600 calibrated radiocarbon years B.P. (Bar-Yosef 2002). Charcoal and other datable substances were rare, but two radiocarbon dates on seeds from locus 4 produced readings of (OxA-742) 12,360 ± 160 and (OxA-743) 12,010 ± 180 b.p. (Hopf and Bar-Yosef 1987), calibrated as 13,403–12,370 B.P. with 1 standard deviation (INTCAL98.calib 4.1).

Figure 2.5
Isolated Mousterian hearth feature (approximately 1.2 m in diameter) in Layer E of the central excavation trench, viewed from above.

Layer C
A Kebaran layer about 2.5 m thick was confined to the entrance of the cave in a matrix that consisted of a loose, reddish, granular silt and clay, locally cemented with calcite and rich in bones. The sediments were largely a mixture of reworked deposits derived from the destruction of the Mousterian layers—especially at the base of Layer C—and sediment accumulation during the Kebaran period. These early parts of the accumulation were formed in an elongated post-Mousterian erosional depression, or trough, located under the drip line at the cave entrance. No datable substances were found in this layer, but the assumed age of the Kebaran, based on comparisons with other dated sites and lithic assemblages, is estimated as 19,000–16,000 cal B.P.

Figure 2.6
Mousterian hearth lenses exposed in the north wall just west of travertine deposits in the central excavation trench, Hayonim Cave.

Layer D

Mostly confined to the central area, this deposit was 0.35–0.55 m thick and composed of an ashy matrix with bone middens and a few well-structured hearths. Radiocarbon dates indicate ages of (OxA-2802) 27,200 ± 600, (OxA-2801) 28,900 ± 650, and (OxA-2805) 29,980 ± 720 b.p. for this layer and its Levantine Aurignacian lithic industry (Belfer-Cohen and Bar-Yosef 1981; Bar-Yosef and Belfer-Cohen 1996). All of the Aurignacian deposit was removed during the first excavation campaign in Hayonim Cave.

Layer E

This layer consisted of predominantly anthropogenic sediments (hearths, ashes, organic matter), but it also contained locally well-defined, clay-rich geogenic units. The sediments exhibited generally reddish and brownish hues and occurred as thick, massive layers. Bone concentrations were rich but localized along the east and west margins of the central area of the excavation, in the upper part of the deep sounding, and laterally toward the cave entrance below the present drip line. The occurrences of travertine stalagmites mirrored this distribution. Elsewhere, phosphatization was extensive. The sediments generally lay horizontally, but the dip increased toward the rear of the cave; inclinations to the north and northwest reflect the presence of one or more sinkholes. The Mousterian industries in Layer E were generally of the Tabun D type, although the uppermost part contained a Tabun C–type assemblage.

Layer F

This layer was predominantly geogenic, consisting of extremely diagenetically altered clay, silica, and quartz silt. Where the layer was most clearly exposed in the deep sounding, some isolated burned areas and numerous pieces of scattered charcoal were identified. The sediments exhibited yellow and brown hues overall. The layer was massive within the cave, with localized laminated units, but at the entrance it was well bedded and increasingly finely laminated with depth. Phosphatization was widespread, as was concomitant opal cement and transformed clay. Calcite was generally absent except in the lower units, where several episodes of calcification and decalcification seem to have occurred. Very few bones were preserved in this layer, consistent with widespread indications of mineral diagenesis in the sediments. In the central area, the sediments were generally horizontal, although they dipped slightly toward a sinkhole in the rear part of the cave. In the deep sounding, the deposit dipped slightly in the upper part, again toward the interior of the cave. Numerous rodent burrows were found in the deep sounding. The Mousterian technology was generally of Tabun D type (Meignen 1998).

Layer G

This layer was partly exposed by the excavation that terminated in 2000. The upper part of the layer was pinkish gray, hard, gritty silt, rich in layered opaline seed coats and with little matrix between the fragments. Below this was a massive unit consisting of mottled brown and yellow-brown silt and seed coats, the latter in much lower abundance. Some charcoal flecks and abundant bird gastroliths were found in this unit, along with traces of millimeter-sized grains of gray silt. No bones were found in this layer. Reddish vein and void coating crystals were observed. All of these observations indicate extensive diagenetic alteration of the sediments and appear to account for the uniform absence of bone. The lithic industries are Acheulo-Yabrudian.

Given the mixed origins and histories of the deposits in Hayonim Cave, we concluded (Goldberg and Bar-Yosef 1998) that interactions between anthropogenic, biogenic (i.e., nonanthropic), and geogenic processes had obliterated any signals of climatic fluctuations. The only exceptions would be the ages of the unconformities, when water activity increased, and the ages of the fallen speleothems in the upper part of Layer E. The latter could be correlated with increasing humidity and precipitation during the final phase of oxygen isotope stage (OIS) 6, or after approximately 135,000–140,000 years ago (following Martinson et al. 1987).

MINERALOGY

Analysis of the microstructures of cave sediments contributes significantly to our understanding of site formation processes. The minerals and organic components that make up the bulk of the sediments are major sources of information in sites, though they are not always well explored. The potential pathways by which minerals may enter a cave include transport by natural agencies such as wind and water and transport in association with the activities of humans and animals. One example is the introduction of wood and other vegetal organics by humans as fuel for fires, as bedding, or as food. Hearths produce ash that is easily identifiable from mineralogical spectra (Schiegl et al. 1994, 1996; Weiner et al. 1995, 2002).

Insoluble silicate minerals such as quartz and clay that are derived from surrounding limestone or dolomite can accumulate essentially in situ during cave formation or the ensuing karstic activities associated with the cave aquifer, even though these minerals are scarce in the bedrock. Authigenic minerals may also be produced in the course of sediment accumulation, a result of some minerals or ions dissolving in certain chemical environments and then reforming into different compounds when the chemical environment changes. A common group of authigenic minerals is the phosphates. The phosphate component generally derives from organic material that breaks down through oxidation or the dissolution of bones. It is thought that wood ash may also be a source of phosphate. A second common group of authigenic minerals is represented by the iron and manganese oxides. Their formation is intimately related to the presence or absence of oxygen in the environment, which in turn is related to the presence or absence of organic matter. In the absence of oxygen, these oxides readily dissolve and migrate by diffusion, whereas in the presence of oxygen they precipitate as insoluble oxides. Perhaps the most common authigenic mineral in caves is calcite, which often forms secondarily within sediment pores or on the sediment surface as a cemented flowstone layer. Stalagmites are also composed of authigenic calcite, although they do not usually form within sediments.

Thermoluminescence, optical luminescence (OSL), and electron spin resonance dating techniques are all directly affected by the mineral assemblages of sediments, because those minerals are important components of the radiation dose from the environment. In some caves, including Hayonim, a major constituent of the sediments was ash produced by man-made fires. The plants that humans bring into caves as food and fuel often contain minerals, including silica in the form of phytoliths and other plant components, compounds that generally preserve for long periods. Thus, some of the mineral components of archaeological layers can be indirect products of human activities. The challenge at Hayonim Cave, as in our previous project at Kebara Cave, was to identify the anthropogenic signals in the cultural layers and to differentiate them from the multitude of independent chemical, biological, and geological processes that inevitably affect sediments. This was the focus of the mineralogical and micromorphological studies at Hayonim Cave.

Figure 2.7
Mousterian hearth areas separated by mineral dissolution fronts in Layer E in the central trench, viewed from above.

More than 2,100 sediment samples were collected and analyzed on-site using Fourier transform infrared (FTIR) spectrometry (Weiner et al. 2002). The analyses were carried out during two weeks of each excavation season. The methods used in the field and in the laboratory have been described in detail elsewhere (Schiegl et al. 1996). The minerals in the sediments of Hayonim and certain other Levantine caves tend to group spatially into two distinct assemblages—the calcite and dahllite (CD) assemblage (Weiner et al. 2002) and the leucophosphite, montgomeryite, variscite, and siliceous aggregate (LMVS) assemblage. The boundary between these two assemblages in certain units of Hayonim Cave is described in chapter 4, because it did not follow the layered stratigraphy. This was especially apparent in the upper eastern part of the section and therefore must have been the result of postdepositional changes (fig. 2.7). Both mineral assemblages presumably formed from the same parent deposit. The presence of abundant lens-shaped hearths and burned clay nodules in association with both mineral assemblages, as well as abundant siliceous aggregates and phytoliths, especially in the LMVS assemblage, is consistent with the bulk of this sediment's having originally contained fresh wood ash. Wood ash contains siliceous aggregates and phytoliths in small amounts (Albert et

al. 2003). That the associated phosphate minerals in both assemblages were not primary minerals but secondary products of dissolution and reprecipitation means that they, too, likely originated in the calcite component of wood ash.

It is significant that wood ash and ash-derived minerals were major constituents of the Mousterian sediments in the central area of Hayonim Cave. The relative proportion of ash decreased toward the cave entrance, and the amount of clay increased. Similar observations were made in Kebara Cave (Weiner et al. 1993) and Amud Cave (Hovers 1998). This phenomenon raises the question of how and why hearth features are preserved, although they are buried in unconsolidated ash. Observations of the Natufian sediments (Layer B) in Hayonim cave were informative in this regard.

In the Natufian layer, the bulk of undifferentiated sediments that filled most of the built-up rooms was composed mainly of a fine, powdery, unconsolidated calcitic ash. The hearths in each of these rounded structures were outlined by small rocks and often had a semirectangular shape. Within these hearths the ash was more consolidated than the surrounding bulk sediment and, judging from its white color in comparison with the gray color of the bulk sediment, was almost free of charcoal. These observations suggest that making fires repeatedly in the same location stabilizes some of the ash calcite that remains in the hearth, though most of the ash produced is dispersed around the hearth. It seems that the practice of making fires in the same place may also account for the preservation of hearth ash in the Mousterian layers, even in the absence of a circle of rocks to enclose and protect the feature. This relative, local stabilization of the ash matrix can persist even after the original calcite dissolves and reprecipitates into another mineral form.

Calcitic ash is relatively unstable chemically, but it was not the least stable mineral present in Hayonim Cave. Land snail shells are composed of another polymorph of calcium carbonate, called aragonite (Lowenstam and Weiner 1989), which is slightly more soluble than calcite. Monitoring the mineralogies of land snail shells demonstrated that in some areas where calcitic ash was preserved, the land snail shells either retained their original aragonitic mineralogy or were transformed in situ into calcite (see also chapter 4). Each of these transformations occurred in proximity to the travertines. At some distance from the travertine, but still within the CD zone, some of the shells had been transformed to dahllite. Despite this second round of dissolution and reprecipitation, the shells retained their original morphology. It is interesting to note that such pseudomorphs of snail shells can be found, whereas in our experience this almost never occurs for bone transforming into another mineral phase (Karkanas et al. 2000). The explanation may be partly related to the small crystal size of bone mineral as compared with that of snail shells.

SITE FORMATION PROCESSES

Paleolithic caves and rockshelters have long been subjects of archaeological and geological research in the Levant. Caves tend to preserve the contexts of material culture in stratigraphic entities. Most of the Levantine caves, however, have such complex stratigraphies that field observations alone cannot unequivocally untangle the depositional processes. Additional information can be obtained from a combination of micromorphology and mineral analyses. But even with increasing attention to the details of depositional and postdepositional processes—whether geogenic, biogenic, or anthropogenic—they are rarely the focus of comprehensive investigation. At Hayonim Cave, as previously in Kebara Cave, we investigated the role of those underlying processes that significantly influenced the archaeological record and its interpretation.

Here we focus on three major issues raised by the Mousterian sequence at Hayonim Cave: erosional features, clay-rich layers, and paleoclimatic implications.

One of the most prominent stratigraphic phenomena in Hayonim Cave was its erosional unconformities. We will not discuss the erosional event that occurred after the Mousterian and accounts for a cultural gap from about 80–100 KYA to the beginning of the Aurignacian occupation (ca. 29–27 KYA). Instead, we focus on the erosion that separated Layers E and F. In the main excavation, this unconformity surface dipped some 50 cm to the north and had a local topographic relief of about 30 cm. In the deep sounding, near the cave mouth, the unconformity surface was expressed as a channel almost 1 m deep. This was surely the result of active water movement under the cave's drip line. Apparently, after the accumulation of Layer F, a period of erosion by water occurred and the sediments were scoured away, especially in the central area and under the drip line. The altered mineral assemblage that testifies to this water activity had a thickness of about 30 cm at the northern end of the deposit and was underlain by normal clays. In the southern part, the thickness was 50 to 60 cm or more, and in the deep sounding at the entrance, the altered assemblage was present throughout the exposed section over a thickness of more than 2 m. This implies that the agents responsible for the erosion and alteration of the mineral phases were much more aggressive near the cave entrance than they were within the cave, presumably because of wetter conditions at the entrance.

It is difficult to determine what sorts of climatic conditions might account for the mineral transformations in the early layers at Hayonim Cave. The only analogous situation of which we are aware is the formation of laterite, which occurs mainly but not exclusively in the tropics, presumably as a function of seasonal high rainfall and elevated temperatures. It is also interesting to note that a similar mineral transformation process appears to have occurred between the C and D layers of Tabun Cave (Margaris 2000). Preliminary dating of the erosional unconformity between Layers E and F in Hayonim Cave to around 200,000 years ago is consistent with the possibility that the same climatic event was responsible for the erosion in Tabun Cave between Layers C and D. Layer C of Tabun Cave has been dated to about 170 ± 17 (165 ± 16) KYA, and the top of Layer D to about 212 ± 22 (196 ± 21) KYA (Mercier and Valladas 2003b; Mercier et al. 1995b, n.d.).

Two rich clay layers were identified in the sedimentary sequence of Layer E of Hayonim Cave. They represent marker horizons that were part of the stratigraphy and not the results of postdepositional processes. As noted earlier, both layers lay more or less horizontally, although they dipped toward the rear (north end) of the cave. Their existence could have been a consequence of local climatic effects that resulted in more clay penetrating into the cave, or they could have been due to less ash or other anthropogenic materials having been introduced into the cave by the human occupants.

From a purely mineralogical point of view, we could envisage three broadly different "local" climatic regimes affecting the accumulation of sediments in Hayonim Layers E and F. The oldest regime produced the erosional unconformity between Layers F and E, which must have been tied to subsidence in the rear of the cave. That is, subsidence created the relief and network needed to drain water associated with the substantial erosion at the top of Layer F toward the back. In addition, it led to the transformation and breakdown of clays. This was followed by a period of extensive travertine formation in a mild hydrological regime (upper part of Layer E). The hydrologic activity became increasingly aggressive, resulting in more LMVS mineral assemblage production. Because the top of Mousterian Layer E was overlain by another erosional unconformity, a fourth change in climatic conditions in the Mousterian must have occurred between Layers E and D.

Dating the Cultural Sequence in Hayonim Cave

Despite the abundant ash remains in the Natufian layer (Layer B), the only samples of datable material there were a few seeds from locus 4 (Hopf and Bar-Yosef 1987). The dates of these seeds reflect the age of the Early Natufian, but only as a range of 15,300 to 13,800 cal B.P. (Calib.4.1; Stuiver et al. 1998). Although formal discussion of the Natufian site situated on the terrace just in front of Hayonim Cave (a.k.a. Hayonim Terrace) is beyond the scope of this chapter, it is worth noting some temporal overlap between the younger Natufian contexts inside the cave and the earliest dates for the Natufian of the terrace (Valla et al. 1991).

No radiometric dates are available for the Kebaran component in Hayonim Cave (Layer C), but on the basis of dates for other Kebaran sites (reviewed by Byrd 1994), as well as typological considerations, Hayonim Layer C is estimated to date between about 19,000 and 16,000 cal B.P. The few available dates for the Aurignacian of Layer D indicate an age range of 27–29 KYA (Bar-Yosef 2000; Belfer-Cohen and Bar-Yosef 1981).

The Middle Paleolithic deposits were dated mainly by TL and ESR techniques on burned flints and ungulate teeth, respectively. Mineralogical studies qualified the calculation of the radiation dose emitted by the surrounding sediments, a parameter essential to the date calculations. This background dose is known as the environmental radiation input and can extend up to 50 cm from the object being dated (Rink 2001). The major sources of the radiation are uranium, thorium, and the radioactive isotope of potassium. These elements are found in different sedimentary minerals. The siliceous aggregates contain relatively large amounts of potassium, and direct measurements of the gamma radiation flux using dosimeters planted in the sediments showed that the LMVS assemblages contained flux about twice as great as that of the CD assemblages (around 1,000 grays per year vs. 500 grays; Mercier et al. 1995a).

This finding had two important implications. First, because the mineral assemblages were transformed in the past, measurements of the flux today might not represent ancient conditions. In a study of diagenesis in another site, Karkanas et al. (2002) concluded that many of the mineralogical changes in the sediments took place at the surface or just below the surface soon after deposition. If this is correct, then the changing mineralogical environments should not be a major problem for dating, because the environmental dose has been the same for almost the entire burial history of the samples chosen for dating. In Hayonim Cave, however, certain boundaries between the two mineral assemblages—CD and LMVS—contained a form of transforming dahllite and hence could have been in a state of flux more recently, an added complication for the dating.

The second implication of the findings was the necessity of identifying the boundaries between the CD and LMVS assemblages for placing the dosimeters, to ensure that they were in the same mineralogical environment as the samples selected for dating. Failing to do this would have produced erroneous dating results. In anticipation of collecting new samples in each succeeding excavation year, we placed many dosimeters at the end of a field season. That is, when suitable teeth or burned flints were exposed, a dosimeter was placed nearby in the sediments, to be retrieved one year later.

It should be noted that the TL and ESR dates for the Mousterian of Hayonim Cave are given as averages. The TL dates are a selection of 77 samples from more than 100 available readings, chosen for reliability on the basis of the mineralogical information reflecting the degree of local diagenesis. The uppermost portion of Mousterian Layer E is undated, and a U-Th date on a fallen stalagmite in the northeast corner of the central excavation area indicates that MP unit 2 of Layer E must be younger than 153 KYA (Mercier et al. n.d.; Rink et al. 2004). Most of the Mousterian deposits, however, date between 130 and 220 KYA, supporting the original contention that the age of the upper part of Hayonim Layer E is globally comparable to Tabun Cave's Layer C, and the lower part of Hayonim Layer E, along with Layer F, is comparable to Tabun Cave's Layer D (Mercier et al. 1995a, 1995b). Finally, certain ambiguities concerning the ESR and TL results were attributed to the effects of postdepositional diagenetic processes on dose rates in Layers F and E, especially in the range dating to the older glacial and interglacial cycles (OIS 5–7).

HAYONIM CAVE: INTERIM SUMMARY OF THE CULTURAL SEQUENCE

Each of Hayonim Cave's stratigraphic units produced a different picture, reflecting differences in lifeways through the ages. The following is a brief overview.

THE NATUFIAN, KEBARAN, AND AURIGNACIAN OCCUPATIONS

The main Natufian occupation (Bar-Yosef 1991; Belfer-Cohen 1988a, 1988b; Munro 2001; fig. 2.4) was marked by the building of a series of rounded or oval rooms from undressed stones obtained in the immediate vicinity of the cave. Their volume and total weight reflect the investment of an incalculable amount of energy in comparison with dwellings of earlier sites in the Levant. The built-up rooms had diameters of 2.0 to 2.5 m, and the width of the joint walls was about 1.0 m. The walls were preserved to a height of 0.60–0.70 m, and no clearly defined entrances were found. Paved floors and built hearths were uncovered in most of the built-up loci designated 4, 5, 7, 8, and 11. Structures such as loci 4 and 5 were paved in the original building phase; this seems to have occurred later in locus 7.

It is difficult to ascertain all the ways in which these loci were used by the Natufian inhabitants. Many of the finds were of a domestic character, but the small diameters of the structures often precluded considering them as dwellings. In certain cases it was clear that special activities had taken place inside; for example,

the evidence from locus 4 showed that it had served, following two occupational horizons, as a small lime kiln and later as a bone tool workshop. A number of incised slabs were incorporated into the paved floors of the lower levels of the same structure. Similar phenomena were observed in loci 8 and 10 (Bar-Yosef and Belfer-Cohen 1998; Marshack 1997). Oddly, such activities are unknown from the domestic structures of Ain Mallaha (Eynan) and Wadi Hammeh 27 (Edwards 1991; Valla 1991).

The presence of hearths in most of the Natufian structures in Hayonim Cave could be interpreted as a sign of domestic activities. In one case (locus 4), two hearths were built one on top of the other, whereas in locus 8 two were placed at opposite sides of the structure, making movement inside the room rather awkward. The most complicated stratigraphy was encountered in locus 4, including two phases of paving and two stages of a built hearth. Some lupine seeds were recovered from the hearths and floors of this room. It was after two paved-floor phases that this room served as a kiln for burning lime (Kingery et al. 1988). Severe charring of the limestone slabs that defined the northern interior of the locus supports this interpretation. No clear evidence for the use of plaster was found in the cave, but a few porous surfaces of stone pestles indicated that some crushing of the burned limestone was done in the cave or nearby. Although some small objects may have been made with the lime plaster (Belfer-Cohen 1991), it seems likely that the pounded lime was transported and used elsewhere, perhaps in building activities on Hayonim Terrace just beyond the cave entrance. The workshop for bone tool production lay above the burnt lime in locus 4, and the final deposits indicated that the room was abandoned and in-filled by collapsed material.

Most of the Natufian loci in Hayonim Cave were abandoned after their initial use and filled by the collapsed upper courses of their walls. Later, graves were dug into them. Yet there were also earlier burials, predating most of the loci just described and tied to the early Natufian phase. These burials included some decorated tools and ornaments (Belfer-Cohen 1988a, 1995). The Late Natufian occupation, on the other hand, consisted mainly of burials, possibly indicating that the cave functioned essentially as a cemetery, although there is evidence for other human activities along the northern cave wall, including three major caches. One cache contained a series of basalt pestles; another, half-worked bovid ribs; and the third, quartzite hammer stones and large quantities of dentalia (tusk shell) beads. Unfortunately, we can only speculate about the nature of the latest Natufian occupation in the cave, because of the disturbance caused in the second to third centuries A.D. when the top of the Natufian layer was flattened and a glass furnace built over it.

The finds from the various Natufian phases are typical of a large base camp in the Natufian "homeland" and include evidence of extensive game processing (Bar-Yosef and Belfer-Cohen 2002; Munro 2001). There are many ground stone tools, and the inventory is typical of the Natufian in general: mortars, pestles, mullers, whetstones, and shaft straighteners. These tools were made from basalt and limestone and were brought into the cave as complete items. Of particular note are the heavy boulder mortars and some decorated smaller ones. Some of the basalt pestles bear evidence of lime pounding as well as grinding of ochre. The chipped stone assemblages incorporate the characteristic tool types, namely, geometric microliths and lunates, predominantly of the Helwan variety in the early phases and the backed variety in the later phases. There are sickle blades and bifacial tools, forerunners of Neolithic tools. All in all, the number of flint artifacts is quite low in comparison with those from other Natufian sites—even Hayonim Terrace immediately outside the cave. This observation suggests to us that the cave was reserved for special purposes to a significant degree.

Also present in the collection are many bone tools of diverse types, some uniquely decorated (Bar-Yosef 2002; Bar-Yosef and Belfer-Cohen 1999; Munro 2001). Most of the bone tools were used for hide working and basketry (Campana 1989). Some of them were produced on-site, as evidenced particularly in locus 4 and more generally by the working debris recovered among the faunal remains (see Munro 2001, among others); such debris included girdle-and-snapped long bone ends, notched or polished pieces, and severed gazelle horn cores. Quite a number of bone pendants were recovered as well, especially from grave contexts (e.g., graves VII, XIII, and XVII), all of which pertained to the early phases of the Natufian occupation. *Dentalium* shell beads were also common among the ornaments. It is interesting that bone beads and pendants were abundant in Hayonim Cave but rarer in other sites. Intersite variability in the character and number of decorative elements is almost certainly related to a combination of site function and group identities within the Natufian as a whole (Belfer-Cohen 1991; Belfer-Cohen and Bar-Yosef 2000).

Our description of the Natufian would be incomplete without some mention of the terrace site in front of the cave. In spite of intensive research at this locality, no Natufian deposit unequivocally links the cave chamber and the terrace as a single site, as was observed for el-Wad Cave and its terrace on Mount Carmel (Garrod and Bate 1937). Published descriptions of the stratigraphic record of Hayonim Terrace seem to indicate that its Natufian occupation dated only to the later part of the Natufian culture period. Its total duration relative to that of the occupation inside Hayonim Cave is unknown (Henry et al. 1981; Valla et al. 1991). The exposed architectural remains in the sediments of Hayonim Terrace indicate less intensive building activity there than inside the cave, although the foundations of several structures were identified. No slab-covered floors or built hearths were uncovered, but there were several deep mortars, one of which was still upright at the time of discovery. Several graves were found on the terrace, the most striking of which yielded a joint burial of adult humans and a young dog (Tchernov and Valla 1997).

The Kebaran remains in Layer C seem to represent several short-term, ephemeral occupations by foragers, perhaps during summer, if only because the area beneath the drip line where the Kebaran-aged materials were concentrated would have been too wet in winter in the later period. The considerable time span of the Kebaran deposits is reflected in the typological changes among the artifacts excavated from this layer (for details see Bar-Yosef 1970, 1991). Core reduction strategies were essentially the same throughout the entire deposit. Typologically, however, there was a shift from bottom to top, with a dominance of curved, retouched bladelets giving way to a dominance of the obliquely truncated backed bladelets that are most typical among Kebaran microliths in general. This industry lacked the microgravette points that appeared earlier in the Kebaran and were found in the upper layer (early Kebaran) of Meged Rockshelter.

The Aurignacian occupation included a dense hunters' midden. A wide variety of activities was conducted on-site, as indicated, for example, by a slab with an engraving of a "horse" and by well-defined, deep hearths (Belfer-Cohen and Bar-Yosef 1981; Marshack 1997) associated with ochre-stained rock slabs (Wreschner 1976 and unpublished observations). In the lithic industry, retouched pieces were numerous and simple debitage products relatively uncommon, suggesting that little blank production took place in the cave. It seems that the Aurignacian foragers generally carried tools in from other places, perhaps including the coastal plain.

THE MOUSTERIAN IN THE LEVANTINE CONTEXT

Most current interpretations of human origins agree that essentially modern-looking humans emerged in sub-Saharan Africa sometime between 300 and 100 KYA, later dispersing across Eurasia. Very early dispersals of anatomically modern humans into Eurasia cannot, however, be correlated with the emergence of Upper Paleolithic cultures, because available evidence for the behavioral changes associated with the Upper Paleolithic greatly postdates the skeletons that bear modern appearance, such as those of the Skhūl-Qafzeh group of Israel (e.g., Klein 1999). These were dated to about 110–90 KYA (Valladas et al. 1988; Mercier et al. 1995b), whereas the earliest Upper Paleolithic cultures or assemblages appeared in the region sometime between 40 and 50 KYA (e.g., Bar-Yosef 2000; Kuhn et al. 2001). Only a few fragments of hominid skeletons have been found in the early Mousterian deposits of Hayonim Cave (Arensburg et al. 1990), and thus we can deal only with questions of hominid behavior and cultural traditions at this site. Both Neandertals and the earliest anatomically modern humans in the Levant used Mousterian technology, and so it is worth exploring in greater depth the nature of variation—or the lack of it—in these industries. Hayonim Cave provides a rare window on the early Mousterian in conjunction with evidence of hominid foraging practices.

The idea of using the long stratigraphic sequence of Tabun Cave for ordering the Middle Paleolithic industry types of the central and southern Levant has been proposed by several scholars (e.g., Bar-Yosef 1989a, 1994, 1996; Copeland 1975; Jelinek 1981, 1982; Ronen 1979) and is adopted here. According to this approach, some variability must be acknowledged among the assemblages used to define each of the pertinent industrial groups (Hovers 1998; Meignen 1998). The causes of such variation are not yet determined (but see Hovers 1997; Meignen and Bar-Yosef 1991). Possible explanations include the economic trade-off between expediency and more systematic core reduction, temporary shortages of raw material, and shifts in the mobility and land use practices of human groups (Kuhn 1995). Several persistent and widely distributed *chaînes opératoires* (operational sequences) are recognized in the Levant. Three general entities, however, dominate a growing number of dated Middle Paleolithic sites and artifact assemblages that have been studied. These entities are as follows, from youngest to oldest.

Levallois industries dominated by the production of short triangular blanks (Tabun B–type)

These industries are characterized by products that were removed mainly from unidirectional, convergent, and recurrent Levallois cores. Typical products are triangular in shape, including remarkable broad-based Levallois points as well as flakes and blades. This type of industry, comprising internal variants (Meignen 1998), was characteristic of Kebara Cave (Meignen 1995; Meignen and Bar-Yosef 1991), Amud Cave (Hovers 1998; Ohnuma 1992), Tor Faraj, Tor Sabiha (Henry 1998a), and Ksar 'Akil XXVIII (Meignen and Bar-Yosef 1992).

Levallois industries dominated by the production of short blanks, mostly oval to rectangular (Tabun C–type)

This type of industry is characterized by large flakes struck from Levallois cores via centripetal or bidirectional exploitation. Triangular points appeared in small numbers in certain horizons such as the top of Layer C in Tabun Cave (Garrod and Bate 1937) and layer XV in Qafzeh Cave (Hovers 1997). Tabun C–type industry was common in Skhūl and Naamé (Fleisch 1970), as well as in Hayonim Cave, upper Layer E. The range of TL dates from 170/150 KYA to 90/85 KYA at Tabun Cave is supported by ESR readings (Bar-Yosef 1995, 1996) and the new TL dates from Hayonim Cave.

Industries dominated by the production of elongated blanks (Tabun D–type)

These industries are composed of products obtained from essentially unidirectionally flaked cores, with limited evidence for bidirectional flaking. Core reduction is often Levallois in form but can be non-Levallois in certain assemblages (Marks and Monigal 1995; Meignen 1994, 2000). The blanks are classified as blades and elongated points. In some cases, the core structure and the presence of crested blades and diagnostic core-trimming elements indicate a reduction strategy based on a volumetric concept resulting in a semiprismatic core geometry close to that of Upper Paleolithic blade industries (Meignen 1998, 2000). Often, Tabun D–type assemblages contain higher frequencies of retouched pieces than do later Mousterian industries. These retouched tools include elongated points, racloirs, and burins. The Tabun D–type industry has been found in a number of sites aside from Hayonim Cave, including Tor Abu Sif and Sahba in the Judean desert (Gordon 1993; Meignen 1994, 1998; Neuville 1951); Rosh Ein Mor and Nahal Aqev 3 in the Negev highlands (Crew 1976; Marks and Monigal

1995); Ain Difla in Wadi Hasa (Clark et al. 1988; Lindly and Clark 1987); Yabrud I (Solecki and Solecki 1995); and Jerf Ajla (Richter et al. 2001; Schroeder 1969) and Douara Cave (Akazawa 1979; Nishiaki 1989) in the Palmyra basin. The Hummalian is a similar industry but is considered to be exclusively non-Levalloisian (Boëda and Muhesen 1993; Copeland and Hours 1983; Meignen 1994).

Tabun D–type assemblages are the earliest Middle Paleolithic industry known in the Levant and correspond to the phase in human evolution in this region that was least known prior to the recent studies at Hayonim Cave. Dating the Tabun D–type industry has proved to be complicated. TL dates in Tabun Cave—the namesake for Tabun D–type assemblages—indicate a time span of 270–180 KYA (Mercier and Valladas 2003b; Mercier et al. 1995b; Rink et al. 2003). Preliminary TL and ESR readings from Hayonim Cave support the findings from Tabun Cave (Mercier and Valladas 2003a; Rink et al. 2004).

Having laid out the context of temporal variation in the character of Levantine Mousterian industries, let us now discuss a number of important features of the lithic assemblages from Layers E and F of Hayonim Cave in particular.

The lithic studies associated with the Hayonim excavations employed the technological approach, following the concept of the *chaîne opératoire,* or operational sequence, an analytical tool developed by social anthropologists specializing in the anthropology of technology (e.g., Lemonnier 1992; Pfaffenberger 1992). Examples of *chaînes opératoires* have been recorded in ethnographic studies, historical documents, and analyses of technologies still in use. In lithic studies, identifying the different steps that define the *chaîne opératoire* reveals the knapping concept employed and the dynamic process of tool making—the kinds of blanks that were made, selected, and used, especially those that ended up being retouched (e.g., Bar-Yosef and Meignen 1992; Boëda et al. 1990; Meignen 1995). Comparing dated archaeological assemblages that exhibit similar *chaînes opératoires* within a known territory may ultimately help to elucidate the histories of loosely defined "social groups" and the effects of mobility patterns on assemblage formation (e.g., Conard 2001 and papers therein; Kuhn 1992a, 1995).

The total number of piece-plotted Mousterian stone artifacts from Hayonim Cave exceeds 25,000. The distribution of lithic artifacts varied with depth, such that Layer E could be subdivided into several units (see also chapter 4, figs. 4.16–4.17). The upper assemblages of Layer E were produced by the Levallois method. The desired blanks were flakes, sometimes of large sizes, obtained mostly by centripetal and unidirectional removals from recurrent Levallois cores. Elongated flakes or blades and a few points also occurred, but always in low frequencies. These assemblages resemble those of Qafzeh Cave, often labeled Tabun C–type (Hovers 1997; Meignen 1998).

The assemblages of lower Layer E and Layer F in Hayonim Cave are characterized by increasing frequencies of elongated products, whether in the form of blades or points. Many of the elongated pieces are retouched on both edges, often forming a point (the so-called Abu Sif point; Bordes 1961). In some cases the points are incurvate and resemble those described by Neuville (1951) for Abu Sif. Some of these elongated products are attributed to the Levallois method, but a portion must instead be assigned to a distinct system of debitage (non-Levallois, if Bordesian terminology is used) called the "laminar debitage system" (Meignen 1998:176). This basic knapping method is closer to what is known from the Upper Paleolithic. The volumetric concept of such cores does not follow the Levallois method, wherein the debitage is organized along the larger face of the block and follows a series of more or less successive parallel splitting planes, with the debitage surface of the core successively prepared and reprepared (Boëda 1995; Meignen 1995). Rather, the concept resembles that which underpinned later industries in which cores, after specific shaping, were reduced in a continuous fashion from one or two striking platforms. In this case, the debitage process follows the maximum length of the block.

Delage's (2001) pioneering study of flint and chert sources in the Galilee demonstrated that most of the raw material used during the Mousterian period at Hayonim Cave was brought from within a 10 km radius of the site, with a few pieces originating in outcrops about 20 km away and other rare raw materials from as far away as 30 to 40 km (Delage et al. 2000).

The contents of Mousterian Layers E and F are most interesting if placed in the context of other studies of Mousterian settlement patterns and inferred degrees of mobility, taking into account observed variation in site sizes and the intensity of occupations, site function, and the exploitation of food resources and raw materials. An important measure of site utilization is the intensity of occupation. This can be measured in relative terms if good information is available concerning dates, sediment volumes, numbers of artifacts (e.g., Bar-Yosef 1983, 1998), and, to the extent possible, seasonal activities and hunting sustainability as interpreted from faunal remains.

Figure 2.8
Meged Rockshelter during excavation season.

The evidence from Kebara Cave, dated at 65,000–48,000 B.P., may provide a scale against which to evaluate the evidence from Hayonim Cave. In the central area of Kebara Cave, where most hearths were located, 1 cubic meter of sediment accumulated over approximately 3,000 years (judging from TL dates) and contained about 1,000–1,200 lithic pieces larger than 2 cm (Bar-Yosef 1998, 2000). Areas near the northern wall of the cave yielded similar quantities of lithics (with a distinct increase in the numbers of cortical elements and cores) and ash, but no hearths. The rates of material accumulation indicate relatively intensive use of this site, especially during the middle part of the late Mousterian sequence (see Speth and Tchernov 2001, 2002, n.d.). Other evidence for intensive human use of Kebara Cave includes the rarity of rodent bones there. In contrast, the same volume of sediment required more than 10,000–15,000 TL years to form in Layers E and F of Hayonim Cave, with fewer lithic pieces and larger quantities of microvertebrate remains. Similarly small numbers of artifacts and high concentrations of microvertebrate remains may reflect ephemeral early Mousterian occupations in Layers 12

and 13 of Sefunim Cave (Ronen 1984), Layer H in Erq el Ahmar Rockshelter (Neuville 1951), and the lower layers at the entrance of Qafzeh Cave (Bar-Yosef 1989b; Hovers 1997).

The situation in the early Mousterian of Hayonim Cave supports the notion of generally slow rates of sediment accumulation and consistently ephemeral use of the site by humans. The differences between the stratigraphic series of Kebara and Hayonim Caves, which are separated chronologically by about 60,000 years, are most apparent in aspects of site structure and related observations discussed in chapter 12. The available evidence leaves us with the conclusion that smaller numbers of people lived in the Galilee per square kilometer in the earlier part of the Mousterian sequence, and far fewer in the Mousterian in general, than during most subsequent culture periods. Much of this case is made from the faunal evidence presented in chapters 5–12, integrated as much as possible with zooarchaeological, mineralogical, and technological findings from Kebara Cave and certain other Mediterranean sites (Meignen et al. n.d.).

Figure 2.9
View of the interior of Meged Rockshelter and the excavation grid. A small spring in the back of the shelter is marked by a large plant growing in the wall just behind the seated person.

MEGED ROCKSHELTER AND ITS EXCAVATION

Meged Rockshelter is situated near the top of a south-facing slope of the Wadi Meged (Kuhn et al. 2004), against a limestone cliff 9.5 m in height (figs. 2.8 and 2.9). Lying 0.5 km northeast of Hayonim Cave, Meged Rockshelter is also 300 m up a steep, rocky incline from the modern wadi bottom. The main exposure of the site is southeast, and during summer the site receives direct sunlight for about half the day. The sheltered area is small, encompassing only about 35 square meters. Massive limestone blocks fallen from the cliff above have created a kind of natural terrace extending several meters beyond the drip line of the shelter. This feature has fostered accumulation of sediments and protected the archaeological deposits from erosion nearest to the cliff (figs. 2.10 and 2.11). Intact archaeological deposits were confined mainly to the sheltered area. The ground descends quickly outside the drip line and consists mainly of colluvial material, judging from findings in a small test pit 5 m west of the main excavation. The matrix in this test unit consisted almost entirely of coarse limestone fragments and lacked fine-grained sediments.

Meged Rockshelter was first investigated in the early 1970s, when a small test trench was excavated in conjunction with the early investigations at Hayonim Cave. Formal excavation of Meged Rockshelter was conducted from 1994 to 1997 (Kuhn et al. 2004), covering a total area of 18 square meters, mostly in one continuous trench. Difficulty of access to the shelter prohibited wet-screening on-site, but all sediments were dry-sieved using 2–3 mm mesh. A sample of approximately 10 percent of the dry-sieved sediment was then carried to the screening operations in front of Hayonim Cave for wet-sieving. The wet-sieved fraction more than anything revealed the effectiveness of the dry-sieving in that very few microliths or identifiable faunal elements were missed in the latter technique.

The stratigraphy was relatively simple within the protected zone of Meged Rockshelter. A surface layer 5 to 10 cm thick consisted of loose, dry, heavily bioturbated sediment that was rich in charcoal from the recent campfires of local shepherds and other occasional visitors. The underlying sediments were prehistoric, reddish in color, and rich in clay (terra rosa), with variable quantities of subangular limestone fragments. The quantity of limestone debris mixed with finer-grained sediments increased away (southward) from

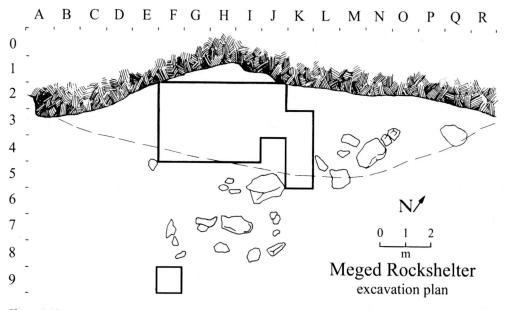

Figure 2.10
Plan view of the 1994–1997 excavations of Meged Rockshelter, including areas of major rockfall.

the shelter wall. The terra rosa and limestone layer, between about 75 cm and 1 m deep, in turn rested directly on a limestone bench; it is possible that another bench or even another shallow rockshelter is buried beneath the slope. The color of the terra rosa became lighter and yellower with increasing depth, apparently reflecting greater carbonate content. In the deepest parts of the trench, sediments were heavily indurated and fully cemented in places. Sediments were also notably harder and more brecciated along the northern end of the back wall of the shelter, especially below a small spring in square J1.

In some excavation profiles, a notable accumulation of relatively large limestone chunks was visible between 200 and 220 cm below the site datum. The zone of large limestone fragments may have been associated with the same event or events that resulted in the deposition of the large limestone blocks already described. It also corresponded roughly to the break between the two main cultural components in the site—a pre-Kebaran, or Upper Paleolithic, phase and an early Kebaran phase (Kuhn et al. 2004). The rockfall might have occurred as a result of an abrupt change in climate or a seismic event. Alternatively, it might simply represent a point at which the gradual undermining of the limestone cliff crossed a critical threshold. There are several other collapsed rockshelters along the same limestone exposure that contains the site of Meged, but no surface indications of prehistoric deposits have been found in the other shelters.

Lithic artifacts, bone, and other archaeological remains were abundant throughout the stratigraphic sequence of Meged Rockshelter, although their densities varied both vertically and horizontally. Bone was less abundant with increasing depth and in the southern part of the site near the shelter's drip line. Although evidence of small-scale bioturbation (root casts, worm castings, and termite burrows) could be found in even the deepest levels, there appears to have been little or no large-scale disturbance of the archaeological sediments by any agent. Most importantly, post-Pleistocene use of the shelter seems to have been minimal; ceramics were rare, for example, and confined to the disturbed uppermost layer.

Because the sediments were relatively uniform through the deposits, the sequence could not be subdivided on sedimentological grounds. Changes in artifact content, however, particularly in the frequencies of microlithic pieces, suggest that a major shift in the nature of technological activities took place at the site. The use of microlith frequencies to establish a "cultural stratigraphy" is justified by the importance of these artifacts in defining the Upper Paleolithic and Epipaleolithic phases in the Levant (e.g., Bar-Yosef 1981; Goring-Morris 1995). Changes in the artifact assemblages were most pronounced in the center of the excavation trench (squares G2–G4 and H2–H4), where the excavation reached its greatest depth. Sediments in squares F2–F4 were somewhat disturbed by the test trench

Figure 2.11
Excavated trench in Meged Rockshelter in 1997, showing general sediment texture and conditions toward the base of the excavation.

from the 1970s, whereas squares in the I, J, and K rows were excavated only to a depth of 200 cm bd.

The sequence at Meged Rockshelter can be divided into three components on the basis of artifact content (Kuhn et al. 2004). In the upper component, from the surface to 200 cm bd, microlithic artifacts accounted for between 60 and 80 percent of the retouched tools. In the lower component, below 215 cm, retouched bladelets accounted for 40 percent of the total or less. These changes occurred at approximately the same depth in all squares, suggesting that the bedding of sediments within the shelter was more or less horizontal. Although the frequency of microlithic artifacts decreased with depth in the sediment column as a whole, there were no significant trends in the frequencies of bladelet tools within either the upper or lower zone. This observation implies that the declining abundance of retouched bladelets with depth actually reflects the existence of two discrete behavioral components rather than a gradual and continuous trend throughout the sediment column. There was a sharp drop-off in the abundance of retouched bladelets in the bottom of the deepest excavation squares, between 245 cm and bedrock, but this was largely a result of natural sedimentation processes. Many larger artifacts at these depths were found directly on top of the limestone shelf that underlay the archaeological deposits. Very small pieces, such as retouched bladelets, would have been more easily swept from the smooth, sloping rock surface by water, wind, or subsequent human activity.

There was no evidence of an occupational hiatus in the stratigraphic series of Meged Rockshelter; the densities of artifactual materials did not decrease within the "intermediate" zone from 200 to 215 cm bd. Nor did this zone seem to represent a behavioral transition between the top and bottom components. Rather, it was likely the product of localized geological mixing of two distinct components. Clay-rich terra rosa soils, when they dry, tend to form deep cracks into which small artifacts deposited on a surface may fall and be incorporated into somewhat older, deeper sediments.

AGE OF THE DEPOSITS

Four accelerator mass spectrometry (AMS) radiocarbon determinations were obtained for Meged Rockshelter (Kuhn et al. 2004). Two very similar uncalibrated dates, $18,065 \pm 120$ and $18,125 \pm 135$ b.p., are from wood charcoal samples from 195–200 cm bd, at the base of the upper cultural component. Two additional determinations, $18,840 \pm 140$ and $20,485 \pm 155$, also on charcoal, represent the middle part of the lower component, at depths of 220–225 cm and 230–235 cm, respectively. Unfortunately, no radiocarbon dates are available for the uppermost 50 cm of the deposit. The existing dates increase in age with depth but do not suggest uniform rates of deposition, and thus it would be inappropriate to project ages for the more recent levels in the shelter. The radiocarbon dates indicate that the occupations at Meged Rockshelter spanned a period of at least 2,500 years, centering on the Last Glacial Maximum, or OIS 2. These dates generally correspond to the end of the Upper Paleolithic and beginning of the Epipaleolithic in the southern Levant (Goring-Morris 1995), a period for which few well-dated sites are known.

THE LITHIC ASSEMBLAGES

The flint artifacts from Meged Rockshelter are patinated but appear to have been made from both nodular and tabular raw materials. Individual nodules acquired by the inhabitants tended to be small, in the range of 8–15 cm. A few very large cores and tools seem to have been made on much larger chunks of a type of raw material that is plentiful in the vicinity of the shelter. Flint-bearing limestones and chalks of both Upper Cenomanian and Lower Turoninan ages (Yanuh and Yirka Formations, respectively) are exposed in the Wadi Meged itself and contain small, rounded and tabular nodules. An even wider variety of flint sources can be found within a 20 km radius of the site (Delage 2001).

The predominant core forms throughout the sequence from Meged Rockshelter are small, carenated or nosed bladelet cores made on pebbles. Despite the similarity in basic core forms, there are some qualitative contrasts between the upper and lower components in Meged Rockshelter. In general, cores from the lower levels appear less well prepared and more irregular than cores from the upper levels. The edges of striking platforms on these cores are more jagged, and the faces of detachment less consistently shaped. This does not appear to reflect simply the extent of reduction, because core masses are statistically indistinguishable.

A more striking contrast between the two cultural components is evinced by the bladelets. Strongly twisted bladelets were present in all levels but were much more common in the lower component. Also, pieces with twisted profiles were always less common as tool blanks than as debitage. Either twisted forms were not selected for retouching or else retouching was used to straighten the profiles of the bladelets. In any case, it appears that twisted profiles were not preferred for retouched tools, even in the lower levels where they were relatively common.

Another point to be made concerns differences in the tool types and retouching of the microliths. The upper component contained a variety of formal microliths, including various types known from Kebaran assemblages more widely. Many of the modified bladelets in the more recent levels display true backing or thick, abrupt retouching. In the lower part of the sequence, most microlithic artifacts were informal, backed or retouched bladelets. True backing was scarce, and fine marginal retouching predominated.

As has been observed for the Upper Paleolithic and Epipaleolithic elsewhere (Belfer-Cohen and Goring-Morris 1986; Goring-Morris 1987, 1995; Goring-Morris et al. 1998), blades and bladelets from Meged Rockshelter vary in more than just their sizes. They were not simply part of a single, continuous core reduction sequence. Somewhat different sets of technological procedures were used in the production of large and small blade blanks: most blade cores were worked primarily along the broad face, whereas the great majority of bladelet cores were exploited along one narrow face (Kuhn et al. 2004).

A small number of tools made of materials other than flint were also recovered from Meged Rockshelter. They included five battered limestone cobbles, probably hammer stones, as well as two fragments of limestone slabs and one large limestone flake that appear to have been used as heavy-duty cutting or chopping implements. Four pieces of nonlocal metamorphic or igneous stone were recovered from the lower component. Two of these are fragmentary hammer stones, and two are small, irregular slabs or palettes showing evidence of abrasion. Although the provenance of the nonlimestone artifacts has not been established, the closest source of igneous or metamorphic stone would have been 40–50 km to the east in the Jordan Rift Valley. No formal grinding implements were found during the excavations, but a possible grinding slab of limestone was found on the surface. A small number of antler or bone points and awls were also found.

THE FAUNAL AND ORNAMENT ASSEMBLAGES

Because the limestone bedrock buffered any natural acids in groundwater, Meged Rockshelter offered reasonably good conditions for bone preservation. The conditions governing bone recovery, however, were less than ideal in the dense terra rosa clays in some parts of the site; the matrix was sometimes cemented with calcite, making it difficult to extract bones fully intact. Meged Rockshelter nonetheless yielded a reasonably large sample of vertebrate and mollusk remains. Roughly 77 percent of the bones came from the upper component, above 200 cm bd. The relatively good condition of the bones was probably a result of rapid burial and protection from the elements by the shelter overhang.

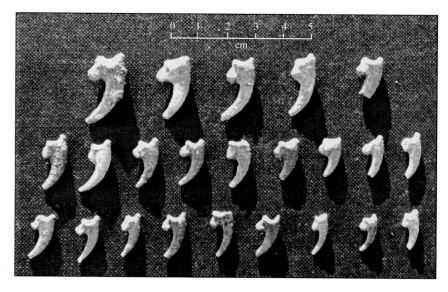

Figure 2.12
Examples of the many raptor (Falconiformes) talons found in Meged Rockshelter. (Image courtesy of N. Munro.)

In addition to the inordinately large numbers of talons of predatory birds (fig. 2.12; see Munro in Kuhn et al. 2004), more than 80 marine and freshwater mollusk shells (Bar-Yosef Mayer in Kuhn et al. 2004) were recovered from Meged Rockshelter. The collection of shell beads was large for a site of this age in Israel, especially in light of the small area excavated. All of the ornamental mollusk species could have been collected from either the Mediterranean littoral or the freshwater lakes and river of the Jordan Valley (e.g., *Theodoxus jordani*). With the exception of dentalia shells, which were always sectioned, the ornamental shells tended to be complete or nearly so. Humans perforated the great majority of them. Some shells show wave wear, small holes drilled by predatory mollusks, or both, suggesting that they were collected as beachcast specimens after the animals inside had died. The shell ornament assemblage is quite similar in content to other early Epipaleolithic assemblages from Ohalo II, Israel (Bar-Yosef Mayer 2002) and Üçağızlı Cave in Levantine Turkey (Stiner et al. 2002), sites discussed for comparative purposes elsewhere in this book.

MEGED ROCKSHELTER: SUMMARY OF THE CULTURAL SEQUENCE

Overall, there was much continuity between the two cultural components at Meged Rockshelter. Typologi-

cally, their large tool assemblages are quite similar, although the earlier levels contained somewhat fewer blade blanks and more blanks that were chunks and cortical pieces. Blank production technology in both components involved similar core forms, the main difference being that the cores from the more recent component appear to have been better prepared and more carefully managed or exploited, whereas twisted bladelets were more abundant in the earlier component. The faunas are even more similar between the two components (chapter 7).

The most obvious difference between the lithic assemblages from the two components at Meged Rockshelter stems from the criterion used to define the components in the first place, the proportions of microlithic artifacts. The percentage value for the earlier component as a whole, 35 percent microliths, is at the lower limit for defining an assemblage as Epipaleolithic. This is not too surprising, considering that these levels date to roughly 20,000 (radiocarbon) years B.P. or earlier, the age conventionally assigned to the end of the Upper Paleolithic and beginning of the Epipaleolithic. The lithic assemblage resembles other, contemporaneous lithic assemblages such as that of Ohalo II (Nadel 1997). The frequency of microlithic artifacts in the upper levels of Meged Rockshelter (about 65 percent) resembles the percentages in other Epipaleolithic assemblages of the Levant, especially those attributed to the early Kebaran (e.g., Hayonim Cave Layer C).

The evidence from Meged Rockshelter suggests that its deposits represent repeated but brief visits to the site by small numbers of people. The sheltered area of the site is small, less than 40 square meters. The absence of preserved constructed features such as formal hearths also suggests that occupations of the shelter were not prolonged. During the Last Glacial Maximum (OIS 2), the Wadi Meged would have been even less well watered than it is today.

Several aspects of the artifact and faunal assemblages from Meged Rockshelter suggest that it might have had a rather specialized function related to the hunting of gazelles and other large animals, at least during the time in which the upper component formed. Supporting this interpretation are artifact assemblages consisting exclusively of chipped stone tools, ornaments and raptor talons, and a few bone points. Although grinding equipment generally was not ubiquitous in the early Epipaleolithic of the Levant (Goring-Morris 1995; Wright 1994), it was present in a number of sites, but not in Meged Rockshelter. Tools for processing seeds and nuts are expected to be more common in residential sites than in task-specific locales. The presence of substantial numbers of shell beads at Meged Rockshelter probably is not diagnostic of site function, although in other Mediterranean Rim sites with deeply stratified culture series their abundance has been correlated most positively with the quantities of large game animals (Stiner 1999; Stiner et al. 2002). This is not to say that there was no residential aspect to the occupations at Meged Rockshelter. In all likelihood, some of these occupations involved complete but small residential groups rather than special task groups (see chapters 5, 7, and 10).

CHAPTER THREE

Experiments in Fragmentation and Diagenesis of Bone and Shell

Mary C. Stiner, Ofer Bar-Yosef,

Steven L. Kuhn, and

Stephen Weiner

TAPHONOMIC ANALYSES OF SKELETAL REMAINS normally succeed by virtue of their clarity, not only in the analytical assumptions employed but also in the classes of information used. In the case of bones and mollusk shells, essential observations should include the chemical and structural diagenetic changes in skeletal specimens and the sediment matrix as well as the patterns of burning damage, fragmentation (specimen size and shape), weathering, and specimen alignments in sediments. Some damage processes act directly on skeletal microstructure and chemistry and have important implications for changes in macrostructure, the latter being what zooarchaeologists generally see. Other processes, such as mechanical loading, act on macrostructure more directly. Finally, the forces that operate on macrostructure may play out very differently on fresh bone than they do on bones that have been chemically altered.

In this chapter we report on our exploration of some of these issues through experiments in bone burning, weathering, and fragmentation. We investigated chemical alteration of bone with the aid of Fourier-transform infrared (FTIR) spectroscopy and geochemical models, following Karkanas et al. (1999, 2000), Schiegl et al. (1996), and Weiner et al. (1993, 2002). We examined mechanical alteration principally in terms of fragmentation in a controlled environment. The most robust results from the experiments served as guidelines for interpreting assemblage formation history and preservation in the Paleolithic layers of Hayonim Cave, with an emphasis on the Mousterian period (chapter 4). The designs, techniques, and samples for the experiments varied considerably and so are presented in the respective sections that follow. First, however, we discuss the differential effects of dissolution versus mechanical processes, even though they were not subjects of experimentation, because they are relevant to the archaeological interpretations to come. In Hayonim Cave, dissolution is likely to have occurred from percolation of water rich in organic acids and hence with low pH (Karkanas et al. 2000; Weiner et al. 1995, 2002). We also offer a brief introduction to our applications of infrared spectroscopy.

BACKGROUND TO THE EXPERIMENTS: CHEMICAL VERSUS MECHANICAL DESTRUCTION

The potential for bias in the preservation of faunal assemblages is a contentious subject in zooarchaeology, not least because our understanding of what the processes responsible for bias should produce in the way of meta- or "macro-" patterns remains limited. Attempts to predict how in situ attrition might distort the contents of archaeofaunal assemblages has until now focused on macrostructural differences—the differences between morphological features composed of compact bone, cancellous or spongy bone, dentin, and enamel—within and between skeletal elements (see also chapters 5 and 10). Bone macrostructure almost certainly relates to some of the many potential causes of element representation bias. However, macrostructure appears to be the sole conceptual basis for nearly all zooarchaeologists' models of in situ attrition (but see Hedges and Millard 1995; Hedges et al. 1995; Lyman 1994). Macrostructural variation is most relevant to locomotor biomechanics in vivo and, in archaeofaunas, to the mechanical breakdown of skeletal elements and fragment sorting. Sources of mechanical damage are diverse and include carnivore gnawing, human processing, trampling and sediment compaction, gravity, and water transport. The relative rates of decay of compact and cancellous bone tissues from dissolution are poorly understood. Because bone is a hierarchical assembly of mineral and organic components, effective models of its resistance to mechanical and chemical forces of destruction must vary accordingly.

Zooarchaeologists' "structural," or "bulk," density standards (grams per cubic centimeter) for mature vertebrate skeletons (e.g., Kreutzer 1992; Lyman 1984; Lyman et al. 1992; Pavao and Stahl 1999) concern mainly the mineral (carbonated hydroxyapatite, also called dahllite) component of bone. Structural density is used as a proxy measure of resistance to destruction at the level of macrostructure. Though quite useful, structural density standards come with some basic liabilities. Fresh bone owes much of its strength to its fibrous composite structure, mainly a combination of collagen fibrils and small apatite crystals (Currey 1984; Weiner and Wagner 1998). Degradation of the collagen component often occurs postdepositionally and results in significant reductions in the mechanical strength of the residual structure. Collagen generally degrades under alkaline conditions in sediments (e.g., calcareous types), which favor the preservation of bone apatite (Karkanas et al. 2000; Weiner et al. 1995). As the fibrous component of fresh bone degrades, a much brittler, more porous structure is left behind (see also Lyman 1994:261). The mineral component, on the other hand, is initially composed of extremely small crystals that are much more soluble than pure hydroxyapatite (Berna et al. 2004). These crystals readily dissolve and, depending upon conditions in the

microenvironment within the bone, may reprecipitate to form a denser and more robust product; otherwise, the dissolved ions are transported out of the bone and the remaining material is more fragile and powdery. Many of the compact bone tissues of large mammals are also riddled with tiny channels, or caniculi. If the bone contains Haversian systems, then larger channels housing blood vessels are also present in the center of each osteon (Currey 1984:30–36). The micropore abundance increases further (in number and area exposed) as the collagen component degrades (Nielsen-Marsh and Hedges 2000a, 2000b).

Chemical processes occur at the microscopic level, the molecular level, or both, implying that microporosity is very important (Hedges and Millard 1995; Hedges et al. 1995). Whereas the microstructure of bone is relatively constant across the vertebrate skeleton, bone macrostructure varies greatly both among and within postcranial elements in mammals (Currey 1984). All bone has the same general composition, roughly 60–70 percent mineral and 30–40 percent organic compounds and water (Currey 1984; Weiner and Wagner 1998). The surface area available to destructive agents, however, is assumed to differ between spongy and compact bone macrostructures, aspects readily visible to analysts. This visual contrast lies at the heart of current disagreements over how in situ bone attrition works. Carbonated hydroxyapatite, the inorganic component of bone, is a calcium phosphate compound that is susceptible to dissolution by acids. Chemical dissolution may be the most common cause of bone attrition in archaeological records worldwide and throughout time, frequently leaving archaeologists with stone artifacts but few or no bones. Dissolution imposes harsh effects on all bone tissues. Compact bone may be less vulnerable to chemical dissolution than spongy bone, but the microporosity of compact and spongy bone, and the rates of loss of the two macrostructure classes in acidic environments, may be more similar than many faunal analysts realize. Mammal teeth fare better simply because the outer enamel layer is denser than bone or dentin and its crystals are much larger (mammal tooth enamel is about 95 percent mineral, Hillson 1986; Wainwright et al. 1976). Much work has been done on what are essentially mechanical effects in bone decomposition and assemblage contents; far less has been done on dissolution effects.

The experiments described in this chapter only begin to address some of these larger issues. We focus specifically on the relative rates of chemical and mechanical decomposition of broad classes of skeletal tissue, mainly but not exclusively compact and cancellous (spongy) bone, and the extent to which differential losses may bias the contents of archaeofaunas. Micro- and macroscale differences in specimen structure represent an essential a priori distinction. *Destruction* refers specifically to the loss of macroscopic structural integrity and thus the recognizability of skeletal remains to the unaided human eye. *Diagenesis* in general refers to all changes that take place after death, but we restrict the use of this term mainly to the molecular transformations that occur during fossilization, heating, and dissolution. Although the principal focus of the experiments was on what happens to bone, mollusk shell was also considered.

FOURIER-TRANSFORM INFRARED SPECTROSCOPY

The infrared spectra of fresh and archaeological bone and mollusk shell provide information on both the mineral and the organic components. FTIR is the method of choice for this research because of its versatility, the portability of its equipment (from laboratory to field and back), and its suitability especially to studies of the composition of biogenic and related compounds (Smith 1996; Weiner and Bar-Yosef 1990; Weiner and Goldberg 1990). In the case of bone, the ratio of the mineral to the organic content can be esti-

mated, as well as the carbonate content and the so-called crystallinity of the mineral. The latter refers to the size and atomic perfection of the crystals. In the case of shells, the infrared spectra provide information about the nature of the calcium carbonate polymorph present, namely, calcite, aragonite, or both.

Sample preparation procedures for FTIR analysis were straightforward (but see chapter 4 on archaeological bone; Surovell and Stiner 2001). The surfaces of specimens to be analyzed were first cleaned mechanically with a scalpel. Representative fractions from each bone or shell were then collected and ground with an agate mortar and pestle. A few tens of micrograms were mixed with potassium bromide (KBr) powder, and a 7 mm diameter pellet was prepared using a Qwik Handipress (Spectratech, Warrington, UK). KBr is transparent to infrared radiation, so only the sample absorbs some of the radiation as it passes through the pellet to the detector. The manner in which the sample absorbs the radiation depends on the atomic structure of the material. We used a Midac Corporation (Costa Mesa, California, USA) spectrometer operated by Grams 386 (for Windows 95) software (Galactic Industries Corporation, Salem, New Hampshire, USA). Many of the analyses in the burning experiments were performed on-site at Hayonim Cave, following Weiner et al. (1993); certain other analyses were performed using essentially the same equipment at the Weizmann Institute of Science (Rehovot, Israel) or at the Department of Anthropology, University of Arizona (Tucson, Arizona, USA).

Peak positions on the wavenumber axis (reciprocal of wavelength in cm) of infrared spectra are used to diagnose compound type. Peak heights reflect approximate quantities of the components in the samples and may be fairly accurate if standardized to a specified baseline, to other peaks, or to both. Note that each compound usually yields more than one absorption band. An expanded spectrum in the 425 to 900 cm[-1] range was used to measure the mineral crystallinity of bone based on the so-called splitting factor (SF), or crystallinity index, following Weiner and Bar-Yosef (1990). The extent of splitting of the absorptions at 603 and 565 cm[-1] reflects a combination of the relative sizes of the crystals in bone mineral and the order of the atoms in the crystal lattice (Termine and Posner 1966). As recrystallization progresses, the two absorption peaks become increasingly separate (fig. 3.1), as measured by an "index of splitting" calculated as the sum of the heights of the two peaks (from a baseline drawn between 495 and 750 cm[-1]) divided by the height of the low point between them (Weiner and

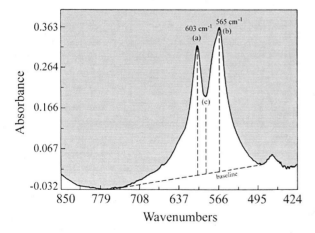

Figure 3.1
Calculation of the bone mineral crystallinity index, or splitting factor (SF = [a + b] / c), from a baseline drawn from approximately 459 to 750 cm[-1] of the infrared spectrum. The sum of the absorption heights at wavenumbers 603 cm[-1] and 565 cm[-1] is divided by the height of the valley between them. (Adapted from Weiner and Bar-Yosef 1990:191.)

Bar-Yosef 1990:189–190). The higher the SF value, the larger and/or more ordered the crystals.

SF is always lowest for fresh bone (about 2.5–2.9), because the carbonated hydroxyapatite crystals formed in living bone are very small (Robinson 1952; Weiner and Price 1986). SF normally increases postmortem, up to 4.5–5.0 under ambient temperatures, as large crystals grow at the expense of small ones—a process known as recrystallization. SF is highest (7.0+) for bone heated to a calcined state. As SF increases with heating, carbonate decreases (e.g., Person et al. 1996; Stiner et al. 1995), because the original dahllite lattice recrystallizes and, in doing so, loses carbonate to form (ideally) hydroxyapatite.

The simpler SF measurement of Weiner and Bar-Yosef (1990) was strongly correlated with Termine and Posner's (1966) measure in our data set ($r = 0.93$, $p < 0.0001$) and thus was assumed to measure the same thing. The ratio of the absorption band of carbonate at 874 cm^{-1} to that of 565 cm^{-1} for phosphate was used to estimate semiquantitatively the relative carbonate contents of the mineral phase (fig. 3.1). This procedure followed Featherstone et al. (1984), who, by using synthetic standards, showed that the ratio of the carbonate 1415 cm^{-1} absorption band to the same phosphate band estimates ($\pm$ 10 percent) carbonate content. We used the 874 cm^{-1} carbonate absorption rather than the 1415 cm^{-1} absorption because the latter was affected by the presence of organic matrix absorption bands in fresh and well-preserved bones.

The insoluble organic matrix fraction of a bone or shell was extracted by dissolving 100–200 mg of powdered sample in 4 ml of 1N HCl at room temperature. The insoluble fraction was separated from the supernatant by centrifugation (14,000 G for 2 min) and then washed twice in the same way with deionized water. The dried pellet was used to obtain an infrared spectrum, as already described, and a carbon/nitrogen (C/N) ratio using a Carlo-Erba (Milan, Italy) model 1108 elemental analyzer.

EXPERIMENTS IN BURNING, FRAGMENTATION, AND MINERAL RECRYSTALLIZATION

A considerable literature exists on how bones are altered by fire and how burning damage to bones and sediments can be reliably identified (e.g., Bellomo and Harris 1990; Brain 1993; Nicholson 1993; Shahack-Gross et al. 1997; Shipman et al. 1984; Sillen and Hoer-ing 1993; Walters 1988; Weiner et al. 1998, 2002). The question of what burned bones reveal about prehistoric human behavior, however, is more complex, requiring an understanding of the contexts in which such damage might occur, including the spatial associations between objects in sediments and how burning and fragmentation of bone relate to the intensity with which humans use domestic space. Much the same can be argued for mollusk shells, especially those of mollusks consumed as food by humans (e.g., Goodale 1957), as well as ornamental shells and the shells of land snails (Stiner 1994, 1999).

In this section we explore the relationships among four phenomena important to interpreting burning damage on bones and shells from archaeological sites (Stiner et al. 1995): visible changes in color; associated changes in mineral and matrix; alterations in susceptibility to fragmentation; and the extent to which sediment may insulate buried bones and other materials from fires built on the ground surface. Our ultimate concerns are with the ways burned bones and shells become part of the fabric of archaeological records inside shelters, with the spatial associations between live fires and the material they damage, and with the influence of fire on differential skeletal preservation. The goal is to establish some physical baselines for the ways in which human activities around a fireplace might be expected to translate into the patterns we see in archaeological deposits hundreds or thousands of years later.

A recurring association between fragment size and the frequency and intensity of burning damage is seen in archaeological sites. Generally speaking, the smallest pieces are the most likely to exhibit burning damage. Riparo Mochi, a rockshelter in northern Italy whose cultural layers span 36,000 to 9,000 years ago (Kuhn and Stiner 1992, 1998a; Stiner 1999), provides one good example. There, median fragment lengths for burned and unburned bones differed consistently for the early Aurignacian through late Epigravettian levels (fig. 3.2), with burned bones seldom exceeding 1.2 cm. These faunal assemblages were retrieved by fine-screening the sediments, and the excavators retained all unidentifiable fragments along with the identifiable pieces. Many of the smaller bones were nearly or fully carbonized, but calcined bones were rare. Generally analogous patterns exist for completely recovered bone from an Epigravettian and a Middle Paleolithic shelter in Latium, Italy (table 3.1). These sites are, respectively, Riparo Salvini, radiocarbon dated to about 12,500 years ago (Avellino et al. 1989;

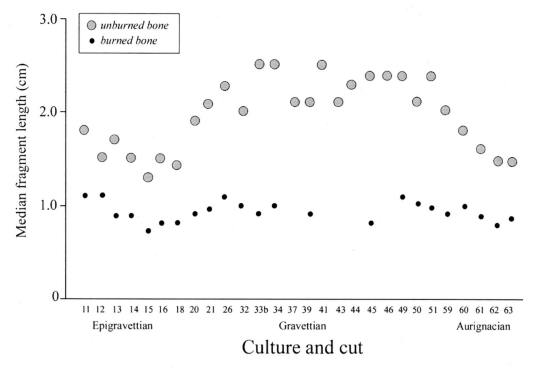

Figure 3.2
Median size differences for burned and unburned large mammal bone fragments from the Middle Paleolithic through late Epigravettian levels of Riparo Mochi, Italy.

Bietti and Stiner 1992), and Grotta Breuil, ESR dated to about 33,000–37,000 years ago (Schwarcz et al. 1990–1991; see also Bietti et al. 1988, 1990–1991). Again, the mean fragment lengths and size ranges for burned bones were significantly smaller than those for unburned bones. The relationship between the incidence of burning damage and bone fragment length held true across the Middle, early Upper, and late Upper Paleolithic shelter assemblages considered.

Because fire figures prominently in the food preparation technologies of most human cultures, it is reasonable to expect that at least some burning of bones stems directly from cooking activities. However, examination of the lithic artifacts from Grotta Breuil and other Paleolithic shelter sites shows that the overall potential of such an explanation for the final state of the bones is quite limited. The percentages of identifiable burned bones and lithic artifacts found in the cultural levels of these caves are rather similar (table 3.2; Stiner 1994). Lithic artifacts—tools, flakes, cores, and debris—were burned (often extensively) as often or more often than were identifiable bones and without regard to raw material type (Kuhn 1995). The lithic data represent another example of the size effect in burned material.

Patterns not unlike those just described were discerned for the Paleolithic levels of Hayonim Cave (chapters 4 and 5), which contained some well-preserved, shallow fireplaces (Weiner et al. 2002). Unburned bones from Hayonim Cave usually exhibited green stick fractures, but the fracture history of the small burned pieces was less clear. Questions about the behavioral and postdepositional circumstances in which these archaeological features and associated material formed prompted our investigations of burning phenomena through controlled experiments. Specifically, how does mineral recrystallization relate to visible changes in bone color as specimens are increasingly heated? To what extent does burning damage alter a bone's strength, long-term durability, recognizability, and preservation potential? Under what conditions might bones buried beneath a fire bed become burned? Finally, why does burning damage to bone in archaeological contexts seldom seem to extend beyond complete carbonization (blackening) of bone? The implications of the experiment results for identifying burning damage, inferring temporal associations between hearth features and artifacts, and understanding humans' use of domestic space in sites are discussed in the final section of this chapter.

Table 3.1
Fragment size data for burned and unburned bones from Grotta Breuil (Middle Paleolithic)
and Riparo Salvini (late Upper Paleolithic/Epigravettian), Italy

Statistic (in cm)	Grotta Breuil Sample		Riparo Salvini Sample	
	Burned ($n = 44$)	Unburned ($n = 347$)	Burned ($n = 124$)	Unburned ($n = 297$)
Minimum length	0.2	1.0	0.4	1.0
Maximum length	3.3	9.5	5.0	8.3
Mean length	1.4	3.0	1.4	2.3
Standard deviation	0.7	1.5	0.6	1.0

NOTE: These are completely recovered faunal assemblages consisting of identifiable and
unidentifiable bones. Statistical significance of length differences between burned and
unburned bone specimens: Grotta Breuil t statistic = 6.636, $p \leq 0.001$; Riparo Salvini t
statistic = 8.731, $p \leq 0.001$.

Table 3.2
Percentages of identifiable burned ungulate bones and burned lithic artifacts in four
Mousterian cave sites, Italy

Site and Layer	Burned Ungulate Bone, % of NISP	% Lithic Artifacts Burned			
		Tools	Flakes	Cores	Debris
Grotta dei Moscerini					
M2	52	42	30	38	44
M3	32	53	39	58	42
M4	7	32	18	40	41
M6	7	41	17	41	32
Grotta Guattari					
G1	*	11	0	3	15
G2	*	5	0	11	11
G4	*	12	3	9	27
G5	*	8	12	9	27
Grotta di Sant'Agostino					
S0	3	11	—	—	—
S1	3	10	13	14	25
S2	6	14	12	14	28
S3	1	12	8	4	16
Grotta Breuil					
Br	2	20	14	16	24

KEY: (*) Burned bone chips were present but relative frequencies are unknown; (—) no
information available.

MATERIALS, METHODS, AND FIRE SETUP

The effects of burning on bone were examined at the
macroscopic and microscopic levels. Bones were
burned to varying extents in controlled fires, and we
subsequently examined their susceptibility to fragmentation (that is, their friability) and, by infrared
spectroscopy, changes in their mineral properties. The
bone samples originated in both modern and archaeological sources. We collected fresh bones for the burning experiments, primarily those of goats and cattle, in
the northern Galilee near Hayonim Cave. The animals

Stages of burning

Figure 3.3
Modern examples of burn color codes 0 through 6 generated under controlled conditions. Light shades on left are cream-colored and represent fresh bone (0) or lightly burned bone (1); light shades on right are light gray to pure white and represent the calcined phase of burning (6, the most advanced). (Reproduced from Stiner et al. 1995.)

had died the previous fall or winter, and their remains were largely skeletonized but still slightly greasy by early summer, when we collected them. Archaeological bone samples were obtained for comparison from the Natufian, Kebaran, Aurignacian, and Mousterian levels of Hayonim Cave. Pairs of archaeological bones from a given excavation unit, one of burned appearance and the other not, were selected on the basis of color from samples of ungulate cortical splinters. The archaeological samples were chosen by a nonarchae-

Table 3.3
Burning damage categories for bone based on macroscopic appearance and color

Burn Color Code	Description
0	Unburned (cream/tan)
1	Slight, localized burning, less than one-half carbonized
2	More than one-half carbonized
3	Fully carbonized (completely black)
4	Less than one-half calcined (more black than white)
5	More than one-half calcined (more white than black)
6	Fully calcined (completely white)

ologist trained to use only our burn color codes (defined later) and darkening of the bone interior as indications of burning damage. The latter criterion was important, because burning damage on bone normally extends deep into the cortex, in contrast to superficial mineral staining (see also Shahack-Gross et al. 1997 for a chemical method of differentiating black bones that were burned from those that were stained by oxides or both burned and stained).

Observations made for the experimentally burned fresh bones included the skeletal element and portion-of-element (defined in appendixes 1.1, 1.2), visible burning intensity, specimen sizes at each step in an experiment, and the proximity of bones to the hottest part of the fire bed. Visible stages of burning were classified by color on an ordinal scale of 0 to 6 (fig. 3.3). The visible grades of burning, described in table 3.3, ranged from unburned (code 0) through intermediate burning stages centering on carbonization (100 percent carbonized, or pure black, = code 3) to the most advanced phase, known as calcination (100 percent calcined, or pure white, = code 6). These color criteria were generally consistent with those used in previous experimental studies of bone burning (Nicholson 1993; Shipman et al. 1984) and with many other archaeologists' informal typologies of burning damage.

Four experimental fires (2, 3, 4, and 5) were made. They were simple arrangements intended to simulate real campfires, and the setup was similar for each experiment. The fireplace was situated on a well-ventilated, dry limestone surface, 2 m from a sheer rock wall. Some fires were built directly on bald rock, whereas later ones were built on a base of sieved cave sediment. Each fire used about 6 kg of local Mediterranean hardwoods—carob, oak, olive, or a mixture thereof—because these would have been the dominant firewood sources in the area during much of the Pleistocene. No piece of wood was greater than 7 cm in diameter, and most were smaller. The peak temperature of each fire was reached within 5 to 10 minutes of setting, and temperatures of up to 900 degrees C were recorded with a thermocouple. (Bellomo and Harris [1990:326] noted similar patterns of heat buildup in their experiments.) Each fire was allowed to burn down naturally, with minimal banking of the coals. Fires were lit in the morning, and if they were not dead by early evening, the coals and contents were collected in a large metal can and allowed to smolder through the night.

Premeasured fresh bones were poured directly into experiment fires 2 and 3. Bones were buried in 1 to 15 cm of soil beneath the beds of fires 4 and 5 (fig. 3.4); a total of 30 liters of fine, dry-sifted sediment, taken from the back-dirt of the Hayonim excavations, served as the matrix in which the bones were buried. The bone material was retrieved from the extinguished fires by gently sieving the mass through a 0.1 cm mesh screen, carefully avoiding further breakage. Bones were then remeasured and classified according to the intensity of visible burning damage, and the recognizability of anatomical elements or fragments was again recorded. To see how burning damage affected bone friability under mild pressure, fragmentation was then induced, first by vigorously shaking the burned bones, sorted by color category, in a cardboard box for 60 seconds and later by having two adult men trample buried bones in situ for several minutes. The extent of fragmentation in the box treatment was measured in terms of median fragment length and the volume of powder (in milliliters) generated for each burn color category.

RESULTS FOR BONE MINERAL RECRYSTALLIZATION AND BURN COLOR

Fresh bone is normally composed of 60–70 percent carbonated hydroxyapatite (dahllite) crystals by weight. At ambient temperatures, the crystals are gradually altered by diagenesis, a process in which large crystals tend to grow at the expense of small ones. This process occurs in vivo and usually continues in the context of

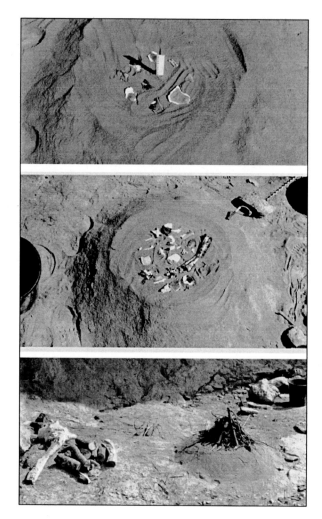

Figure 3.4
Experiment setups for fires 4 and 5. Screened sediment from the Mousterian layer of Hayonim Cave served as matrix, and either a combination of fresh bones (some fragmented, some whole) and ostrich eggshell (top) or pieces of newly fractured natural flint (middle) were buried at set increments in the sediment base. Heat sensors were placed at certain depths in the mound and in the active coal bed. The fire was set on a pyramid of fine wood with a grass center (bottom) and was then fed larger hardwood logs to raise the temperature rapidly to about 900 degrees C.

fossilization or of weathering over a few months or years. High-temperature diagenesis, by contrast, is nearly instantaneous. At temperatures below about 650 degrees C, the changes that occur in the crystals of heated bone are probably similar to those occurring more slowly at low temperatures (e.g., during weathering), but above that temperature, solid state recrystallization takes place (Shipman et al. 1984).

As we explained earlier, part of the information provided by the infrared spectra of bone relates to the degree of crystallinity of the carbonated hydroxyap-

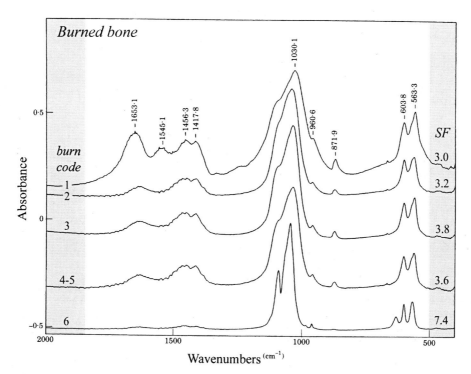

Figure 3.5
Infrared spectra of modern burned bones. Burn color codes appear on the left, and splitting factors (SF) on the right. Note that the collagen peak (1653 cm^{-1}) still prominent in the burn code 1 sample is greatly reduced in all of the other samples, which were heated to higher temperatures, and that the fully recrystallized sample burned to code 6 contains almost no carbonate, as evidenced by the absence of the carbonate peaks at 1456 cm^{-1}, 1417 cm^{-1}, and 872 cm^{-1} on the wavenumber axis. (Reproduced from Stiner et al. 1995.)

atite (see also Shipman et al. 1984 on the use of x-ray diffraction for a similar purpose). This is a function of the extent of splitting of the two absorptions at 603 and 565 cm^{-1} (Termine and Posner 1966) and reflects a combination of the relative sizes of the crystals and the extent to which the atoms in the lattice are ordered. The higher the splitting factor (SF) value, the larger and more ordered are the crystals. Hence, the crystallinity index is always lowest for fresh bone (i.e., SF = 2.5–2.9) and highest for calcined (or highly fossilized) bone (i.e., SF about 7.0), when using the convention defined by Weiner and Bar-Yosef (1990).

Figure 3.5 shows infrared spectra for modern bones burned to color codes 1 through 6. The progressive splitting, or separation, of absorption peaks at around 603 and 565 cm^{-1} can be seen at the right end of the wavenumber axis. Figure 3.6 plots the changes in splitting factor (SF) and carbonate (CO$_3$) values as functions of visible changes in burn color on the bones from the experiment fires. The SF values increase and CO$_3$ values decrease as the extent of burning damage intensifies. The differences in SF and CO$_3$ content for

burning color values 1–3 are minimal, however, in comparison with the changes associated with more advanced burn color codes 4–6, representing partial and then complete calcination. It is easy to identify the calcined phase in bones by infrared spectroscopy, because the original carbonated hydroxyapatite lattice recrystallizes and in doing so loses carbonate to form hydroxyapatite.

Analyses of the 1N HCl–insoluble matrix fractions of the modern goat and cow bones from the experiment fires (table 3.4) indicated that collagen was still preserved in bones with a burn color code value of 1. This was apparent from the infrared pattern (DeNiro and Weiner 1988a) and from a C/N ratio of around 3.0 (for a related C/N study, see Brain and Sillen 1988). Bones burned to color values 2–5 all yielded insoluble organic fractions with infrared spectra characteristic of pyrolyzed material. However, fully calcined bone (color code 6) had no insoluble matrix; the HCl solution used to dissolve the mineral was clear, presumably because the entire organic matrix was destroyed by heat.

Table 3.4

Summary of 1N HCl–insoluble fraction results by burn color code for bones from experiment fires

Burn Color Code	Infrared Spectrum Reading (Observed Color)	C/N ratio
0 (fresh bone)	Collagen (cream)	3.03
1	Collagen (yellow)	3.00
2	Pyrolyzed matrix (partly black)	n.d.
3	Pyrolyzed matrix (completely black)	4.36
4–5	Pyrolyzed matrix (black and white)	4.20
6	No insoluble fraction	—

KEY: (n.d.) Not determined; (—) no insoluble matrix.

RESULTS FOR BONE FRAGMENTATION AND BURNING INTENSITY

The great strength of fresh bone is lost to some extent in diagenesis, presumably due mainly to the loss of the collagenous matrix. Because burning, too, causes a loss of the organic matrix, the strength of burned bone should also be reduced. Though we can expect that as a rule, burned bone is more likely than fresh bone to crumble, the question here is about the shape of the damage continuum.

Bone samples representing the full range of burn phases were generated in the experiment fires. Their susceptibility to fragmentation as a function of burning intensity was observed in each of three ways: as a direct product of heat alone, in the form of spontaneous cracking, without added pressure (see also Nicholson 1993); as a product of the vigorous agitation of fragments sorted by burn color inside a box for 60 seconds; and as a product of the trampling of premeasured whole bones buried beneath the cooled fire bed. Agitating burned specimens in a box permitted complete recovery of all fractions, including fine powder, whereas the trampling experiments better typified the effects of foot traffic on shallowly buried burned bones.

Upon burning, large, isolated cracks appeared spontaneously in some of the whole bones, especially in long bones with compact tubes (fig. 3.7). Heated teeth tended to display the same series of color changes noted previously but also tended to explode into many small, irregular fragments (fig. 3.8).

The stark contrast in brittleness between burned and unburned bones was clear from visual inspection of

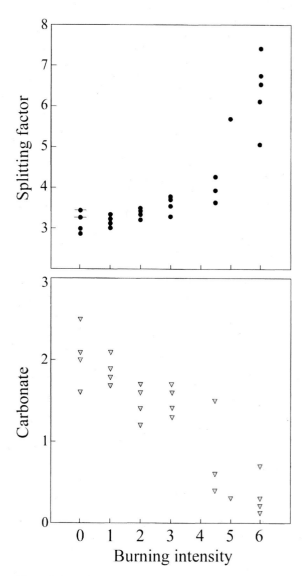

Figure 3.6

Splitting factor and carbonate (CO₃) values for experimental bone by burn color code. A filled circle with a dash represents the SF for the slightly weathered exteriors of otherwise fresh bone specimens. (Reproduced from Stiner et al. 1995.)

the material and from the fragmentation data. Agitation or trampling quickly reduced the burned material to smaller sizes and some of it eventually to powder (table 3.5). Unburned fresh bones buried under hearths were hardly affected by trampling. Figure 3.9 shows a composite of results for induced bone fragmentation across the experiments. The data reveal a monotonic, nonlinear decrease in median fragment length for the agitated or trampled bones as burning increased from code 0 to codes 5–6 (fig. 3.10). Con-

Figure 3.7
Spontaneous cracking of a whole goat radius-ulna that was shallowly buried (less than 5 cm) beneath the active coal bed.

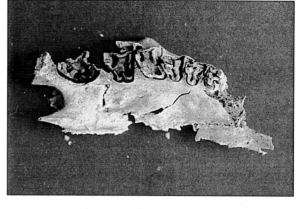

Figure 3.8
Fire-exploded teeth in a goat maxilla, contrasted with the more stable morphology of the bone palate in the absence of mechanical pressure.

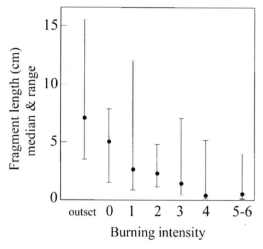

Figure 3.9
Composite illustration of bone fragment size reduction (medians and ranges) as a function of burn code and pressure-induced fragmentation (agitation in a box or foot trampling). The data represent a compatible set of results from bone samples from fire experiments 2 and 5. "Outset" represents fresh bones selected for the experiments. (Reproduced from Stiner et al. 1995.)

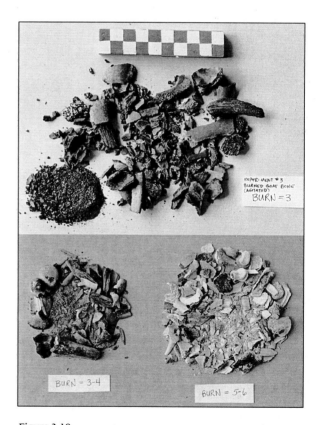

Figure 3.10
Differing degrees of fragmentation and pulverization from agitation in a box or from foot trampling for (top) carbonized bone burned to code 3 and (bottom) partly carbonized and partly calcined bone burned to codes 3–4 and the mainly calcined bones burned to codes 5–6.

cordantly, bone identifiability declined, initially with respect to skeletal element and ultimately with respect to the recognizability of bone tissue itself. Agitation of burned bones in code class 3 resulted in approximately 6 percent fine powder by volume (table 3.5). The same amount of agitation produced fractions of 10 percent and 11 percent powder by volume for burned bones in code classes 4 and 5–6, respectively. With extreme burning, or calcination, we might expect bones to be nearly or completely reduced to a powder by com-

paction and trampling. Instead, carbonized bones were extensively fragmented but usually were still recognizable as bone.

Table 3.5

Bone fragment size as a function of burn code before and after agitation and trampling

Burn Color Code	Mechanical Action	No. Observed	% of Volume as Powder	Median Size (cm)	Size Range (cm)
Bones buried 1–4 cm below fire bed, trampled half only (fire 5)					
0	(Outset) fresh	14	—	7.0	3.5–15.5
0	Trampled	6[a]	—	5.0	1.6–7.8
1	Trampled	11	—	2.7	0.9–12.0
2	Trampled	27	—	2.3	1.2–4.8
3	Trampled	87	—	1.5	0.5–7.0
Bones placed partially in fire, later agitated in box (fire 3)[b]					
3	(Outset) selected	80	—	2.0	0.4–6.4
3	Agitated	741	6	0.4	0.2–5.8
Bones placed directly in fire, later agitated in box (fire 2)[c]					
0	(Outset) fresh	74	—	3.5	1.2–9.2
3	No agitation	6a	—	2.9	2.6–3.4
4	No agitation	32	—	3.1	0.4–5.6
5	No agitation	55	—	2.2	0.4–7.0
6	No agitation	41	—	1.4	0.4–7.0
4	Agitated	228	10	0.4	0.2–5.1
5–6	Agitated	536	11	0.6	0.2–4.0

[a] Sample was exceptionally small.

[b] Burning to color code 3 was produced by placing the bones partly in or atop the fire; a range of carbonized fragments was selected from the burned material, measured, and then agitated in a box. (—) Powder not recovered for trampled samples.

[c] Bones were placed in the center of the fire; this produced advanced burning damage only.

The rates at which bones were altered by heat are also important from an archaeological perspective. Because the effects of heat are nearly instantaneous, the transformations and increased susceptibility to fragmentation caused by burning may take place within the same time frame as human activities at a site. This is less true of weathering and not at all true of fossilization, as we discuss later.

RESULTS FOR BURNING OF BURIED BONE

The next question was whether bones buried below a fire could be burned, and if so, what extent of burning was possible. In some of the experiment fires, bones were placed at various depths in sieved dry sediment and the fires built on top. Although bones were buried as deep as 15 cm below the fire, only those specimens in the first 5 cm were much affected by heat from the fire (table 3.6). What is more, these shallowly buried bones were burned only to the point of carbonization (burn code 3). The contrast between the maximum extent of burning of buried and unburied bone is shown in figures 3.11 and 3.12. Goat ribs buried 5 cm below the fire bed were visibly altered, whereas bones buried 10 cm deep were largely unmarred by heat. Interestingly, the heads of some goat ribs that protruded upward through the soil were locally burned in a way that resembled roasting damage observed in ethnographic contexts (e.g., Gifford 1977; Gifford et al. 1980; Yellen 1977). Tests for randomness in the anatomical positions of localized burning would be the most obvious way to address the possibility of roasting damage at the assemblage level.

The buried bones underwent corresponding alterations in color and mineralogical characteristics (table 3.7). The SF and carbonate values for buried bones burned to color codes 1–3 are analogous to values for similarly burned bones that had been exposed to fire on the ground surface. We were unable to induce calcination in bones buried under any depth of soil, despite the fact that our experiment fires were comparatively hot (maximally 900 degrees C; see also Bennett 1999). This is one potential explanation for the paucity of calcined bone in archaeological sites; another is that calcined bones are more easily destroyed by crushing, due partly to their greater inherent fragility.

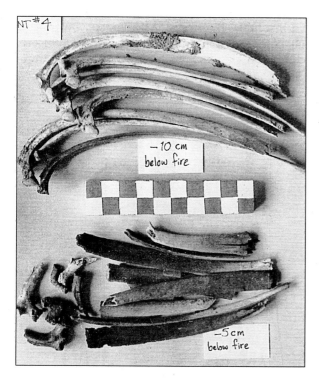

Figure 3.11
Differential burning on goat rib elements buried at 5 cm and 10 cm below the coal bed in fire 4. (Reproduced from Stiner et al. 1995.)

Figure 3.12
Contrasting conditions of partial goat and cow bones calcined (white) within the red coal bed (in the absence of mechanical pressure) and carbonized skull and mandible fragment (black) buried about 5 cm below the fire bed.

Table 3.6
Numbers of bones in each burn category by depth below experimental fires 2, 4, and 5

Depth Buried in Sediment (cm)	No. Buried	No. in Burn Color Code						
		0	1	2	3	4	5	6
0, in fire bed	134	—	—	2	4	32	55	41
1–4	142	5	11	34	92	—	—	—
5	17	2	2	4	9	—	—	—
10	8	5	2	1	—	—	—	—
15	4	4	—	—	—	—	—	—

NOTE: The numbers of specimens buried 5 to 15 cm below the experimental fire beds were small, primarily because large, whole goat bones were used.

INFRARED SPECTRA OF FRESH AND BURNED MOLLUSK SHELLS

The mineral component of most fresh mollusk shells (gastropods and bivalves) found in archaeological sites, including the marine and terrestrial species in our study area, is composed primarily of the carbonate mineral aragonite (e.g., Sabelli 1980; Wilbur 1964). Aragonite is an orthorhombic form of calcium carbonate ($CaCO_3$), but it possesses a crystal structure different from that of calcite, the other common polymorph with the same chemical composition (Parker 1994). Aragonite also has a higher specific gravity and less marked cleavage than calcite. In addition, aragonite is relatively unstable, and it may dissolve and reprecipitate as calcite during diagenesis. Heat from fire transforms the aragonite of mollusk shells to calcite almost instantaneously. Under sedimentary conditions, buried shells may be transformed to calcite via diagenesis, and in certain other instances they may be transformed to carbonated apatite, a situation of interest to research on skeletal preservation (chapter 4).

Aragonite and calcite are easily distinguished by FTIR spectroscopy. Aragonite, whether from inorganic or biological sources, displays characteristic infrared peaks at the wavenumbers 1083, 858–860, 713, and 700 (cm⁻¹) (fig. 3.13). Calcite displays characteristic infrared peaks only at the wavenumbers 874–876 and 713 (cm⁻¹). FTIR spectroscopy cannot, however, distinguish between slow diagenesis and rapid, heat-induced alteration. Visible discoloration of mollusk shells—specifically, the transition from a white or cream color to a dull true gray—is more helpful on this question in archaeological contexts, a distinction supported by the experimental evidence. In nearly every Pleistocene snail shell sample from Hayonim Cave tested by FTIR ($n = 38$), a true gray color corresponded to a calcite signal, and a white-cream color, to aragonite. In only two cases for which burning damage was deemed possible but uncertain on the basis of color, the infrared spectrum indicated calcite. However, additional tests are needed to prove that all of the gray calcitic shells in Hayonim Cave were burned.

Table 3.7

Mineralogical and color properties of bones burned while buried beneath fire 5

Splitting Factor	CO_3 (874/565)	Burn Color Code	Visible Color
3.01	0.15	1	Yellow/brown
3.12	0.15	1	Yellow/brown
3.09	0.19	2	Yellow/gray
3.31	0.19	2	Brown/gray
3.26	0.17	3	Black
3.18	0.19	3	Black
3.50	0.13	3	Black

NOTE: The maximum degree of burning that could be achieved on buried bones was carbonization (code 3).

Many of the terrestrial gastropod shells (*Levantina spiriplana*) in Hayonim Cave appear to have been burned by fire. Such snails are known to inhabit caves and crevices (fig. 3.14), and they are common in cracks in the rocks of Hayonim Cave and its vicinity. Although food refuse abounded in the Paleolithic layers, there is no indication that these snails were consumed as food. After the snails died, their shells fell into the sediments below and were burned inadvertently by human-built fires (see also Stiner 1994:171–174, 1999).

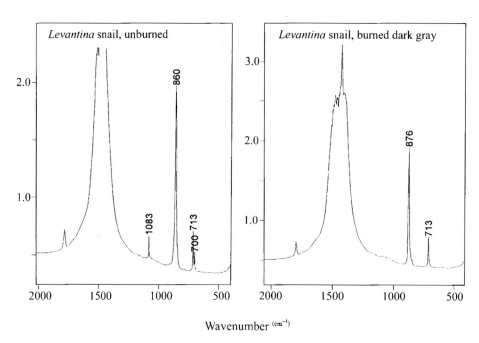

Wavenumber (cm⁻¹)

Figure 3.13

Left: Infrared spectrum for unburned shell of the modern land snail *Levantina spiriplana*, dominated by aragonite and displaying peaks at the wavenumbers 1083, 858–860, 713, and 700 (cm⁻¹). Right: Infrared spectrum for shell of the same species, burned to a dark gray color, in which the original aragonite has been transformed by heat to pure calcite and displays peaks only at the wavenumbers 874–876 and 713 (cm⁻¹).

Figure 3.14
Live *Levantina* snails grouped in a small crevice just outside the entrance to Hayonim Cave.

OVERLAPPING SIGNATURES OF BURNING, WEATHERING, AND FOSSILIZATION IN BONE

A second, related group of observations from the experiments on bone concerns potential overlap between the signatures of recrystallization caused by heat and those arising from weathering and fossilization. Shipman et al. (1984) have shown clear-cut relationships between bone discoloration, microscopic morphology, crystalline structure, and shrinkage due to heating in modern bone samples. A central implication of their study was that burned bones should be recognizable in the archaeological record by these same criteria. Our comparisons of archaeological and modern bone expose a few contradictory observations regarding the identification of burning damage by infrared spectroscopy and, probably, x-ray diffraction techniques. These contradictions offer some additional

insights into the nature of fossilization, weathering, and heat-induced transformations of archaeological bone.

Bone weathering normally results from exposure to wind, sun, and freeze-thaw cycles (Behrensmeyer 1978; Lyman 1994; Lyman and Fox 1989). It can also result when bones lying on the surface wick up ground water (Trueman et al. 2004). The smooth cortex of fresh bone degenerates fairly quickly under these conditions, cracking, splitting, and eventually flaking away from the outside inward. Visible weathering in the form of rectilinear cracking and exfoliation seems to occur only if bones are exposed to the atmosphere, especially where solar exposure is great. The rate of weathering can vary enormously with local conditions and even within a single locality.

SF values for a variety of weathered bones from two arid environments show consistent patterns (Stiner et al. 1995). Dog and goat bones exposed for approximately two years in Israel and cow bones exposed in a controlled setting for exactly nine years in New Mexico (USA) span the lightest and severest degrees of weathering; only the cow bones, some of them semi-protected by vegetation, were severely deteriorated. The exteriors of the bones that possessed only slightly weathered surfaces showed SF values significantly higher than values for the interior portions of the same bones (fig. 3.15). The interior SF values for these bones were lower in every case tested (table 3.8) and were generally equivalent to values for genuinely fresh bone (see also Weiner and Bar-Yosef 1990). SF values for the outer cortex ranged between 3.15 and 3.43, regardless of exposure time and degree of exfoliation, and the infrared spectra indicated some reduction of collagen content. The SF results suggest that the greatest microscopic transformations caused by weathering take place rapidly, apparently within the first year or two of exposure, and then begin to stabilize. These microscopic changes may therefore signal the onset of the process, but they do not correspond to the full trajectory of macroscopic degradation that zooarchaeologists associate with weathering damage. The changes in modern bone crystallinity caused by weathering partly overlap with those caused by heat between burn color codes 0 and 3 (i.e., up to complete carbonization).

The crystal lattice of bone mineral may also change if bones are buried in sediments for long periods of time. Indeed, some degree of alteration is to be expected in any archaeological or paleontological setting in which water percolation occurs (Weiner and Bar-Yosef 1990). Thirteen pairs of apparently burned and unburned bones, found together in Natufian or

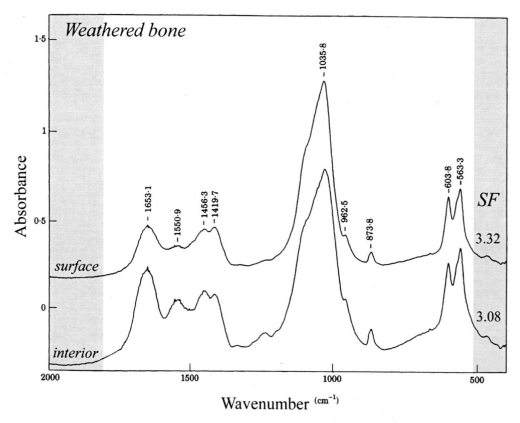

Figure 3.15
Infrared spectra of the surface and interior of a lightly weathered modern dog calcaneum, exposed for
approximately two years. Note the reduction in collagen (peak 1653 cm⁻¹) in the surface of the compact
bone wall in comparison with its interior only a few millimeters away. (Reproduced from Stiner et al. 1995.)

Paleolithic strata of Hayonim Cave, were analyzed for
splitting factor, carbonate ratio, and HCl-insoluble
fraction. The specimens in each pair presumably had
been subjected to roughly the same diagenetic history,
because they were found within the same 50 × 50 × 5
cm unit. The samples were selected conservatively on
the basis of color and with the knowledge that mineral
and organic stains might masquerade as burning. Surprisingly, the infrared data revealed no significant differences in SF or carbonate content between the
apparently burned and unburned archaeological samples (table 3.9). The results were remarkably uniform;
all specimens had SF values between 3.0 and 4.0,
apparently the usual condition for Paleolithic bones
from Hayonim Cave. The infrared spectra—at least
with respect to SF and carbonate content—did not differentiate between the mineral phases of apparently
burned and unburned archaeological bones, although
the same infrared data keyed to color codes readily distinguished the two states in modern bones. Moreover,

we found no correspondence between infrared readings and the age of the cultural material in our sample, in general agreement with Sillen's (1981) earlier
findings on calcium/phosphate ratios for Natufian and
Aurignacian samples.

It is conceivable, particularly in light of the foregoing observations about weathering, that diagenetic
processes can achieve the same effects in unburned
bones buried for thousands of years that moderate
burning achieved instantly in our experiment fires. If
this is correct, then another question arises. The apparently unburned bones from these strata all had rather
low SF values, ranging between 3 and 4. Why had the
crystallinity of the burned bones not continued to
increase with time and, in doing so, maintain the
expected difference in SF values between bones of
burned and unburned appearances?

One possible explanation for the similarities in SF
and carbonate content is that the darkened archaeological bones were not actually burned. To address this

Table 3.8

Splitting factor (SF) values for the surfaces and interior portions of modern weathered bones

Sample	Exposure Time (Years)	SF, Surface	SF, Interior
Goat bone, Israel[a]	<1	3.26	2.88
Goat bone, Israel[a]	<1	3.43	2.98
Dog calcaneum, Israel[b]	~2	3.32	3.08
Dog phalanx, Israel[b]	~2	3.36	3.11
Cow phalanx, New Mexico[c]	9	3.15	2.90
Cow femur, New Mexico[c]	9	3.27	2.95

[a] Specimens were skeletonized but somewhat greasy when collected for analysis; the bones had been exposed to full sun and were from individuals that had perished the previous winter in the northern Galilee, Israel.

[b] Two bones from one individual, collected from parkland outside Jerusalem, Israel; estimated time of exposure was two years.

[c] Samples from a nine-year control study conducted by Stiner in Albuquerque, New Mexico, USA. The specimens were semi-protected by vegetation throughout the nine years of exposure, but the degree of deterioration varied somewhat with the amount of plant cover; both specimens are severely weathered.

question, the 1N HCl–insoluble organic matrix fractions were extracted from the darkened bone of each pair, and their C/N ratios (analyses kindly performed by Dan Yakir of the Weizmann Institute) and infrared spectra were analyzed (following DeNiro 1985; DeNiro and Weiner 1988a, 1988b). Collagen, an important component of fresh bone, is burnable. Well-preserved bones in geological sediments may retain fair amounts of collagen or its combustion products, most of which show up in the HCl-insoluble fraction (Masters 1987). Table 3.9 shows the carbon/nitrogen (C/N) ratios of the 1N HCl–insoluble fractions. All of the C/N values are for the "burned" specimen of each archaeological bone pair from Hayonim Cave; generally, no insoluble organic phase is recoverable from unburned bones. All of the C/N values are above the range that is characteristic of collagen (i.e., 2.9–3.6; DeNiro 1985), and table 3.9 shows that the 1N HCl–insoluble fractions of the so-called burned bones from Hayonim Cave have essentially the same ratios as those calculated from infrared spectra. The spectra for this group are quite different from those for unburned bone tissues in different diagenetic states (fig. 3.16).

Table 3.9

Splitting factors (SF), carbonate (CO$_3$) ratios, and 1N HCl–insoluble fractions for apparently burned and unburned archaeological bone pairs from Hayonim Cave, Israel

Excavation Square	Cultural Affiliation	Unburned		Burned		Insoluble Fraction, Burned Only	C/N Ratio, Burned Only
		SF	CO$_3$ (874/565)	SF	CO$_3$ (874/565)		
F27d	K	3.62	0.18	3.45	0.16	0.8	4.11
J25b	N	3.35	0.17	3.07[a]	0.20	Absent	—
C27a	K	4.00	0.12	3.66	0.15	2.1	4.65
J25d	N	3.19	0.16	3.72	0.15	0.6	n.d.
K26a	N	3.31	0.17	3.26[a]	0.14	Absent	—
E28b	K	3.23	0.18	3.33	0.16	2.0	4.39
H18c	M	3.21	0.18	3.78	0.13	0.3	n.d.
M25c	N	3.21	0.19	3.39	0.18	4.8	4.98
K18d	A/M	3.18	0.19	3.44	0.18	1.0	7.90
K26b	N	3.31	0.18	3.82[a]	0.12	Trace	n.d.
D26d	K	3.50	0.18	3.39	0.20	1.0	4.84
C27b	K	3.26	0.19	3.39	0.18	2.3	4.54
G27b	M	3.35	0.18	3.30	0.21	2.1	5.25

NOTE: Bones were classified as burned or unburned strictly on the basis of specimen color. All samples are cortical specimens from ungulate long bones. Pairs of fragments were obtained from the same Paleolithic levels, subsquares, and 5 cm cuts to control for variation in preservation conditions within the cave deposits..

KEY: (N) Natufian cultural affiliation; (K) Kebaran; (A) Aurignacian; (M) Mousterian. (n.d.) Not determined.

[a] Only trace amounts of 1N HCl–insoluble fraction.

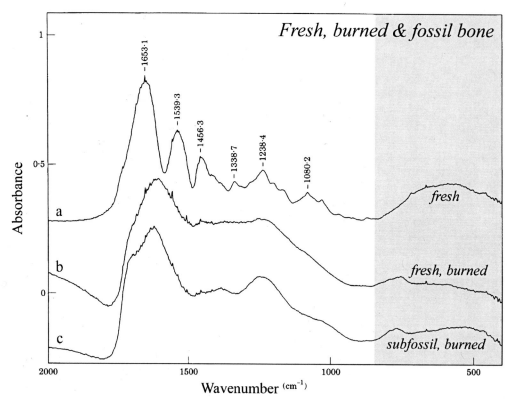

Figure 3.16
Infrared spectra of the 1N HCl–insoluble fractions of (a) modern fresh goat bone, (b) modern bone burned to color code 3, and (c) a fossil bone of burned appearance from square F27d in Hayonim Cave at 370 cm bd. The fresh bone spectrum is essentially that of collagen, characterized by the Amide I band at 1653 cm^{-1}, Amide II at 1539 cm^{-1}, and the proline absorption at 1456 cm^{-1}. The modern and fossil burned bone spectra are similar to each other and very different from the spectrum of fresh bone. The burned bone spectra can also be easily distinguished from the spectra of the insoluble fractions extracted from unburned fossil bones (see DeNiro and Weiner 1988a). Because the spectra represent only the collagen extracted from the samples, their contours do not resemble the spectra for collagen plus mineral content shown in the other figures. (Reproduced from Stiner et al. 1995.)

At least 11 of the archaeological bones classified as burned on the basis of color were in fact burned. The color criterion, so readily discernible by the naked eye, worked reasonably well. As a cross-check, the HCl-insoluble procedure could well be used as an additional criterion for identifying burned bones in archaeological deposits. Although visually based assessments of burning appear to have been correct most of the time (11 of 13), we note that two other bones identified as burned on the basis of color did not yield insoluble fractions; either these specimens were erroneously identified as burned or the burning damage was sufficient only to change their color but not their internal matrixes. Shahack-Gross et al. (1997) developed an alternative technique that is considerably more reliable on a specimen-by-specimen basis, but its systematic application to the large sam-ples in the Hayonim study was impractical because of its labor intensiveness.

Another explanation for the similarities in SF and carbonate content of burned and unburned archaeological bones gains some credibility here. A threshold seems to separate the consequences of low-temperature diagenesis of all sorts (mild fire exposure, weathering, and fossilization) from the changes that occur with calcination at high temperatures. Nonetheless, the signatures of crystallinity of bones altered by weathering, burning, and fossilization partly overlap. Infrared spectrometry effectively measures heat-induced changes in the crystallinity of modern bone mineral, with results analogous to those obtained by Shipman et al. (1984) using x-ray diffraction. The two methods do not monitor exactly the same physical parameters (Ziv and Weiner 1994), but they can be expected to produce related readings of modern and fossil bones. We found that the HCl-insoluble techniques we used, along with observations of macroscopic changes in internal bone color, diagnosed burning damage on archaeological bones more reliably than does x-ray diffraction.

DISCUSSION

Our experiments addressed topics in the fragmentation and diagenesis of skeletal tissues, including burning, weathering, and fossilization, with a focus on phenomena most likely to arise in shelter sites. The burning experiments compared the molecular and macroscopic signatures of damage in fresh, weathered, and partly fossilized bones. They also explored the conditions under which progressive levels of burning might occur in archaeological sites and the ways in which burning damage changes bones' crystal structure and susceptibility to fragmentation. Some bones buried down to 5 or 6 cm below the experiment fires were carbonized, but calcination occurred only with direct exposure to live coals. Infrared spectroscopy revealed that marked changes in crystallinity accompany the macroscopic transformations in color and friability of modern, fire-altered bone. There also was a monotonic, nonlinear decrease in mean fragment length across the six color categories when samples were agitated or trampled and a concordant decline in bone identifiability, first with respect to skeletal element and ultimately with respect to the recognizability of bone tissue itself.

The identification of burning damage on archaeological bone is a separate issue: molecular signatures of recrystallization in modern burned bones partly overlap with recrystallization caused by weathering after only one to two years of exposure in an arid setting and by partial fossilization of archaeological bones over the long term. Although infrared and x-ray diffraction techniques effectively describe heat-induced changes in modern bone mineral and are important aids for modeling diagenetic processes, these techniques do not reliably identify burning damage to archaeological bones. Readily visible color phases, augmented by HCl-insoluble fraction data, proved much more effective and economically feasible for the latter purpose (but see Shahack-Gross et al. 1997).

The findings on heat damage to bone and shell help qualify the behavioral implications of the stratigraphic associations between artifacts and hearth features in sites and the intensity of the use of space by human occupants. Burned bones are more fragile or brittle than unburned bones, and their mechanical strength varies nonlinearly with the extent of burning. In addition, bones buried in sediments beneath a fire may be burned by that fire, implying that bone deposition and bone burning can represent separate events during the formation of archaeological sites. Evidence of burning simply indicates that fires and bone debris were in close proximity to each other at some moment in the past; such evidence does not in itself prove that they were contemporaneous, although they might have been. The extent of damage to buried bones and other materials from fire is considerably less than damage to those directly exposed to red coals: bones were easily calcined in open campfires fed with Mediterranean hardwoods, but bones buried just a few centimeters below the surface were burned only to the point of carbonization.

Because burning makes bones more susceptible to fragmentation, many of the identifiable features of extensively burned bones may be lost soon after deposition if mechanical disturbances occur. With the application of pressure or agitation in our experiments, the greatest decline in the identifiability of bone (*sensu* the definitions provided in appendix 1) occurred between burn color codes 0 and 3, a gradient that contrasts the great strength of fresh bone with the fragility of completely carbonized bone. As burning damage approaches complete calcination, agitated bones may be reduced to ever finer fragments or even a fine powder. Burned bones in archaeofaunas generally are smaller than unburned bones because burned bones are more easily broken. As fragile as burned bones may become, however, it is not heat alone that breaks them, but also pressure (see also Knight 1985). Trampling is probably the foremost, or most immediate, cause of breakage in cultural contexts, especially if people use a place for prolonged periods or visit it repeatedly. This kind of situation is especially likely to arise in natural shelters and in masonry rooms (Schiffer 1983).

Beyond the basic question of whether hominids used fire, some of the greatest potential of burning data in human behavioral studies lies in the realm of site structure and spatial analysis. The archaeological signatures of burning considered here result from the interaction of heat, burial, and pressure, the latter commonly arising from foot traffic in an occupied space. Patterns of bone fragmentation and burning intensity may therefore inform us about the intensity of human activities, if qualified by data on sedimentation rates. One may wonder where cooking behavior, one of the presumed reasons for using fire, fits into all of this. Certainly the coincidence of fireplaces and food debris may represent behavioral associations. But cooking and extensive burning of bone seldom represent adjacent steps in site formation processes, since burning generally is not the purpose of cooking. (Exceptions may be the burning of shells of tortoises [chapter 5] and edible mollusks and of isolated bone ends that protrude from butchered meat.) Much of the burning damage in Hayonim Cave discussed in this chapter and in chapter 5 may have occurred postdepositionally, relating foremost to the reuse of certain locations within the site.

Finally, weathering can rapidly induce changes in bone mineral crystallinity in a manner resembling the effects of fire. In situ diagenesis, associated with fossilization, may also lead to analogous effects on bone matrix over the course of many years. Infrared and x-ray diffraction techniques applied to bone in modern controlled settings are valuable tools for understanding and modeling the character of diagenetic progressions. These techniques, however, are probably inappropriate for confirming burning damage on archaeological bones. Although these criteria certainly diagnose burning damage in modern circumstances, many things can happen in the interval between bone discard by prehistoric humans and excavation by archaeologists. If anything, archaeologists' color-based methods, supplemented by HCl-insoluble fraction analyses, are more reliable for identifying burning damage in archaeofaunas.

Many of these findings are applied in chapter 4 to the problem of patchy bone distributions in the Mousterian layers of Hayonim Cave. They are called upon as well in the analyses of macroscopic damage in chapter 5.

CHAPTER FOUR

Bone, Ash, and Shell Preservation in Hayonim Cave

Mary C. Stiner, Steven L. Kuhn,

Todd A. Surovell, Paul Goldberg,

Amy V. Margaris, Liliane Meignen,

Stephen Weiner, and

Ofer Bar-Yosef

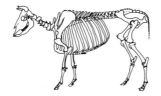

THAT BURIED BONES AND WOOD ASH can dissolve away raises the question of whether such materials were once present in archaeological sediments that lack them today. In this chapter we are concerned with differential preservation among excavation units in the Mousterian layers of Hayonim Cave—and not simply with how dissolution might distort our perceptions of site structure and human disposal behavior but also with what can be done to control for its effects. Flint artifacts were ubiquitous in the Mousterian layers of Hayonim Cave, whereas bones and the visible traces of fire displayed very uneven distributions. Distinguishing human and other mechanical phenomena from preservation effects requires information on the sedimentary environment of the Mousterian layers as well as on the condition of the bones in question.

Two materials common to archaeological sites, animal bone and wood ash, are rich in phosphate and carbonate minerals, respectively. Bone mineral is composed of the calcium phosphate mineral dahllite, also known as carbonated apatite (McConnell 1952), whereas calcite is a major component of fresh wood ash. The mineral compounds in vertebrate skeletons, most mollusk shells (mainly aragonite), ostrich eggshells (calcite), and wood ash are all susceptible to dissolution and recrystallization in acidic sedimentary conditions. Dissolution removes minerals, which may reprecipitate elsewhere. Heating and fossilization also promote recrystallization of bone mineral, the crystals of which can grow and become more orderly postmortem (Hedges and Millard 1995; Shipman et al. 1984; Stiner et al. 1995). The processes of dissolution and recrystallization require water, a powerful and common solvent, and the presence of organic compounds that together may lower the pH of the surrounding sediments (Hedges and Millard 1995; Hedges et al. 1995; Karkanas et al. 2000; Shahack-Gross et al. 2004; Weiner and Bar-Yosef 1990).

Because mineral constitutes the bulk of bone, substantial loss of mineral reduces a bone's visibility and identifiability in sediments. Bone dissolution from pH reduction inevitably occurs when organic matter is oxidized (Berner 1971; see Berna et al. 2004 for a detailed analysis and discussion of bone solubility and recrystallization). The presence of calcite in sediments, on the other hand, tends to stabilize pH at around 8.0 and thus minimizes or halts dissolution. Once it begins, however, dissolution seldom proceeds at a gradual pace—a variety of studies indicates that dissolution is promoted by subtle changes in sediment chemistry and may occur very rapidly (Hedges et al. 1995;

Karkanas et al. 2000; Weiner and Bar-Yosef 1990; Weiner et al. 1993). The completeness of bone destruction by dissolution depends upon acid concentration and the degree of solution recharge relative to the mass of the subject material.

Dissolution and reprecipitation should leave characteristic mineral traces in their wake. For example, wood ash contains a small fraction of highly resistant mineral components, siliceous aggregates and phytoliths, that persist in sediments after carbonates and phosphates have been flushed from them (Albert et al. 1999; Schiegl et al. 1994, 1996). The presence of calcite, dahllite, or both in sediments indicates conditions conducive to good preservation of bones and wood ash (Schiegl et al. 1996; Weiner et al. 1993), regardless of whether these compounds originated from biological or geological sources. When calcite or dahllite is absent from the sediments and other, less soluble phosphate minerals are present (Schiegl et al. 1996), one may conclude that any bones once present would not have been preserved (Weiner et al. 1993). These points also apply to mollusk shells, since aragonite is even less stable than calcite.

To understand the formation history of the Mousterian layers at Hayonim Cave and of the faunal assemblages within them, we began by examining the condition of wood ash residues and other minerals in the sediments using infrared spectroscopy. We then compared these findings with infrared data on bone mineral condition and macroscopic observations of bone characteristics such as fragmentation, the ratio of porous to compact tissue types, and the presence of abrasion. The results of these analyses pointed to yet another process, mechanical disturbance in the strata column, the scale of which was evaluated in part from the proportion of intrusive (post-Mousterian) artifacts. Distinctive features of this study included the large scale at which infrared analysis of bones and sediments was conducted and the systematic cross-referencing of the results of the mineralogic and macroscopic data. The results allowed us to isolate the variables most suitable for addressing site formation processes and faunal preservation in this Mediterranean cave.

ARCHAEOLOGICAL DISTRIBUTIONS

Artifacts and faunal remains were prevalent in the Natufian (Layer B), Kebaran (C), Aurignacian (D), and Mousterian (E) layers of Hayonim Cave (Bar-Yosef 1991; Belfer-Cohen and Bar-Yosef 1981; Tchernov 1994). Layers F (also Mousterian) and G (Acheulo-Yabrudian), though rich in stone artifacts, lacked bones almost entirely. Layer E was about 2.4 m thick in the central excavation trench and represented tens of thousands of years of deposition (see chapter 2). The study sample came from the central excavation trench (see fig. 1.2), where the Mousterian layer was overlain by Aurignacian and, to a lesser extent, Natufian deposits. There, the bedding in Layer E was generally horizontal, spanning excavation squares I–K 20–24.

Bones and flint artifacts were particularly abundant between 400 and 470 cm below datum (bd) in the central trench (fig. 4.1). They were present in smaller numbers between 300 and 400 cm bd and between 470 and 520 cm bd. In contrast to the roughly continuous horizontal scatter of lithic artifacts, a diagonal swath nearly devoid of bones bisected the central trench area from top to bottom (fig. 4.2). The boundaries of this peculiar feature were quite sharp. Bones were most abundant under calcareous breccia shelves near the cave walls, where flowstones and localized calcium saturation shunted seeping water toward the center of the cave. Isolated "hearth areas" rich in charcoal and recognizable ash lenses (fig. 4.3) were common where bones were abundant and generally absent where bones were rare.

The "bone-poor" zone did not lack bone entirely—bones were just conspicuously few in comparison with nearby units. Below 470 cm bd, the positions of the bone-poor and "bone-rich" zones shifted somewhat, but the contrasting distributions of bones and stone artifacts persisted. The orientations of the long axes of piece-plotted stone artifacts (fig. 4.4) in the sediments indicated no alignment biases of stone artifacts in the bone-rich versus the bone-poor zone. This was in contrast to the orientations of the bones (fig. 4.5), which were different in the two zones; bones in the bone-poor zone were oriented most randomly of all. Thus, there is no evidence of hydrological sorting or removal of archaeological materials, refuting the possibility that the bone-poor swath represented an erosional channel. Localized dissolution might account for the markedly uneven distribution of bones in Layer E. If so, it occurred before the formation of the Aurignacian layer (D), which was uniformly rich in bone (Belfer-Cohen and Bar-Yosef 1981).

TWO HYPOTHESES FOR THE BONE DISTRIBUTIONS IN LAYER E

Differences in bone and flint artifact distributions in Hayonim Cave are easily appreciated from the maps of piece-plotted materials (fig. 4.1). These observations

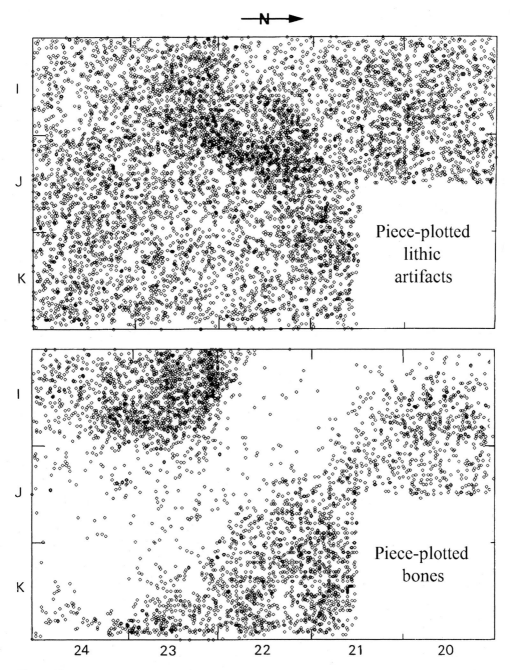

Figure 4.1
Spatial distributions of piece-plotted lithics (top) and bones (bottom) in Mousterian sediments from 300 to 540 cm bd in the central trench, viewed from above. Horizontal spatial units are square meters. The mild clustering apparent in the lithics distribution is due only to a small concentration of artifacts encountered at the boundary of Layers E and F.

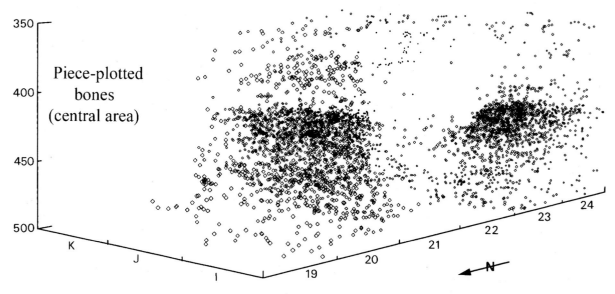

Figure 4.2
Three-dimensional view of the distribution of piece-plotted bones from Mousterian Layer E in the central trench. Horizontal spatial units are square meters; vertical increments are centimeters.

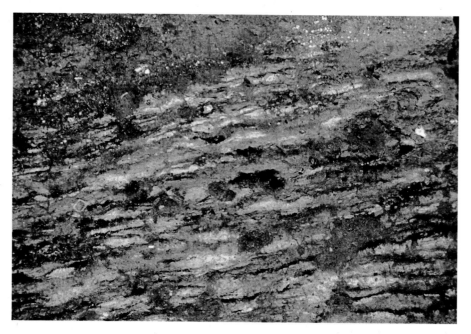

Figure 4.3
Detail of the Mousterian hearth lenses exposed in the north wall just west of travertine deposits in the central excavation trench.

raised two contrasting hypotheses and sets of test implications about the cause of the bone distributions in the Mousterian layer:

1. The heterogeneous distribution of bones was the result of spatially discrete disposal behavior by humans that evidently was not practiced for stone

artifacts; that is, bones never were deposited in some units but were preferentially deposited in others (as they were in Kebara Cave, Bar-Yosef et al. 1992; Speth and Tchernov n.d.). If the preservation environment did not vary among units, then the condition of the few faunal specimens present in the bone-poor units should be about the same as

LITHIC ORIENTATIONS

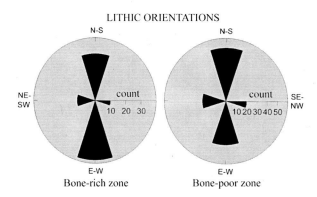

Bone-rich zone   Bone-poor zone

Figure 4.4
Orientations of piece-plotted lithic artifacts in the bone-rich and bone-poor zones of Mousterian Layer E of the central trench. Frequency distributions are shown according to four potential axes: N-S, E-W, NE-SW, and SE-NW.

BONE ORIENTATIONS

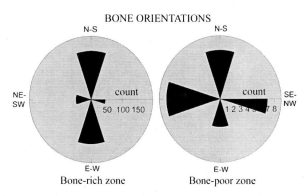

Bone-rich zone   Bone-poor zone

Figure 4.5
Orientations of piece-plotted bones in the bone-rich and bone-poor zones of Mousterian Layer E of the central trench. Frequency distributions are shown according to four potential axes: N-S, E-W, NE-SW, and SE-NW.

the condition of those in the bone-rich units. The sediments in bone-poor units should contain appreciable amounts of calcite, dahllite, or both.

2. The heterogeneous distribution of bones was the result of differential preservation over a scale of centimeters. Bones were once present in virtually all units but were dissolved locally in some and preserved in others. Whatever caused the near disappearance of bones in the bone-poor units operated almost exclusively in this spatial domain for tens of thousands of years. The few specimens in the bone-poor swath or zone could be in semialtered states as remnants of a nearly completed dissolution process. Sediments of the bone-poor units should not contain dahllite or calcite but should retain the more stable (i.e., insoluble) decomposition products of these parent minerals.

INFRARED AND MACROSCOPIC METHODS

Two distinct analytical scales were employed to evaluate the two hypotheses. At the molecular scale, we used Fourier-transform infrared (FTIR) spectroscopy to focus on the condition of generated minerals, judged according to independently established diagenetic sequences. FTIR analysis is well suited to research on carbonate and phosphate minerals (Weiner et al. 1995) because it reflects the strengths and arrangements of molecular bonds and hence the relative orderliness of crystal structure (Smith 1996). Abundant calcite or dahllite in sediments is taken to indicate good preservation conditions for bones and ash because these minerals are relatively unstable and dissolve easily; dissolution removes mineral from bone, just as it removes carbonates and phosphates

from the surrounding sediments. Recrystallization involves molecular reorganization of the bone's mineral component but is not necessarily independent of dissolution (Hedges et al. 1995). Although these diagenetic processes can be induced by many factors, water of neutral-to-low pH is essential for dissolution and recrystallization (Karkanas et al. 2000; Schiegl et al. 1994; Shipman et al. 1984; Stiner et al. 1995; Weiner and Bar-Yosef 1990; Weiner et al. 1993; Ziv and Weiner 1994). As explained earlier, advanced diagenesis is indicated by the prevalence of relatively stable decomposition products and an abundance of resistant silica components in sediments (Karkanas et al. 1999, 2000; Schiegl et al. 1996; Weiner et al. 2002). Thus, although the state of bone mineral preservation was the primary issue here, the results of FTIR spectroscopy for bones were compared with FTIR data for the surrounding sediments, the latter of which determined the preservation environment (chapter 2; Weiner et al. 2002). We examined sediment content and bone condition in relation to bone abundance by weight across multiple spatial units in the central trench.

The second analytical scale was macroscopic. At this scale we looked at variation in bone damage characteristics such as fragment length (cm), burning, and abrasion, in conjunction with skeletal tissue type, bone abundance per excavation unit ($50 \times 50 \times 5$ cm), and specimen identifiability. Mechanical and chemical decomposition should have consequences for most or all of the damage characteristics. Quantitative units for the analyses of bone were number of specimens (NSP), number of identified specimens (NISP), number of unidentifiable specimens (NUSP), and total bone weight (g) per $50 \times 50 \times 5$ cm excavation unit.

More data on macroscopic damage to bones, relating mainly to human behavior, are presented in chapter 5.

The FTIR bone sample was primarily from a SW-NE transect of the central trench, oriented to crosscut the maximum horizontal variation in bone abundance (fig. 4.6). Bones in the FTIR sample were also examined for macroscopic damage characteristics.

INFRARED SPECTROSCOPY OF BONES AND SEDIMENTS

Bone mineral recrystallization can be measured more or less directly using the infrared index known as the splitting factor (SF) (Termine and Posner 1966). Probable loci of high dissolution can be identified from the absence in the sediments of the two relatively unstable phosphate minerals most indicative of preservation—dahllite and calcite (Weiner and Goldberg 1990; Weiner et al. 1993). Phosphate in cave sediments may originate in biological sources (e.g., guano, organic refuse) or as decomposition by-products of fresh wood ash. Our method does not distinguish primary and secondary sources of phosphates—it recognizes only their presence or absence in the form of phosphate-containing minerals. Regardless of origin, the presence of calcite and dahllite in sediments is consistent with bone preservation, a relationship well demonstrated for Kebara Cave (Weiner et al. 1993). The decomposition of calcite and dahllite may follow one of several pathways, resulting in products as diverse as taranakite, montgomeryite, leucophosphite, and variscite, among others (Karkanas et al. 2000; Schiegl et al. 1996). In addition, the more resilient siliceous aggregates and phytoliths may be the only surviving remnants of wood ash in extreme cases (Schiegl et al. 1994).

In our analysis, more than 2,000 sediment samples and more than 600 bone samples were subjected to infrared spectroscopy, using two Midac Corporation (Costa Mesa, California, USA) Fourier-transform infrared spectrometers operated by Spectracalc (for Ms-Dos) or Grams 386 (for Windows 95) software (Galactic Industries Corporation, Salem, New Hampshire, USA). Sediment samples were taken throughout Mousterian Layers E and F and analyzed mainly on-site. Nearly all of these sediments contained ash derivatives of some kind, albeit in varied concentrations and states of preservation (Weiner et al. 2002). Preparation of sediment samples for FTIR analysis is straightforward, because the matrix is examined only for rank-order abundance of key minerals, most importantly carbonates, phosphates, quartz, clay, manganese and iron oxides, and silica compounds. In this case, each sediment sample was ground in an agate mortar and

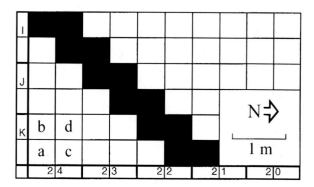

Figure 4.6
Transect placement in the central trench. Twelve 50 × 50 cm horizontal units, each about 5 cm deep through a total depth of 1.2 m, were oriented to crosscut the maximum variation in bone abundance.

pestle. An extract of a few tens of micrograms of this was mixed with a few tens of milligrams of potassium bromide (KBr) powder, ground again, pressed into a 7 mm diameter pellet using a Quik Handipress (following Weiner et al. 1993), and then inserted into the sample chamber of the infrared spectrometer.

The FTIR analyses of bone were performed at the University of Arizona (Tucson, USA), owing to the emphasis on quantitative variation in hundreds of bone spectra and the more elaborate system needed to standardize the pellet preparation procedure and control analyst-induced error. Experimentally based refinements in pellet production effectively kept SF measurement error to no more than ± 0.05 (for a complete technical description, see Surovell and Stiner 2001). Altered bone may be mineralogically heterogeneous, necessitating the use of relatively large, homogenized bone samples. Thus, the standardization of measures is an absolute necessity for the comparison of analytical results between studies or even between samples. Intensive grinding of bone samples for FTIR analysis can result in a net decrease in splitting factor; if grinding is overintensive, then SF measurements may respond more to variation in sample preparation than to the differences in bone mineral crystallinity it is intended to monitor.

We developed a set of standards for calibrating infrared crystallinity measures and a protocol for sample preparation (Surovell and Stiner 2001). After the surface of a specimen was mechanically cleaned, one piece approximately 10 mm in diameter and at least 3–5 mm thick was homogenized by pulverization for 15 seconds in a Wig-L-Bug ball mill (Crescent), using a steel capsule and mortar ball. This powder was sifted through nested 45 and 63 μm mesh screens to limit the particle size range. The powder fraction collected within this size range was added to powdered KBr at a ratio of 1 percent bone to 99 percent KBr. The mixture was returned to the Wig-L-Bug mill for an additional 30 seconds of regrinding and blending. A 50 mg extract of the fully homogenized mixture was then pressed into a pellet in the manner already described. The ball mill's steel capsules and ball bearings were cleaned between sample preparations, first by manual scraping with a small metal spatula, then by vigorous scrubbing with a test tube brush in a sudsy Alconox and tap-water solution, and finally by immersion in a fresh Alconox solution in a jewelry-grade gem sonicator run for one hour. The clean steel vials were then rinsed in distilled water and air dried under a heat lamp.

The FTIR analysis of bone employed four infrared peak indexes: the dahllite crystallinity index called the splitting factor (SF) and the carbonate (876 cm^{-1}), water (3500 cm^{-1}), and, if present, calcite (712 cm^{-1}) peaks, each standardized to the 563 cm^{-1} phosphate peak. Although not a direct measure of dissolution, SF is useful for tracking diagenesis in general (see chapter 3). Recrystallization of buried bone may occur gradually over many years, as part of fossilization, or through rapid transformations caused by weathering over a few months to decades. In contrast, high-temperature diagenesis is nearly instantaneous, especially above 650 degrees C, when solid-state recrystallization occurs (Shipman et al. 1984; Stiner et al. 1995). Much of the bone diagenesis in the Mousterian layer of Hayonim Cave seems to have taken place at ambient or low burning temperatures, exceptions being those few bones calcined by fire. Because we wanted to know about ambient temperature recrystallization of bone mineral, burned and obviously weathered bones were excluded from the FTIR analysis on the basis of visual inspection; darkened specimens of any sort were avoided.

MACROSCOPIC ANALYSES OF BONE

The macroscopic study of bone damage included 15,492 identifiable skeletal specimens (NISP) recovered from throughout the central trench between 300 and 540 cm bd. The great majority of these specimens were fragmented. In addition, 3,768 unidentifiable specimens (NUSP) from the transect area (see fig. 4.6) were examined. The FTIR bone samples (>600) were also unidentifiable fragments, raising the total NUSP obtained from the transect to more than 4,400, although the macroscopic qualities of the FTIR sample set were analyzed separately. The macroscopic variables considered were maximum fragment length in cm; presence of burning damage; identifiability to element and taxon or body size class (otherwise considered "unidentifiable"); tissue macrostructure, ranging from spongy to compact bone and "stony" enamel; and presence and intensity of abrasion damage.

RESULTS FOR SEDIMENT MINERALOGY IN RELATION TO BONE ABUNDANCE

Prior work in Hayonim and Kebara Caves (Schiegl et al. 1996; Weiner et al. 1993) had identified diagenetic sequences that remove calcite and dahllite, form montgomeryite, leucophosphite, and variscite, and leave behind siliceous aggregates in the sediments. Some of the chemical pathways observed in these two sites might have been unique to the geology of the region (compare, for example, Karkanas et al. 1999, 2002). The most important distinction, however, was one common to many caves—the presence of calcite or dahllite in sediments versus that of decomposition products known to derive from these parent minerals under reactive conditions. Figures 4.7 and 4.8 show certain 10 cm thick cuts in which sediments of the bone-poor swath lacked calcite or dahllite but in which highly decomposed mineral phases were abundant. Calcite and dahllite were well preserved on either side of this diagonal feature. Bone abundance by bulk weight (g), also shown in figures 4.7 and 4.8, clearly corresponds to the distribution of intact dahllite and calcite in the sediments, as does the distribution of terrestrial snail shells. FTIR analysis of 38 snail shell specimens confirmed that all of the unburned shells were composed of aragonite, a particularly soluble carbonate mineral discussed in chapter 3, again indicating favorable preservation conditions on either side of the bone-poor swath.

Figure 4.9 displays the percentages of sediment samples dominated by each of six key minerals relative to bone abundance for all of the units from 300 to 470 cm bd. There is a significant positive spatial relation between the mineral compositions normally conducive to ash and bone preservation and the observed quantities of bone in the central trench ($n = 521$, Spearman's $rho = 0.54$, $p < 0.001$), in spite of the fact that point-plotted sediment results were compared with bone NISP by gross $50 \times 50 \times 5$ cm excavation units. Calcite is the first compound to decline as bone abundance declines, followed by dahllite. The other minerals in the diagenetic cascade, representing conditions unfavorable to bone and ash preservation, predominate where bones are least abundant.

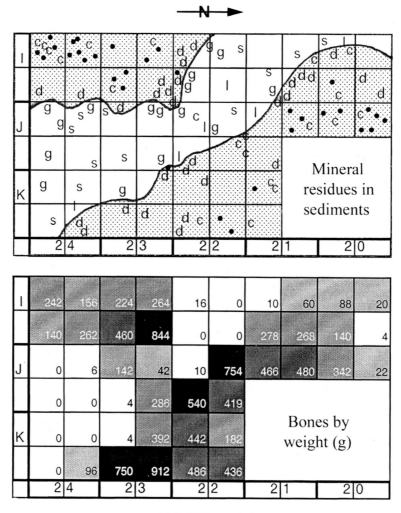

440-449 cm bd

Figure 4.7
Distribution of dominant minerals in sediments (top) relative to screen-recovered bone by weight in grams (bottom) between 440 and 449 cm bd in Mousterian Layer E of the central trench. Stippled shading (top) indicates sediments containing as their major components calcite (c) and dahllite (d)—good preservation conditions for faunal remains. Sediments containing poorly preserved wood ash and/or bone residues are indicated by the presence of montgomeryite (g), leucophosphite (l), and siliceous aggregates (s) as their major components—poor preservation conditions for faunal remains. Solid dots represent preserved land snail shells. Dissolution fronts appear as solid lines. Mapping of the fronts was performed interactively on-site on the basis of visible color changes and point-specific FTIR sediment analysis. This resulted in some clustering of samples on and around the dissolution boundaries. The number of sediment samples was 84. (Data on bone weights are missing for three subsquares in J–K 21.)

The uneven distribution of two types of anthropogenic materials in the central trench—faunal remains and wood ash—is therefore best explained by geochemical dissolution, not by selective deposition by Middle Paleolithic humans. Dissolution did not affect the flint artifacts, the chemical composition of which (mainly microcrystalline SiO_2) is quite resistant to common acids.

BONE MINERAL CONDITION RELATIVE TO SPECIMEN SURFACE AREA

Diagenetic processes that involve water could be mediated somewhat by variation in specimen surface area relative to mass. The macroscopic surface area of any skeletal specimen increases with fragmentation, while the density of the constituent fragments remains constant. Susceptibility to diagenesis could therefore

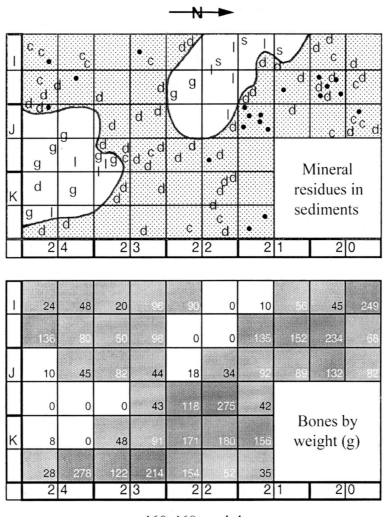

460-469 cm bd

Figure 4.8
Distribution of dominant minerals in sediments (top) relative to screen-recovered bone by weight in grams (bottom) between 460 and 469 cm bd in Mousterian Layer E of the central trench. Symbols are as in figure 4.7. The number of sediment samples was 83.

increase with fragmentation. In our analysis, bone specimen length served as a proxy (and decidedly imperfect) measure of surface area, since nearly all of the bones were fragmented to some degree. The analysis, applied mainly to the excavation units rich in bone, helped to qualify our perceptions of preservation. A negative relation between fragment length and SF was expected if diagenesis was enhanced with increasing surface exposure relative to bone mass.

The FTIR bone data weakly fit the expectation ($r = 0.27$, $p = 0.0001$), meaning that some diagenesis may have occurred to the well-preserved bones, at least in the form of recrystallization at the molecular level. However, its extent was quite limited overall, since the relation, though statistically significant, explains less

than 10 percent of the observed variation. Beyond noting this observation, we find it difficult to evaluate the data, because we do not know what to expect in the way of relative rates of macrostructure loss from dissolution.

Another way of testing for diagenetic effects relative to macrostructural differences is to compare mean SF values for two distinct bone tissue structures, the spongy (cancellous) and compact types, consistent with issues raised in chapter 3. Table 4.1 shows that SF is only slightly higher for spongy bone fragments than for compact bone fragments, given a potential range of 2.8 to around 6; other FTIR indexes remain constant between the two bone tissue classes. None of these ratio differences is significant. Slightly higher mean SF values for smaller specimens in general and for spongy bone specimens in particular might indicate that the surviving bones from the central trench were exposed to mild diagenesis, but this is not proved. Although dissolution cannot be measured directly by our techniques, bone crystal changes relative to exposed surface area of specimens suggest that very mild diagenesis occurred. The macroscopic condition of specimens nonetheless appears to be very good overall.

Diagenesis affects bone crystallinity at a scale much finer than that of the visible macrostructure. Because the crystals themselves have submicron dimensions (Robinson 1952; Weiner and Price 1986), this might be the critical scale for these effects, rekindling a point raised early in chapter 3. The compact bone tissues of large mammals are riddled with tiny canaliculi. If the bone contains Haversian systems, then larger channels housing blood vessels are also present in the center of each osteon (Currey 1984:30–36). Micropore abundance also increases as the collagen component breaks down (Nielson-Marsh and Hedges 2000a, 2000b). Bone macrostructure varies greatly both among and within postcranial elements in mammals, on account of porosity, but the microstructure and composition of bone are relatively constant across the vertebrate skeleton (Currey 1984)—roughly 65 to 70 percent mineral and 30 to 35 percent organic compounds and water (Currey 1984; Weiner and Wagner 1998).

RESULTS FOR BONE CONDITION IN RELATION TO BONE ABUNDANCE

Sediment mineralogy provides clear evidence for bone dissolution in a restricted area of the central trench yet very good preservation elsewhere. These findings call

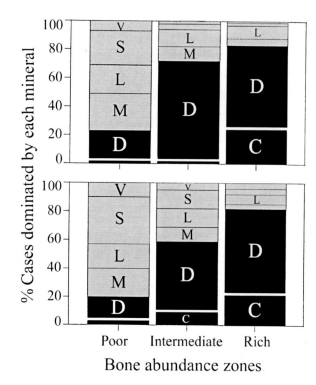

Figure 4.9
Percentage of the total number of sediment samples from each of three bone abundance zones that were dominated, by weight, by (V) variscite, (S) siliceous aggregates, (L) leucophosphite, (M) montgomeryite, (D) dahllite, (C) calcite. All minerals except dahllite and calcite represent secondary products of diagenesis. Black shading indicates relatively unaltered carbonates and phosphates, and light shading, the consecutive stages of diagenetic alteration of the parent minerals, from bad to worse.

for information about the condition of the few bone specimens present in the areas where preservation was generally poor. Toward that end, we looked for indications of bone diagenesis in relation to bone distribu-

Table 4.1
Splitting factor (SF) and other FTIR index means for compact and spongy bone specimens

Bone Tissue Type	SF[a]		Carbonate: Phosphate		Calcite: Phosphate		Water: Phosphate	
	Mean	SD	Mean	SD	Mean	SD	Mean	SD
Compact ($n = 479$)	3.2	0.2	0.2	0.1	.01	.02	.64	.14
Spongy ($n = 75$)	3.4	0.3	0.2	0.1	.02	.04	.61	.12

NOTE: Compact bone is dense and thick-walled. Spongy bone is composed of a material that is equally dense at the microscopic level but thin-walled at the level of macrostructure, and it possesses a higher ratio of surface area to mass once the thin cortical covering is breached. N = number of faunal specimens observed.

[a] The potential range of variation for recrystallization (SF) at low temperatures is roughly 2.8-4.5.

420-469 cm bd

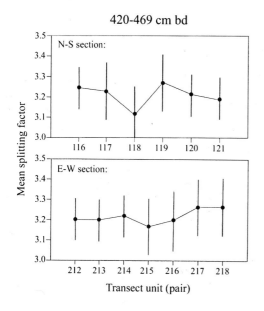

470-539 cm bd

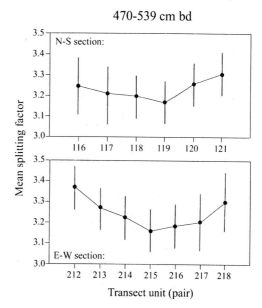

Figure 4.10
Mean splitting factors (SF) for FTIR-sampled bones from transect subsquare pairs, viewed N-S and E-W, above and below 470 cm bd. Vertical bars are standard deviations.

tions in space by conducting an FTIR analysis of bone fragments from the transect of the central trench (fig. 4.6) over a vertical range of 420 to 540 cm.

Figure 4.10 compares mean SF values for bones from transect unit pairs running N-S and E-W, above and below 470 cm bd (data in table 4.2). Little variation exists in bone mineral crystallinity in relation to bulk bone abundance (F ratio = 1.588, $n = 636$, $r^2 = 0.005$, $p = 0.205$). Oddly, mean SF is consistently lower (i.e., closer to that of fresh bone) where bones were least abundant and ash was most diagenetically altered. To the extent that bone mineral preservation varied among excavation units, SF values are the reverse of what was expected from sediment mineralogy and bone abundance. If different at all, the bones in the bone-poor zone were in *better* condition than preservation chemistry predicted.

Mean SF declines only slightly at the lower boundary of faunal assemblages in the central trench, varying between 3.2 and 3.3 before bones disappear below about 520 cm bd. There is no evidence for gradual vertical or horizontal transitions in bone condition, an observation that supports our perception of sharply bounded "dissolution fronts." Isolated specimens elsewhere in Hayonim Cave have SF values as high as 4.8 and partly decomposed appearances, but they are very rare. SF values for bone samples within the transect never exceed 4.5.

We also examined the relations of fragment size, burning damage, and skeletal tissue representation (spongy bone, compact bone, and tooth enamel) to spatial variation in bone abundance and sediment mineralogy. Presumably, the proportion of delicate to robust (dense) skeletal structures should decline with poorer preservation conditions. However, table 4.3 shows the relative frequency of fragile (spongy) bone, compact bone, and highly resistant (stony) tooth enamel to have been about the same across the bone-poor, intermediate, and bone-rich zones. What is more, mean specimen lengths and burning frequencies bear no consistent relationship to bone abundance across the three zones (table 4.4).

In contrast to the results for sediment mineralogy, few spatial correlations exist between the most readily visible aspects of bone specimens and bulk bone abundance. The only exception, as we discuss in the following section, is the somewhat higher degree of abrasion on bones from units where the total quantity of bone was small.

WHAT WAS THE ORIGIN OF THE BONES IN THE BONE-POOR ZONE?

Some macroscopic observations help to resolve the question of the origin of the bones in the so-called bone-poor zone of Layer E. Figure 4.11 compares the

Table 4.2

Splitting factor (SF) statistics for faunal specimens from transect subsquare pairs in Mousterian Layer E of the central trench

Excavation Grid Unit or Units	Unit Code	420–469 cm bd			470–539 cm bd		
		N	Mean SF	SD	N	Mean SF	SD
North-south view							
I24a, c	116	55	3.246	0.2	48	3.246	0.3
I24b, I23d	117	60	3.229	0.3	63	3.211	0.3
J23a, c	118	33	3.112	0.3	59	3.199	0.2
J23b, J22d	119	35	3.273	0.3	61	3.166	0.2
K22a, c	120	46	3.214	0.2	62	3.259	0.2
K22b, K21d	121	59	3.188	0.2	55	3.306	0.2
East-west view							
K21d	212	33	3.201	0.2	29	3.370	0.2
K22a, b	213	49	3.200	0.2	55	3.271	0.2
K22c, J22d	214	39	3.222	0.2	60	3.226	0.2
J23a, b	215	41	3.168	0.3	58	3.160	0.2
J23c, I23d	216	44	3.200	0.3	62	3.185	0.2
I24a, b	217	51	3.266	0.3	54	3.204	0.3
I24c	218	25	3.265	0.3	30	3.300	0.3

NOTE: Pairings increase sample sizes in this comparison.

severity of mechanical abrasion on FTIR-sampled bones, ranked into four classes (no abrasion to severest abrasion) relative to total bone weight per excavation unit. The degree of abrasion is always highest for specimens from units with the lowest weights of total recovered bone. A Spearman's *rho* statistic has a value of −0.38 ($n = 451$, $p < 0.001$) for abrasion intensity (a specimen-specific characteristic) against total bone weight (for a 50 × 50 × 5 cm volumetric unit). This relationship is significant in light of the differing sampling strategies and units of comparison for the two variables. Considerable tumbling or trampling is required to produce visible abrasion damage on bone. The fact that abrasion damage was most advanced in the bone-poor excavation units suggests that these specimens were subjected to greater mechanical disturbance. The more diverse inclinations of bone specimens in the bone-poor units, summarized in table 4.5, lend support to this conclusion. Whereas the majority of the piece-plotted specimens lay horizontally in the central trench, the proportion of nonhorizontal pieces was about twice as high in areas of low to moderate quantities of bone than in areas where bone was abundant. The bone and lithic orientations discussed earlier (see fig. 4.5) tell a related story.

Another perspective on the nature of bioturbation in the Mousterian layer can be gained from the frequency with which articulated bones and teeth occurred in it.

Articulated faunal specimens were rare in the Mousterian deposits overall, a condition typical of Paleolithic sites and therefore not surprising in the case of Hayonim Cave. It is interesting, however, that there is no clear relation between bone abundance, as measured by either unit or total sample size (NISP), and the incidence of articulated specimens (fig. 4.12, $n = 54$, $r = 0.161$, $p = 0.309$); the Mousterian assemblages are quite uniform in this regard. The number of articulated specimens here refers to all potentially separable specimens that remained in articulation up to the time of excavation, including teeth still seated in a bone armature and bone elements connected at natural joint surfaces (e.g., snake vertebrae, ungulate phalanges). Regrouping the data according to rank-order differences in bone abundance (fig. 4.13) reveals a slightly greater incidence of articulated specimens in

Table 4.3
Percentages of mineralized skeletal tissue classes relative to bone abundance in Hayonim Cave

Skeletal Tissue Class	Bone Abundance Zone		
	Poor (%)	Intermediate (%)	Rich (%)
Central trench			
Spongy (e.g., horn core, antler, vertebrae)	25	21	22
Fine spongy (e.g., carpals, tarsals, tortoise bone)	41	48	48
Compact (e.g., mandible, long bones, phalanges)	27	24	23
Stony (e.g., tooth enamel, ostrich eggshell)	7	7	7
Total NISP	(516)	(2,112)	(9,804)
Transect only			
Spongy (e.g., horn core, antler, vertebrae)	14	23	22
Fine spongy (e.g., carpals, tarsals, tortoise bone)	57	44	50
Compact (e.g., mandible, long bones, phalanges)	22	22	22
Stony (e.g., tooth enamel, ostrich eggshell)	7	11	6
Total NISP	(113)	(436)	(3,334)

NOTE: Tissue categories are generalized; they ignore known heterogeneity in total element structure, because the question is about relative, surface-mediated differences in resistance to decomposition on an ordinal scale. Percentages are calculated from NISP for large mammal, tortoise, and ostrich eggshell remains.

KEY: Tissue codes, from least dense to densest (following Wainwright et al. 1976), are as follows:

Spongy — Mostly spongy structure with thin compact bone covering. Macrostructure is relatively open, ratio of surface area-to-volume is relatively high, and thus this tissue class is the least dense of the four types.

Fine spongy — Mostly a fine sponge structure with thin compact bone covering.

Compact — Mostly compact or dense bone structure, but some spongy portions may also be present (e.g. epiphyses), and tissue is relatively dense.

Stony — Highly crystalline mineral structure, exceptionally dense, with a low surface-to-volume ratio.

Table 4.4
Fragment length statistics for specimens from the bone-poor, intermediate, and bone-rich zones in Mousterian Layer E

Abundance Zone	300–419 cm bd			420–469 cm bd			470–539 cm bd			All Depths		
	NISP	Mean (cm)	SD	NISP	Mean (cm)	SD	NISP	Mean (cm)	SD	NISP	Mean (cm)	SD
All specimens (burned and unburned)												
Bone-poor	61	3.7	2.0	132	3.1	1.7	124	2.7	1.4	351	3.1	1.7
Intermediate	160	3.6	1.7	265	3.2	2.0	1,242	2.8	1.7	1,698	3.0	1.8
Bone-rich	1,199	3.5	2.1	6,499	3.5	2.0	2,635	2.7	1.6	10,367	3.3	2.0
Burned specimens only												
Bone-poor	11	3.0	1.1	15	2.5	1.5	25	2.4	1.1	59	2.6	1.2
Intermediate	31	3.0	1.8	36	2.5	1.5	293	2.2	1.0	366	2.4	1.2
Bone-rich	208	3.0	1.3	1,119	3.0	1.7	479	2.2	1.3	1,815	2.7	1.6

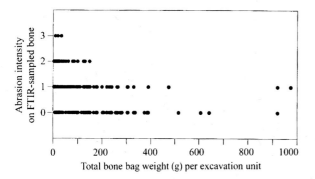

Figure 4.11
The intensity of abrasion damage on FTIR-sampled bone specimens relative to total screen-recovered bone bag weight (g) for volumetric excavation units in Mousterian Layer E of the central trench. Zero indicates no abrasion; 3 indicates the severest abrasion.

Table 4.5
Inclination results for piece-plotted bones from Mousterian Layer E

Inclination	No. Observed	% by Bone Abundance Zone	
		Rich	Poor
Horizontal	3,040	86	75
Inclined	392	8	15
Vertical	262	6	9

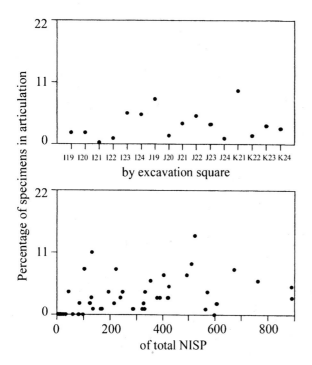

Figure 4.12
Percentages of articulated faunal specimens in Mousterian Layer E of the central trench, by 1 × 1 m horizontal excavation unit and relative to total NISP.

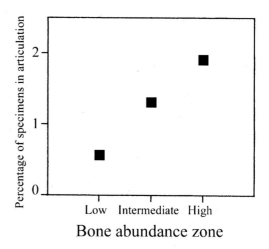

Figure 4.13
Percentages (of total NISP) of articulated faunal specimens in Mousterian Layer E of the central trench, by bone abundance zone.

the bone-rich areas of the Mousterian excavations. These differences are subtle, however, on the order of only 1.5 percent of total NISP, and are not statistically significant ($n = 54$, Spearman's *rho* $= 0.066$, $p > 0.1$). The skeletal articulation data indicate that the degree of sediment disturbance in Layer E was similar (i.e., relatively uniform) across vertical and horizontal units

of the central trench, consistent with the results presented earlier.

The most likely sources of mechanical disturbance in Hayonim Cave were human traffic and small fossorial animals (toads, snakes, and rodents), whose burrow trails were observed in some units at the time of excavation (fig. 4.14). Burrows and micromorphological evidence of trampling raise the possibility of time-averaging within the Mousterian strata series. If the dissolution hypothesis presented earlier is generally correct, but the few bones remaining in the bone-poor zone are in good rather than bad condition, then those bones could represent recent material introduced into older

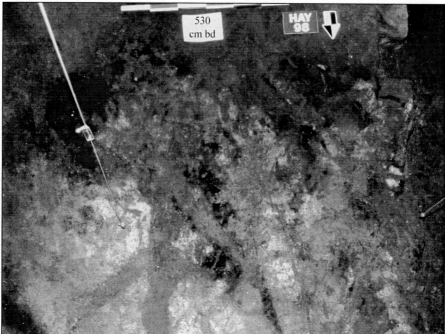

Figure 4.14
Rodent burrows crosscutting Mousterian hearth lenses (top) and features (bottom), damaging them but not destroying their visibility overall.

layers via bioturbation in combination with gravity. The visibility and quantitative importance of these effects for zooarchaeological analyses will be proportionately greatest in the units where the initially deposited bone was already lost by dissolution (fig. 4.15). The Aurignacian and Natufian layers just above the Mousterian in Hayonim Cave would have been ready sources of younger bone. New introductions would not have suffered the same fate as the Mousterian bones if the chemical environment had already stabilized. The lack of an empty swathe in the distribution of Natufian and Aurignacian bones in Layers D and B in the central trench confirms this interpretation.

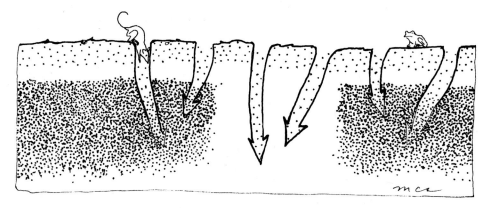

Figure 4.15
Scenario of the differential visibility of bioturbation effects in units preserving large quantities of original bone versus units in which original bone was lost by dissolution. Gravity tends to move younger material downward into older, bone-rich and bone-depleted units as animals make and modify burrows.

EVALUATING THE EXTENT OF BIOTURBATION

The great contrasts in bone preservation among units of Layer E in the central trench can be used to gauge the extent of time-averaging in the Mousterian sediment series. The potential effect of bioturbation on the chronological integrity of the cultural strata is a critical consideration, because the Mousterian layer is thick and resulted from a long occupation history. Attempts to scale the time-averaging effects of sediment disturbance are becoming common in paleontology (see Flessa et al. 1993; Kowalewski et al. 1998; Olszewski 1999), but seldom is the approach taken to archaeological sites (but see Grave and Kealhofer 1999). Gauging the extent and scale of bioturbation effects is feasible for many caves, because the inhabitable space tends to be limited, and cultural debris may be concentrated and superimposed in relatively small areas. The probability that time-coherent sedimentary series will form in caves is therefore high, though by virtue of the same conditions, the possibility for the mixing of younger and older material is also considerable.

Downward migration of younger stone artifacts into Layer E can be evaluated from the relative frequencies of Mousterian and post-Mousterian tool types. Time-diagnostic artifacts generally make up higher proportions of Epipaleolithic industries than is true for Mousterian industries, something that would amplify rather than suppress signals of potential stratigraphic mixing in the case of Hayonim Cave. Table 4.6 shows that post-Mousterian artifacts constituted about 3 per-cent of all diagnostic tools in the upper section of the Mousterian layer (300–419 cm bd), about 2 percent in the middle section (420–469 cm bd), and 0 percent in the lowest section (470–539 cm bd). Downward mixing could also have occurred in the lower sections of the Mousterian layer, but its aftermath cannot be detected from the stone tools. The artifact proportions nonetheless indicate that downward movement of younger material could account for as much as 3 percent of the total assemblage and that vertical migration usually occurred over short distances, since there is a clear decline in the frequency of post-Mousterian tools with depth.

It should be possible to evaluate from the faunal remains whether or not bioturbation occurred in the deeper portion of the vertical Mousterian series of Hayonim Cave, because the excavation units rich in bone flank those that are poor in bone throughout the sediment column of Layer E. The initially deposited bones probably dissolved soon after burial, consistent with the proposal of Karkanas et al. (2000) for diagenesis rates in caves in general, meaning that all of the specimens in the bone-poor units could be intrusive. Table 4.7 compares the frequencies of identified bone (NISP) in the bone-rich and bone-poor units, subdivided into three vertical segments; bone-intermediate units were not considered. The contrasting bone frequencies indicate a potential maximum addition of 2 to 5 percent (averaging 3 percent overall) of younger bone to the faunal assemblages of any vertical segment. An average of 3 percent downward migration of bones (and presumably artifacts) is not a great deal of mixing, at least for layers that retain original bone in large quan-

Table 4.6
Intrusive Epipaleolithic and Upper Paleolithic formal stone artifacts in Mousterian Layer E

Square	300–419 cm bd			420–469 cm bd			470–539 cm bd		
	pM	M	% pM	pM	M	% pM	pM	M	% pM
J19	2	160	1.2	0	49	0	0	32	0
J20	2	105	1.8	5	150	3.2	0	50	0
J21	3	60	4.8	2	70	2.7	0	28	0
J22	14	344	3.9	2	101	1.9	0	128	0
J23	7	279	2.4	5	275	1.8	0	256	0
J24	9	272	3.2	0	328	0	0	263	0
All units	37	1,220	2.9	14	973	1.4	0	757	0

NOTE: Data are for the J row of grid units only. Percentages of post-Mousterian artifacts (% pM) were calculated by dividing pM by the total number of diagnostic artifacts for all periods in a given sediment volume unit. Because later Paleolithic industries tend to contain higher proportions of time-diagnostic elements than do Mousterian industries, this analysis provides a *maximum estimate* of the contribution of younger, intrusive material to the subject assemblage. The proportion of intrusive material might have been less in actuality.

KEY: (pM) Post-Mousterian artifacts; (M) Mousterian artifacts.

Table 4.7
Inferred intrusive bone in the sediment column of Mousterian Layer E

Depth (cm bd)	Bone-Poor Zone (NISP)	Bone-Rich Zone (NISP)	% Intrusive Bone
300–419	61	1,199	5
420–469	132	6,499	2
470–539	124	2,635	5

NOTE: Percentages of intrusive bone were calculated by first subtracting bone-poor NISP from bone-rich NISP and then dividing bone-poor NISP by the "adjusted" bone-rich NISP. It was assumed for the sake of argument that all bones in the bone-poor units were intrusive. Units containing intermediate quantities of bone (in grams) were excluded from the comparison.

tities. Of course intrusive bones could constitute a high proportion of total bone where they penetrate units previously emptied of bone by dissolution. But intrusive bones constitute only a very minor proportion of the total bone in the dense Mousterian bone beds in Hayonim Cave. We conclude that the chronological integrity of the Hayonim cultural sequence in the central trench is essentially intact, but mild time-averaging occurred throughout the Mousterian series.

SEDIMENTATION, COMPACTION, AND HUMAN OCCUPATION INTENSITY

A final set of tests concerned the possibilities of sediment compaction from dissolution and variable rates of sediment accumulation relative to human presence in the cave. We compared lithic artifact abundance and sediment chemistry throughout the Mousterian sediment column of Layers E and F, squares I–K 20–24 (see fig. 1.2). The distribution of piece-plotted Mousterian lithic artifacts is shown for the J-row section of the central trench in figure 4.16. Although lithics were generally ubiquitous, artifact densities varied throughout the vertical column (fig. 4.17). One explanation for the fluctuations in artifact density is sediment compaction due to diagenesis (dissolution). Where the carbonate or apatite fraction of the sediments was little altered, sediments should have been preserved in most of their original volume, and artifact densities might be lower as a result. In this scenario—"hypothesis A"—evidence of mineral dissolution is expected where artifact densities are high. If this hypothesis is correct, then artifact density should be negatively correlated with the proportion of the sediment samples that was dominated by carbonate or apatite.

An alternative explanation for the variation in artifact densities in the sediment column—"hypothesis B"—is that sediments from external sources accumulated at different rates at different times. When large

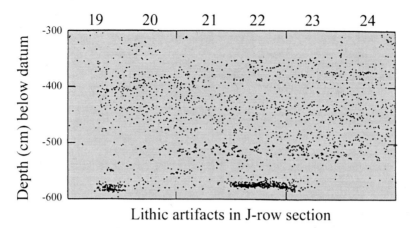

Figure 4.16
Distribution of piece-plotted lithic artifacts in the J-row section of the Mousterian sediment column (Layers E and F) of the central trench.

quantities of sediments were entering the cave, artifact densities would be lower. If sediment accumulation slowed relative to artifact accumulation, then artifact densities would be higher. According to this hypothesis, artifact density should be negatively correlated with the proportion of sediment samples dominated by quartz, the best proxy for allochthonous sediments, since it is not a major component of limestone.

The analysis began with a comparison of FTIR results for sediment chemistry across narrow vertical bands through the central excavation area that were characterized by high and low artifact densities. Then the *transitions* between the high- and low-density zones were considered in order to eliminate any signal due to long-term trends in the sediment column as a whole; the focus was instead kept on localized shifts from top to bottom. Table 4.8 indicates that the presence of calcareous sediments did not correspond to artifact densities. A low-density zone in the sediment column exhibited the extreme value for proportions of calcite/apatite, but high-density zones displayed high to intermediate proportions as well. These results suggest, contrary to hypothesis A, that the sediments were relatively unaltered overall and that little compression, if any, had taken place in the central trench area. The results for quartz fit somewhat better with hypothesis B. The two low-density zones in the comparison showed relatively large proportions of quartz, suggesting substantial inputs of external sediments—but although one high-density zone contained little quartz, as predicted, the other exhibited the highest levels of quartz of all. Clay could have originated from a variety of immediate and external sources, so it is an ambiguous signal, but data on clay were included for comparative purposes.

Another way to look at the questions of compaction and variable sediment input is in terms of contrasts in the calcite/apatite and quartz contents of adjacent sets of high- and low-density artifact zones (table 4.9). The results are similar to those from the first analysis. In only one of five transitions considered does the change in artifact density correspond as expected to the change in the condition of carbonate/apatite in the sediments—that is, there is less correspondence than would be expected by chance. For quartz, only two of the five "transitions" are as predicted by hypothesis B. Neither reduction in sediment volume nor variable rate of sediment input was the dominant influence on variation in artifact density in Mousterian Layers E and F. Instead, a behavioral signal is strongly suggested, reflecting the frequency and duration of human occupation of the cave. Noncultural factors, though present, were weaker, with the magnitude of external sediment input being somewhat higher than dissolution-induced compaction. Neither of these two factors acted globally in the deposits, however.

Some subtle, localized examples of external sediment input and compaction via mineral diagenesis and leaching were apparent in the central trench, particularly at the margins of the area from which the main bone sample derived. This was the zone of poorest bone preservation, where one could see some short, dipping "surfaces" in the distributions of the lithics around the point where bones disappeared in section view. There, sediments were altered locally and some compaction likely occurred. The amount of "deflection" was small, however, just 15 to 20 cm at most, and thus was unlikely to have had a major influence on artifact density. These effects were more apparent

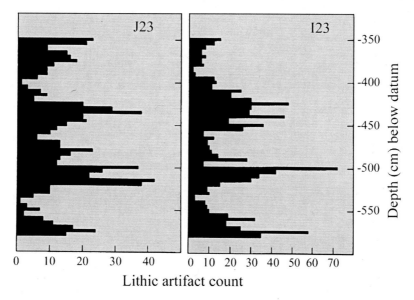

Figure 4.17
Examples, from 1 × 1 m squares J23 and I23, of vertical variation in lithic artifact counts throughout the Mousterian sediment column (Layers E and F) in the central trench.

DISCUSSION

In Hayonim Cave, calcite and dahllite were common in the Mousterian sedimentary units that also contained large quantities of bone. Conversely, decomposition products derived from calcite and dahllite dominated the units in which bones were scarce. The macroscopic condition of preserved bones was very good across the board, and only mild diagenetic activity in the form of recrystallization was indicated for the assemblages. The strong spatial agreement between the preservation conditions indicated by sediment mineralogy and by bulk bone and land snail shell abundances points to dissolution as the primary cause of the patchy bone distributions in the central excavation area. The hypothesis of an anthropogenic cause for unevenness in the bone distributions is refuted (in contrast with the situation in Kebara Cave; see Bar-Yosef and Meignen n.d.).

One might have expected, on the basis of these observations, that the few bones recovered from the "bone-poor" units would have been less well preserved. Contrary to expectation, bone mineral crystallinity (SF) and the ratio of fragile to resistant skeletal tissue types varied remarkably little across the bone-rich and bone-poor zones. The physical state of bone specimens was actually slightly better where bones were least abundant. Taken alone, the latter finding might seem to support the idea that the heterogeneous bone distributions in Mousterian Layer E resulted from variable disposal by prehistoric humans, habits that evidently did not apply to stone artifacts. The robust indications of preservation environment obtained from sediment mineralogy analysis prohibit this conclusion. Instead, the higher intensity of abrasion damage on bones from the bone-poor units, along with the presence of younger, intrusive artifacts in those units, argues for infrequent, small-scale bioturbation.

The observation that similar proportions of downward-migrating bones occurred in the three segments of the Mousterian sediment column is consistent with the notion that the initially deposited bones dissolved before any of the Upper Paleolithic and later archaeological deposits formed. Getting a grip on the issue of dissolution makes it possible to gauge the extent of stratigraphic mixing and, in doing so, to evaluate the chronological integrity of the faunal series. Burrowing animals and penecontemporaneous trampling appear to have moved younger materials downward in the sediments on occasion. This effect might also have been greatest where sediments were decalcified. The small proportions of intrusive material indicate that

Table 4.8

Sediment mineralogy across high and low lithic artifact density zones of Mousterian Layer E

Depth (cm bd)	Artifact Density	% Calcite/ Apatite	% Quartz	% Clay
375–405	Low (185/m³)	26.0	45.5	32.5
465–500	Low (238/m³)	55.4	50.5	17.8
430–455	High (469/m³)	54.9	23.9	5.6
500–520	High (500/m³)	41.3	47.5	35.0

NOTE: Percentages represent those sediment samples of the total number analyzed in which calcite or apatite was the first or second most abundant mineral.

Table 4.9

Sediment mineralogy in transitions between high and low artifact density zones of Mousterian Layers E and F

Transition (Depth, cm bd)	Artifact Density	Carbonate Preservation	Presence of Quartz	Hypothesis A	Hypothesis B
280–350 vs. 350–420	+	–	+	Yes	No
350–420 vs. 420–465	+	=	–	No	Yes
420–465 vs. 465–495	–	–?	+	No	Yes
465–495 vs. 495–530	+	+?	+	No	No
495–530 vs. 530–600	–	–	–	No	No

KEY: (+) Increases with depth; (–) declines with depth; (=) no change; (Yes) hypothesis is supported by the data; (No) hypothesis is not supported by the data.

bioturbation effects in Layer E were relatively unimportant, at least in the many units where bone preservation was good. A detailed view of these data therefore identifies the excavation areas that yielded samples most suitable for paleoeconomic research.

Dissolution is also important for distinguishing sediment compaction from differential rates of artifact and sediment accumulation. Lithic artifacts were abundant throughout the cultural layers of Hayonim Cave, yet artifact densities varied in the sediment column of Layers E and F, at least in the central trench area. The vertical variation in artifact densities is not explained by sediment compaction from mineral diagenesis or by variable rates of sedimentation. Instead, it reflects mainly the frequency and relative durations of the human occupations of the cave. Additionally, we found no consistent relation between infrared data for bone on a vertical gradient and the age of the cultural material in our sample, nor between the Mousterian, Kebaran, and Natufian layers. These observations are in general agreement with Sillen's (1981) earlier find-

ings on calcium/phosphate ratios for Natufian and Aurignacian samples from Hayonim Cave. The results, however, argue against the general possibility of using levels of apatite recrystallization for relative dating purposes (Sillen and Parkington 1996).

Our findings reflect more generally on current hypotheses about the timing and rates of dissolution in sediment formation histories. The crisp boundaries between favorable and unfavorable preservation environments for calcite and dahllite—and thus for ash and bones—in Hayonim Cave document the existence of major dissolution fronts in the Mousterian layers. Evidence for transitional specimen states was sought at the boundaries between the bone-rich and bone-poor zones, following visible changes in sediment color in the walls and floor of the excavation. Partly altered bone specimens did exist, but they were exceedingly rare. It was remarkably difficult to detect "transitional" bone specimens in the Mousterian deposits of Hayonim Cave, despite the obvious juxtaposition of favorable and unfavorable chemical environments in the sediments, as observed in their mineralogy.

Why should this have been so? The mineral decomposition processes seem to have operated at the nanometer or molecular scale. All-or-nothing preservation situations seem to have prevailed in the Mousterian layers of Hayonim Cave, as they probably did in other sites as well. Changes in mineral composition appear to have been rapid and thorough while the chemical conditions were reactive. Karkanas et al. (2000) probably are correct in their suggestion that once started, phosphate and carbonate dissolution proceeds rapidly until a new chemical stability field (equilibrium) is reached. The problems of preservation and phosphate diagenesis may be especially acute for Mediterranean caves and are perhaps somewhat less severe for colder areas of Europe.

Combining macroscopic and FTIR observations provides a more comprehensive picture of the taphonomic history of faunal remains in Hayonim Cave, a relatively complex situation that is to some degree typical of Mediterranean and other middle-latitude limestone caves. Identifying "preservation zones" in sites greatly simplifies a host of other assumptions about the quality of faunal data for paleoeconomic studies. This strategy proved to be effective and economical in our study. Differential skeletal body part representation or fragment refitting could not have resolved these questions about in situ bone loss in the Mousterian layers. Finally, some indicators of bone condition clearly disagree with those concerning the sedimentary preservation environment and are bound to mislead investigators if time-averaging effects are not also considered.

CHAPTER FIVE

Vertebrate Taphonomy and Evidence of Human Modification

ESTABLISHING THE AGENTS OF BONE COLLECTION and modification is not the most challenging problem of this chapter, although doing so is essential to the dietary analyses to come. The data presented here demonstrate the consistent imprint of human economic behavior on the Wadi Meged faunas from the early Mousterian through the Epipaleolithic period. The quality of bone and tooth mineral preservation is generally quite good where bones occurred in the deposits (fig. 5.1), though the assemblages are not altogether free of attritional effects. The absence of evidence for bone collectors and modifiers other than humans—at least for the macrofaunal remains—is unusual but not unknown for Mediterranean shelter sites (Stiner 1994).

The greater challenge in the taphonomic analyses was to assess in situ attrition, which might have biased the contents of assemblages and therefore archaeologists' perceptions of food transport behaviors. In this regard the focus was on density-mediated attrition of teeth, compact bone parts, and spongy bone parts of ungulates; macrodamage from weathering, burning, gnawing, humans' tools, and fractures across vertebrate groups (reptiles, birds, ungulates, and carnivores); and comparisons of carcass processing techniques and intensity within and between Paleolithic periods. The results described in this chapter clear the way for questions treated in later chapters on prey acquisition, carcass transport, and processing decisions of Paleolithic foragers in the Wadi Meged. The analyses were focused on the Mousterian, pre-Kebaran, and Kebaran faunas. For related taphonomic studies of Hayonim Cave, see Munro 2001 on the Natufian and Rabinovich 1998 on the Aurignacian.

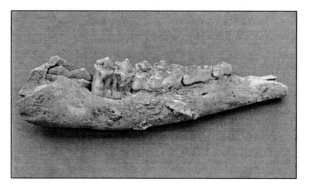

Figure 5.1
Mandible of a subadult fallow deer from Mousterian Layer E of Hayonim Cave with fragile components intact.

METHODS AND VARIABLES

The piece-plotted and screen-recovered fractions from Hayonim Cave were combined for the taphonomic analyses (see chapter 1). The main counting unit for the study of macrodamage was the number of identified specimens (NISP). The study of density-mediated skeletal attrition employed a subset of NISP—portion-of-element counts—which in some analyses was converted to the derived variable MNE, or minimum number of elements (see appendix 1). *Portion-of-element* refers to specimens that possess unique, irreducible features from which the original number of skeletal elements can be estimated. Several portions were considered per element type (see appendix 2). MNE was estimated for each skeletal member of a given taxon in an assemblage, based on the most common morphologically unique "portion" or feature for that element. Bone and dental elements still in articulation at the time of excavation were counted as separate specimens to avoid the conflating effects of variable postdepositional disturbance (Stiner 1994:69–73). For example, the mandible containing teeth in figure 5.1 and the articulated radius-carpal set in figure 5.2 each yielded a NISP count of six, because mechanical reworking of sediments *could* easily have separated them under other circumstances. Many of the analyses were divided according to vertebrate body size classes. For the ungulates, the small body size class included mountain gazelles *(Gazella gazella)* and the rare roe deer *(Capreolus capreolus);* the medium body size class included mainly fallow deer *(Dama mesopotamica)* along with some wild pigs *(Sus scrofa),* wild goats *(Capra aegagrus),* and wild asses *(Equus* cf. *hemionus);* the large body size class included mainly aurochs *(Bos primigenius)* along with horses *(Equus caballus),* red deer *(Cervus elaphus),* and an occasional hartebeest *(Alcelaphus buselaphus).*

The question of in situ bone attrition centers on what might be missing from an assemblage, relative to a complete animal model, as a result of destruction on-site. In situ attrition can result from mechanical or chemical processes. Obviously, skeletal parts that never reached a campsite have implications very different from those of skeletal parts lost through decomposition on-site or pilfering by carnivores. One of the more interesting advantages of working with shelter faunas is that the bones of herbivorous vertebrates usually must be transported into these places by other agents; ungulates in arid ecosystems may seek out rock overhangs as protection in harsh weather, but seldom do they die there in large numbers. The fact that humans

Figure 5.2
Transversely fractured distal radius of an artiodactyl with five carpal elements in anatomical connection and for which total NISP is six. In this case, all elements were recorded separately, along with information about the elements in articulation.

often take large herbivore carcasses apart to make them easier to transport to shelter camps adds yet another layer of information about the decisions of prehistoric consumers. In situ attrition can erase or distort the patterns of anthropological interest, however, and this issue must be addressed first.

Prey remains in Paleolithic sites tend to be highly fragmented, so much so that the damage can be quite characteristic, offering insights into carcass handling behaviors. Unfortunately, most kinds of damage also work against the identifiability of skeletal elements. The burning experiments described in chapter 3 clarified some of the mechanical (especially trampling) and thermal conditions that might normally occur around small campfires, as well as the forces that might further modify surface refuse or bones that were shallowly buried in sediments. Chemical decomposition was indicated for Mousterian Layer E of Hayonim Cave but seems to have had a more random effect with respect to body part preservation. The issue of differential skeletal destruction was considered in terms of biases in the relative proportions of spongy epiphyses versus compact shafts of ungulate limb bones, of bones versus teeth, and of developing versus fully mature tooth enamel or bone. In this chapter and again in chapter 11 I take up the subject of differential loss of element portions or of entire identifiable elements with the help of several other techniques.

The analysis of superficial macrodamage in the Wadi Meged archaeofaunas centered on positive indications of bone modification. Several types of macrodamage (see appendix 2) were cross-referenced in order to infer assemblage formation histories (following Behrensmeyer 1991; Lyman 1994:452–465; Stiner 1994:153–157). Weathering damage reflects mainly rates of burial and intensity of exposure to solar radiation (chapter 3; Behrensmeyer 1978; Lyman and Fox 1989). Gnawing damage by carnivores or rodents refers to the activities of nonhuman resident or visiting species. Burning, though usually a human signature in shelter sites, is mainly about fire-centered activity and spatial patterns of hearth rebuilding and somewhat less about the immediacies of cooking meat (chapter 3). Tool marks and impact damage are also diagnostic of human activities and, in some cases, informative about carcass processing routines. Many of the macrodamage criteria used here have been reviewed extensively elsewhere (e.g., Stiner 1994). Specific applications of these criteria are discussed where appropriate in the following sections.

A practical challenge to the analysis of the Meged Rockshelter faunas (Kebaran and pre-Kebaran) arose from the calcite concretions that covered many of the bones (chapter 2). Faunal specimens were extracted from a matrix rich in limestone rocks and hard clays, and many were broken during excavation as a result. Observations of fine superficial damage—especially cut marks—were hindered by the presence of mild to moderate concretions, the downside of an otherwise favorable preservation environment for the mineral constituents of the bones and mollusk shells. The concretions were seldom severe or thick enough to prohibit observations of gnawing damage at Meged Rockshelter, a type of damage that tends to significantly reshape natural bone surfaces. Observing cut marks was more problematical for the faunas from Meged Rockshelter; tool marks tend to be small and subtle in all of the Paleolithic periods represented in the Wadi Meged and many other Mediterranean sites. Consequently, a systematic comparison of cut mark frequencies for Meged Rockshelter with those for the assemblages from Hayonim Cave was not attempted. Processing evidence in the forms of burning damage, fractures, and heavy tool marks is readily visible in the Meged Rockshelter faunas, and in these aspects the assemblages are directly comparable with those from Hayonim Cave.

DENSITY-MEDIATED SKELETAL ATTRITION

A Comparison of Teeth and Cranial Bone

Teeth resist virtually all decomposition factors better than do bones, because they are more mineralized. All forms of mature bone are composed of no more than about 70 percent mineral, whereas fully developed tooth enamel is composed of about 95 percent mineral (Currey 1984; Hillson 1986; Wainwright et al. 1976). Teeth and bones therefore represent the most profound contrast in mineral density in mammalian skeletons. The first approach used in this study for evaluating the relative rates of loss of skeletal tissues capitalized on this contrast by comparing cranial MNE estimates based on a variety of dental and bony cranial "landmarks." The comparison was confined strictly to the heads of ungulates, in order to circumvent some of the biases in postcranial body part representation that might have arisen from differential transport by Paleolithic hunters.

Figure 5.3 presents a model of the expected rates of attrition of cranial bones versus teeth (following Stiner 1994:99–103). In this application, the slope of the line expresses the loss of compact bone features relative to the loss of teeth. Bony portions of the skull include the petrous, incisive, occipital condyle, mandibular condyle or coronoid process, anterior horizontal ramus, and so on (appendix 2). The strength of the correlation between bone-based and tooth-based MNE counts is also of interest with respect to consistency in preservation among stratigraphic units or assemblages. Of course the two estimates must be autocorrelated to some extent, because bones and teeth occur together in life. Assuming an intercept of zero, a slope of 1.0 represents perfect preservation of the two classes of skeletal material. A slope of less than 1.0 indicates that bone has suffered relative to teeth; a slope of 0.50, for example, means that compact bony parts of the cranium are preserved, on average, only half as often as dental parts.

Figure 5.4 (right) compares the highest bone-based MNE to the highest tooth-based MNE for the crania and mandibles of small, medium, and large ungulates in the seven early Mousterian assemblages (vertical units 1–7) from Hayonim Cave (see appendix 4). MNE is used in lieu of MNI because all of the bone portions and tooth elements considered occur as pairs in the natural skeleton; n refers to the number of paired observations. Deciduous teeth were included in the counts to the extent that their state of development or wear posed no risk of double-counting individuals. The observed slope of 0.832 and a squared multiple *r* of 0.888 indicate that loss of identifiable bone relative to identifiable teeth was quite limited in the Mousterian assemblages. Only 11 percent of all variation in bone-based and tooth-based estimates of ungulate cranial and mandibular MNE might be explained by density-mediated attrition. Adherence of the points to the line of the slope is strong, meaning that the relation is consistent for ungulates of different body sizes and among the seven Mousterian stratigraphic units. These results are remarkably similar to those obtained for Paleolithic faunas in Italian caves (Stiner 1994:101).

Essentially the same degree of preservation is evidenced for gazelles from the early Natufian deposits of Hayonim Cave (fig. 5.4; Natufian data from Munro 2001). Bone preservation was somewhat more compromised in the Kebaran layer of Hayonim and in the assemblages from Meged Rockshelter. The slope in figure 5.4 (left) declines to 0.559 for the Kebaran and pre-Kebaran ungulates (r^2 = 0.831) when the Natufian case is removed.

BONE SURVIVORSHIP VERSUS BULK BONE DENSITY

The next question concerns how different bone macrostructures of the cranial and postcranial skeleton might have fared relative to one another. Although all mammal bone is composed of about 70 percent hydroxyapatite, bulk density values (usually expressed on a scale of 0 to 1) for spongy and compact bone macrostructures differ. *Bulk density* (*sensu* Lyman 1984)

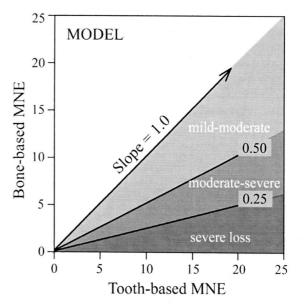

Figure 5.3
Modeled relations of the relative attrition of cranial and mandibular bone versus teeth. The slope of the line expresses the rate of loss of compact bone (about 70 percent mineral) relative to that of teeth (enamel is about 95 percent mineral). The strength of the correlation is of interest for assessing consistency in bone and tooth preservation among stratigraphic units. Assuming an intercept of zero, a slope of 1.0 corresponds to ideal preservation of these two classes of skeletal material; a slope of less than 1 indicates that bone has suffered relative to teeth.

is mainly an expression of variation in the bulk or concentrated mass of bone specimens that is semi-independent of bone volume, because some bones contain

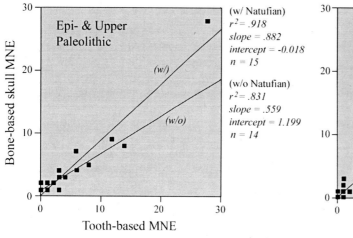

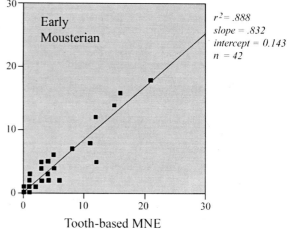

Figure 5.4
Comparison of highest bone-based MNE to highest tooth-based MNE for crania and mandibles of small, medium, and large ungulates in the Wadi Meged assemblages. Left: Epipaleolithic and Upper Paleolithic with and without Natufian gazelles. Right: Seven Mousterian stratigraphic units.

more open pore spaces than others. Pores introduce weaknesses, and measurements of bulk density are meant to reflect differences in bone resistance to destructive forces, mainly at the level of macrostructure (Currey 1984). This proxy for bone strength interests zooarchaeologists mainly because of the great importance of skeletal macrostructure in taxonomic and anatomical identifications. Changes in shape that arise from fragmentation must also affect inherent resistance to mechanical loading.

In the rest of this section I address potential biases in skeletal part representation at the level of portions-of-elements throughout the ungulate bony skeleton using photon densitometry measurements (Lyman 1984, 1994). This approach has grown out of innovative early work by C. K. Brain (1969, 1981) and L. R. Binford (1978, 1981) for the economic analysis of ungulate faunas. Its advantages center on its attention to the anatomy as a whole, achieved through comparisons at the level of portions-of-elements.

In this study, limb end (epiphyseal) and shaft features such as foraminae and attachment scars were used to estimate limb element MNEs (fig. 5.5), and all fragments, including shaft splinters, were examined. For the skull, only the bony portions were applied in the comparisons with postcranial MNEs, because tooth enamel is so much denser than bone of any kind. The portion-of-element categories tend to be hierarchical, because fragment sizes vary and a specimen may contain more than one portion suitable for estimating MNE (appendix 2). However, some portions-of-elements nearly always yielded higher counts than others, possibly because of their greater inherent resistance to mechanical destruction. Countable portions should be independent of fragmentation effects as much as possible, so small, compact features were favored for this purpose.

Many of the portions listed in appendix 2 coincide with Lyman's photon densitometry scan sites (1994: 234–250). Calculation of portion-of-element survivorship followed the reasoning of L. R. Binford (1978) in that MNE counts were standardized relative to a complete anatomical model to obtain MAU (minimum animal unit) values (observed MNE/ expected MNE for one whole skeleton). MAU was then normed to the highest MNI value for the subject taxonomic group in the assemblage, in order to estimate the relative survivorship of various skeletal parts. The analysis included many axial as well as appendicular skeletal features, permitting a thorough examination of density-mediated effects across the ungulate skeleton.

Figure 5.5
Examples of portions-of-elements representing limb shafts used to count elements (MNE): (a) distal diaphysis of the scapula at its narrowest section; exterior and interior views of nutrient foraminae of (b) humerus, (c) femur, and (d) tibia. Examples are from cervids, but the criteria also apply to wild bovids.

The parameters for macrostructure density of skeletal portions in table 5.1 are for ungulate species other than those that lived in the Pleistocene Levant, but the reference taxa possess similar body masses and structures. Pronghorn (*Antilocapra,* from Lyman 1994) serves

Table 5.1

Average structural density values for skeletal portions of antelope, deer, and bison, as determined through photon densitometry

Element	Portion	Lyman's Scan Site Label	Antelope (Antilocapra)	Deer (Odocoileus)	Bison (Bison)
Mandible	Anterior horizontal ramus	DN1	n.d.	0.55	0.53
	Anterior-mid horizontal ramus	DN3	n.d.	0.55	0.62
	Mid horizontal ramus[a]	DN4	n.d.	0.57	0.53
	Posterior horizontal ramus	DN5	n.d.	0.57	0.53
	Mandibular condyle	DN7	n.d.	0.36	0.49
Atlas	Complete or nearly complete	AT2	0.13	0.15	0.91
	Anterior articulation[a]	AT3	0.32	0.26	0.34
Axis	Anterior-ventral articulation[a]	AX1	0.13	0.16	0.65
Other cerv v	Pre-/post-zygopophysis[a]	CE1/2	0.12	0.19/0.15	0.37/0.62
Thor vert	Body (centrum)	TH1	n.d.	0.24	0.42
	Dorsal spine[a]	TH2	n.d.	0.27	0.38
Lumb vert	Pre-zygopophysis[a]	LU1	0.15	0.29	0.31
	Post-zygopophysis[a]	LU2	0.11	0.30	0.11
Rib	Head (proximal end)	RI2	n.d.	0.25	0.35
	Proximal diaphysis (if long)[a]	RI3	n.d.	0.40	0.57
Sacrum	Anterior body[a]	SC1	0.11	0.19	0.27
Innominate	Acetabulum (all, ac-il, ac-is, ac-pub)	AC1	0.14	0.27	0.53
	Iliac body[a]	IL2	0.33	0.49	0.52
	Ischial blade	IS1	0.28	0.41	0.50
Scapula	Distal end (or subset thereof)[a]	SP1	0.27	0.36	0.50
	Distal diaphysis (narrowest section)[a]	SP2	0.10	0.49	0.48
	Proximal rim (or subset thereof)	SP5	0.21	0.28	0.17
Humerus	Proximal end (or subset thereof)	HU1	0.06	0.24	0.24
	Diaphysis or fragment with foramen	HU3	0.25	0.53	0.45
	Distal end (or subset thereof)[a]	HU5	0.33	0.39	0.38
Radius	Proximal end (or subset thereof)[a]	RA1	0.26	0.42	0.48
	Diaphysis or prox attachment scar[a]	RA3	0.57	0.68	0.62
	Distal end (or subset thereof)	RA5	0.34	0.43	0.35

Continued on the next page

as the model for mountain gazelle, American deer (*Odocoileus,* Lyman 1994) for Mesopotamian fallow deer, and American bison (*Bison,* Kreutzer 1992) for aurochs, a robust form of extinct wild cattle that had a moderate thoracic hump. The best possible matches were sought between species-specific sources of bulk density parameters and the archaeofaunal species, although none is ideal. In defense of their application, I note that the differences in bone density distributions in the anatomy of pronghorn and American deer are minor (collective $n = 129$, $r^2 = 0.724$, $p < 0.0001$), which should also be true for the differences between gazelle and fallow deer, the most common ungulates in

the Wadi Meged series. The differences between bison and pronghorn ($r^2 = 0.260$, $p < 0.0001$) and between bison and deer ($r^2 = 0.229$, $p < 0.0001$) are greater, which might also have been true for the larger Mediterranean ungulates of the Pleistocene. Only common ungulate species were considered in the analysis of in situ attrition. Bone specimens falling into the small ungulate category were grouped with those identified as *Gazella,* medium artiodactyl ungulates were grouped with *Dama,* and large artiodactyl ungulates were grouped with *Bos* (appendixes 5–8). The combined medium-size ungulate group may have contained scant remains of red deer, goat, or wild pig (see chapter 7).

Table 5.1, continued

Element	Portion	Lyman's Scan Site Label	Antelope (Antilocapra)	Deer (Odocoileus)	Bison (Bison)
Ulna	Proximal end (or subset thereof)[a]	UL2	0.26	0.45	0.69
Femur	Proximal end (or subset thereof)[a]	FE1	0.16	0.41	0.31
	Diaphysis or fragment with foramen[a]	FE4	0.33	0.57	0.45
	Distal end (or subset thereof)	FE6	0.27	0.28	0.26
Tibia	Proximal end (or subset thereof)	TI1	0.18	0.30	0.41
	Proximal diaphysis	TI2	0.26	0.32	0.58
	Diaphysis or fragment with foramen[a]	TI3	0.48	0.74	0.76
	Distal end (or subset thereof)[a]	TI5	0.29	0.50	0.41
Calcaneum	Distal end (or subset thereof)	CA1	0.29	0.41	0.46
	Diaphysis	CA2	0.55	0.64	0.80
	Proximal end (or subset thereof)[a]	CA3	0.50	0.57	0.49
Astragalus	Complete	AS1	0.39	0.47	0.72
	Middle	AS2	0.48	0.59	0.62
	Distal end (or subset thereof)[a]	AS3	0.57	0.61	0.60
Metacarpal	Proximal end (or subset thereof)[a]	MC1	0.33	0.56	0.59
	Diaphysis (long)	MC3	0.57	0.72	0.69
Metatarsal	Proximal end (or subset thereof)[a]	MR1	0.47	0.55	0.52
	Diaphysis (long)	MR3	0.57	0.74	0.67
Metapodial	Distal end (or subset thereof)[a]	MC6/MR6	0.44	0.51	0.53/0.48
Phalanx 1	Proximal end (or subset thereof)	P11	0.24	0.36	0.48
	Distal end (or subset thereof)[a]	P13	0.45	0.57	0.48
Phalanx 2	Proximal end (or subset thereof)	P21	0.23	0.28	0.41
	Distal end (or subset thereof)[a]	P23	0.30	0.35	0.46
Phalanx 3	Proximal end (or subset thereof)[a]	P31	0.25	0.25	0.32

SOURCES: Lyman 1982, 1984 for pronghorn and deer; Kreutzer 1992 for bison.

NOTE: Only portions for which control data on bulk density are available are listed. "Or subset thereof" means that this area of the element can be subdivided further, usually into morphologically unique anterior, posterior, medial, and lateral segments.

[a] Densest portion for this kind of element.

Table 5.2 presents Spearman's *rho* statistics for the relation between portion-of-element densities and their survivorship in the Natufian (data from Munro 2001:143–144), Kebaran, pre-Kebaran, and seven Mousterian assemblages from Hayonim Cave and Meged Rockshelter. The results are presented by ungulate body size group. Significant correlations between survivorship and structural density are apparent for the Kebaran assemblages and, in the case of deer, several of the Mousterian assemblages as well. Density-mediated attrition potentially explains some of the observed variation in portion-of-element representation in the Wadi Meged faunas. However, linear regressions of these data reveal buckshot-style scatters of points (fig. 5.6) and poor adherence to the lines plotted through them (table 5.3). In only four cases can density explain a substantial amount of the variation in skeletal survivorship: the Hayonim Kebaran fallow deer assemblage (31 percent), the Meged Kebaran gazelles (24 percent), the fallow deer in MP unit 7 (34 percent), and aurochs in MP unit 6 (21 percent). Skeletal density explains less than 20 percent of skeletal survivorship in all the other cases. The causes of attrition in the Wadi Meged assemblages are unclear. It is unlikely, judging from such evidence, that in situ attrition added significant biases in skeletal part representation (see also chapter 11 on prey body part profiles).

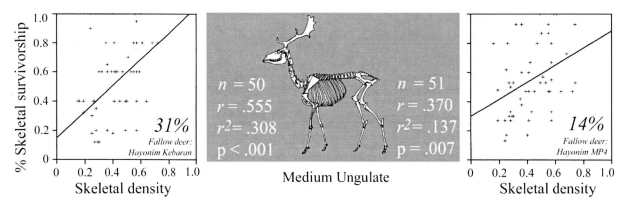

Figure 5.6
Two examples of the degree to which skeletal density parameters may explain observed variation in skeletal survivorship in medium-size ungulates from Hayonim Cave as revealed by linear regression analysis: 31 percent in the case of the Kebaran (Layer C), and 14 percent in Mousterian (MP) unit 4. Values of 14 percent and lower are typical for small, medium, and large ungulates of Mousterian Layer E.

Table 5.2
Spearman's correlation statistics for the relation between bulk density and survivorship of element portions for ungulates in the Wadi Meged assemblages

Stratigraphic Unit/Culture	Small Ungulates			Medium Ungulates			Large Ungulates		
	n	rho	p	n	rho	p	n	rho	p
Hayonim Natufian	17	0.546	<u><.05</u>	—	—	—	—	—	—
Hayonim Kebaran	47	0.367	<u>.03</u>	50	0.538	<u>**<.001**</u>	27	0.219	>.1
Meged Kebaran	45	0.276	<u>**.0007**</u>	43	0.264	.1	—	—	—
Meged pre-Kebaran (UP)	44	0.492	<u>**.001**</u>	38	0.508	<u>.03</u>	—	—	—
Hayonim MP 1	27	0.064	>.1	14	0.407	>.1	—	—	—
Hayonim MP 2	30	0.435	<u>.03</u>	18	0.433	<u>.06</u>	—	—	—
Hayonim MP 3	47	0.087	>.1	48	0.399	<u>**.008**</u>	15	0.206	>.1
Hayonim MP 4	49	0.063	>.1	51	0.378	<u>**.007**</u>	42	0.3663	<u>.06</u>
Hayonim MP 5	48	0.179	>.1	47	0.438	<u>**.002**</u>	23	0.245	>.1
Hayonim MP 6	47	0.128	>.1	44	0.223	>.1	20	0.464	<u>.04</u>
Hayonim MP 7	21	0.538	<u>.07</u>	24	0.664	<u>**.003**</u>	—	—	—

SOURCE: Results for the Natufian are from Munro 2001:143–144, an analysis that focused on the appendicular skeleton but included shaft-based portion counts.

NOTE: *n* is the number of portion types for which observed archaeological MNE is greater than zero. No comparable data are available for the Aurignacian of Hayonim Cave.

KEY: (—) Sample too small for comparison. An underscore indicates that the correlation is mildly significant; an underscore plus boldface, that the correlation is very significant.

BIASES IN THE REPRESENTATION OF LIMB BONE ENDS VERSUS SHAFTS

A related point concerns the relative representation of diagnostic end (epiphyseal) and shaft portions of ungulate long bones in the faunal assemblages. This sort of comparison is not a particularly clean test of density-mediated attrition, because it embodies some conflict-ing perceptions about the utility of MNE counting procedures. The Wadi Meged data offer insights that, while not completely settling this debate, help us to prioritize the issues, especially since we already know that density-mediated attrition exerted relatively minor effects on most of the archaeofaunal assemblages.

Two different expectations (survivorship ratios) for the relative attrition of ends versus shafts of long bones

Table 5.3
Regression statistics for assemblages that display significant relations between the bulk density of bone element portions and survivorship

Stratigraphic Unit	n	r	r^2	p	% Survivorship Explained by Skeletal Density
Gazelles					
Hayonim Kebaran	17	.3197	.1022	.03	10
Meged Kebaran	44	.4918	.2419	.0007	24
Hayonim MP 2	30	.3900	.1521	.03	15
Hayonim MP 7	21	.4015	.1612	.07	16
Fallow deer					
Hayonim Kebaran	50	.5553	.3084	<.001	31
Meged pre-Kebaran (UP)	38	.3573	.1276	.03	13
Hayonim MP 3	48	.3807	.1450	.008	14
Hayonim MP 4	51	.3701	.1370	.007	14
Hayonim MP 5	47	.4367	.1907	.002	19
Hayonim MP 7	24	.5822	.3389	.003	34
Aurochs					
Hayonim MP 4	42	.2965	.0879	.06	9
Hayonim MP 6	20	.4591	.2107	.04	21

NOTE: n is the number of portion types for which observed archaeological MNE is greater than zero.

have been forwarded or implied by other authors. The first is a maximum potential differential of 1:2 to 1:3 for spongy versus compact bone feature survivorship, following Lyman's (1984, 1991, 1994) photon densitometry approach and Lam et al.'s (1999) unadjusted BMD_1 measurements using computed tomography (Stiner 2004). An alternative view is a maximum differential in end and shaft representation of 1:8 for spongy versus compact bone feature survivorship, implied by feeding experiments involving captive and wild hyenas (Capaldo 1997; Marean and Spencer 1991; Marean et al. 1992), adjusted computed tomography measurements of the density of limb shafts (Lam et al. 1998, 1999 BMD_2 criteria), and Marean and Kim's (1998) refitting study of the Mousterian ungulate faunas from Kobeh Cave in Iran.

Figure 5.7 plots the relationship between end-based and shaft-based MNE counts for the major limb bones of common small, medium, and large ungulates from all seven Mousterian units of Hayonim Cave combined. The limb elements are the scapula, humerus, radius, femur, tibia, calcaneum, and metapodials. Only the more common of two potential ends of an element is considered, since this should be enough to "find" the elements in fragmented material. The overall cor-

relation between end-based and shaft-based MNE counts is significant ($n = 21$, $r^2 = 0.472$, $p = 0.0006$; Spearman's $rho = 0.808$ for rank-ordered data, $p = 0.0001$), and a slope of 0.423 (intercept = 6.50) indicates that shafts are somewhat underrepresented relative to the more common end of the elements considered. There are some major outliers to the distribution: of the three body size categories distinguished in figure 5.7, small ungulates are the deviant group, including one outlier that has been omitted from the graph. In contrast, the relation between end-based and shaft-based MNE counts for medium and large ungulates combined is much tighter, with the counts of shafts and the harder end in nearly perfect agreement ($n = 14$, $r^2 = 0.935$, slope = 0.933, intercept = 1.61, $p = 0.00001$; Spearman's $rho = 0.964$, $p = 0.0001$).

Are the small ungulate remains less well preserved than the remains of the larger ungulates? The analyses of density-mediated attrition presented in the previous section and those presented in chapter 4 indicate that the answer is no. Figure 5.8, which compares the density values for limb shafts and the denser end of the limb in the same range of elements of modern ungulates, indicates a consistent relation when the

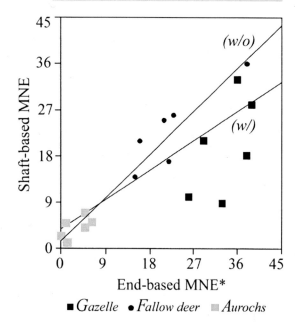

(w/o gazelles)	(w/ gazelles)
$r^2 = 0.935$	$r^2 = 0.472$
slope = 0.933	slope = 0.423
intercept = 1.61	intercept = 6.50
n = 14	n = 21

Figure 5.7
Comparison of end-based and shaft-based MNE counts for the major limb bones of small, medium, and large ungulates from all Mousterian units combined.

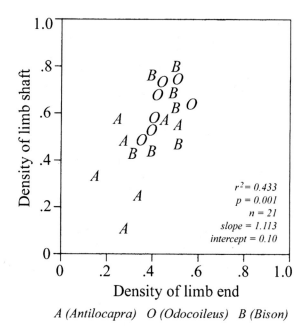

A (Antilocapra)   O (Odocoileus)   B (Bison)

Figure 5.8
Comparison of density parameter values for shafts and the denser end of the same limb elements in *Antilocapra* (= small ungulate), *Odocoileus* (= medium ungulate), and *Bison* (= large ungulate). The correlation is very weak if the *Antilocapra* data are removed ($n = 14$, $r_2 = 0.271$, $p = .06$).

three body size groups are combined (appendix 9). The shafts are slightly denser than the ends of these bones, according to photon densitometry standards, but the slope is close to the ideal value of 1. However, the values for *Antilocapra*, the small ungulate referent, tend to fall on the lower end of the slope, and in fact there is a poor correlation for the data set if *Antilocapra* is removed ($n = 14$, $r^2 = 0.271$, slope = 0.904, intercept = 0.218, $p = .06$). The points for *Antilocapra*, *Odocoileus* (the medium ungulate referent), and *Bison* (the large ungulate referent) each form tight clusters along the end-density *(x)* axis (0.2–0.6) but are more scattered along the shaft-density *(y)* axis (0.1–0.8). This suggests that end-based counts can yield more consistent results among limb elements and among ungulate species (Rogers 2000; Stiner 2002b). The results do not support the idea that shaft-based MNE counts are superior to or more reliable than end-based counts (*contra* Marean and Kim 1998). The points for *Antilocapra* are the most scattered of all with respect to the

shaft-density axis. Thus, the anomaly for the small ungulates may arise from the unique skeletal design, especially of the femur and humerus, of light-framed ungulates and perhaps additionally from size-dependent problems in recognizing foraminae and other important shaft features of small ungulates during identification.

Table 5.4 summarizes comparative data on MNE counts for shafts and the more common ends of limb elements for Pleistocene shelter faunas of diverse origins in Mediterranean Italy. All of these assemblages were completely recovered, and the types of collectors ranged from Middle and Upper Paleolithic humans to spotted hyenas. The assemblages collected by hyenas exhibit extensive gnawing damage typical of den contexts in general (Stiner 1994). No systematic differences exist among the assemblages created by the disparate agents.

End-based MNE counts therefore should not signal systematic undercounting of the original number of elements in an assemblage, because one end generally tends to be as hard as the shaft in most limb bones (see also chapter 11). The anomalies observed for small ungulates cannot be generalized to data on the medium-size and larger ungulate species of the types that dominated Mousterian faunas in Eurasia. On the

Table 5.4

Limb MNE counts from unique shaft features and epiphyseal features of the more common of two ends in assemblages from three Italian Mediterranean caves and Mousterian Layer E of Hayonim Cave

Element	Riparo Salvini[a] MU Shaft	Riparo Salvini[a] MU End	Grotta Breuil[b] MU Shaft	Grotta Breuil[b] MU End	Buca della Iena[c] MU Shaft	Buca della Iena[c] MU End	Buca della Iena[c] LU Shaft	Buca della Iena[c] LU End	Mousterian Layers in Hayonim Cave (All Units Combined) SU[b] Shaft	SU[b] End	MU[b] Shaft	MU[b] End	LU[b] Shaft	LU[b] End
Scapula	1	5	0	1	0	0	1	2	28	39	25	21	4	5
Humerus	6	8	4	4	1	2	4	5	9	33	14	15	7	5
Radius	1	3	3	3	2	3	4	5	21	29	21	16	5	1
Femur	3	7	2	3	1	1	1	3	10	26	14	15	2	0
Tibia	5	7	5	4	2	3	6	6	18	38	26	23	7	5
Humerus/femur	(4)	—	—	—	—	—	—	—	—	—	—	—	—	—
Total MNE	20	30	14	15	6	9	16	21	86	165	100	90	25	16

SOURCE: Data for Italian sites are from Stiner 1991.

NOTE: All cases are from completely recovered assemblages in which every fragment was examined. The humerus and femur could not always be distinguished on the basis of foraminae in the highly fragmented material from Riparo Salvini.

KEY: (SU) Small ungulate, (MU) medium ungulate, (LU) large ungulate.

[a] Collected by Epipaleolithic humans exclusively.

[b] Collected by Middle Paleolithic humans exclusively.

[c] Collectors were principally or exclusively spotted hyenas (Stiner 1994).

other hand, the observations presented here provide good reason for considering *all* unique landmarks on bone elements during data collection, as many zooarchaeologists routinely do (e.g., Bar-Oz and Dayan 2002; Brain 1981; Binford 1978; Bunn and Kroll 1986, 1988; Delpech 1998; Morlan 1994a, 1994b; Stiner 1991a, 1994, 1998a; Todd and Rapson 1988), rather than the rigid privileging of shaft portions-of-elements over others for estimating limb MNE.

MACRODAMAGE PATTERNS

Weathering damage from atmospheric exposure and gnawing damage from carnivores or rodents are exceptionally rare (generally less than 1 percent of NISP) in the Wadi Meged assemblages (tables 5.5 and 5.6). Faunal remains in the central trench of Hayonim Cave were protected by the roof of the cave and perhaps also by rapid burial. Because rodent burrows were fairly common in some parts of the Mousterian excavation (chapter 4), some evidence of rodent gnawing was expected, but there were few examples. The near absence of carnivore gnawing damage is unusual for Mediterranean Middle and early Upper Paleolithic cave sites (cf. Speth and Tchernov 1998 on Kebara Cave;

Stiner 1994 on Italian sites) but is not unique (e.g., the late Mousterian site of Grotta Breuil, Italy). Only in the Natufian layer of Hayonim Cave was exposure-related weathering and rodent damage somewhat more common (Munro 2001:35).

Information about human activities comes from common fracture forms, tool marks, and burning damage on bones and teeth. The prevalence and uniform distributions of burning damage on the macrofaunal remains together constitute one of several important signatures of human activity for the Wadi Meged series. Burning damage is prevalent in all of the assemblages, usually ranging between 8 and 30 percent of NISP for common prey taxa (tables 5.7 and 5.8). Calcined bone was rare in Hayonim Cave and only slightly more common in Meged Rockshelter. Burning shows few if any biases in terms of body parts in the ungulate remains, but there are anatomical biases in the tortoises that may relate to the way people roasted them, as I discuss later. Otherwise, burning damage varies little in its distribution or intensity among taxa or across assemblages. This observation is exemplified in figure 5.9 for tortoises and ungulates, following the burning stage criteria discussed in chapter 3. Similar rates of burning are seen on the bones of other small game animals. The bones in Meged Rockshelter

Table 5.5

Incidence of weathering damage by stage in the Wadi Meged assemblages

Assemblage	Weathering Frequency (NISP)					Total Observed NISP	% Weathered
	Stage 0	Stage 1	Stage 2	Stage 3	Stage 4		
Hayonim Kebaran	2,981	16	4	1	0	3,002	<1
Meg E. Kebaran	2,077	4	2	1	0	2,084	<1
Meg pre-Kebaran (UP)	625	0	0	0	0	625	0
Hayonim MP1	158	3	0	0	0	161	2
Hayonim MP2	352	1	0	0	0	353	<1
Hayonim MP3	1,851	6	10	3	0	1,870	1
Hayonim MP4	8,879	15	12	2	3	8,911	<1
Hayonim MP5	3,175	4	0	0	0	3,179	<1
Hayonim MP6	2,343	5	2	1	0	2,351	<1
Hayonim MP7	269	1	0	0	0	270	<1

SOURCE: Weathering stages generally follow Behrensmeyer 1978.

assemblages are burned and calcined somewhat more often than those from Hayonim Cave, perhaps because the rock shelter contained much less habitable space, making reburning of bone debris more likely.

Small campfires fueled by Mediterranean hardwood can attain high temperatures (900 degrees C) and may carbonize bones buried below the coal bed, whereas calcination seems to occur only within the coal bed itself (chapter 3). Given the prevalence of fire features and ash products in the sediments of Hayonim Cave, the charring patterns on the bones must at least partly reflect accidental reburning of discarded material. Burning damage was more common on unidentified specimens than on those that could be identified to taxon, body part, or both (fig. 5.10, table 5.9). Burned specimens also tended to be smaller than unburned specimens (fig. 5.11, table 5.10). In Hayonim Cave, roughly twice as many unidentified specimens as identified ones were burned in Mousterian units 1–4, although no such differential existed in units 5–6. With greater fragmentation comes a loss in specimen identifiability, making burned bone more common in the "unidentified" fraction of the assemblages (chapter 3). As I have argued in earlier publications on this subject, the frequency as well as "the intensity of burning damage refers as much or more to relationships between successive events as it does to the activities comprising any one depositional event. . . . Burning on bones and stones as a function of fragment size may reflect, for example, differing intensities of activity

Table 5.6

Incidence of gnawing damage in the Wadi Meged assemblages

Assemblage	% Carnivore Gnawing	% Rodent Gnawing	Observed NISP
Hayonim Kebaran	<1	<1	3,002
Meg E. Kebaran	<1	0	2,084
Meg pre-Kebaran (UP)	<1	0	625
Hayonim MP1	0	<1	161
Hayonim MP2	0	<1	353
Hayonim MP3	<1	<1	1,870
Hayonim MP4	<1	<1	8,911
Hayonim MP5	0	0	3,179
Hayonim MP6	<1	0	2,351
Hayonim MP7	0	0	270

area maintenance relative to rates of sediment buildup" (Stiner 1994: 149).

Certain fracture forms also point to humans as the principal or sole modifiers of the Wadi Meged macrofaunas. Impact and transverse fractures—the latter being relatively clean breaks oriented perpendicularly to the main axis of limb and vertebral elements—are diagnostic of human influences, not so much from simple presence or absence as from their occurrence in high frequencies. Only fractures with old edges were considered. The consistently "clean" appearances of these breaks were also critical to excluding the possibility of carnivore modification.

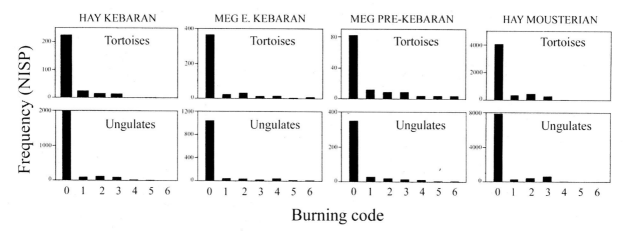

Figure 5.9
Frequency distributions (NISP) of burning by intensity on tortoise and ungulate remains in the Kebaran, pre-Kebaran, and Mousterian assemblages. (HAY) Hayonim Cave; (MEG) Meged Rockshelter. Burning codes are (0) unburned, (1) lightly carbonized, (2) more than one-half carbonized, (3) fully carbonized, (4) lightly calcined, (5) more than one-half calcined, and (6) fully calcined.

So-called green bone (a.k.a. green-stick) fractures are characteristic of postmortem damage, but they occur only while the natural fibrous composite built during life remains intact (e.g., Currey 1984). Transverse green-bone fractures on large mammal remains in Paleolithic sites reflect the practice of sectioning carcasses for transport, although they probably were not confined to this activity alone. Transverse breaks are produced by a shearing action. They differ from the nearly perfectly squared or planar breaks commonly found on fossilized bones (which are by then composed almost exclusively of mineral) in that short hinges and mildly undulated break surfaces are present on transversely broken fresh bone. Thus, transverse fractures on fresh bone are consistent with the tendencies of green-bone fracture morphology in general, but they are much more restricted in length and orientation than the classic long, spiraling fractures seen in compact bone.

Humans are not the only species that can create "shear," or transverse, breaks on fresh bone (for raptors, see Hockett 1991; for coyotes and cougars, see Schmitt and Juell 1994; Stiner 1994), but humans tend to generate them on limb bones and vertebrae in high frequencies. These types of fractures were uncommon (4–11 percent of total NISP) on ungulate bones modified by a variety of modern predators and by those in Pleistocene large carnivore dens in Italian caves (Stiner 1994). Transverse fractures were also rare (4–6 percent) on small animal remains in coyote scats from Chaco Canyon and on juvenile ungulate bones fed upon by American cougars in the Guadalupe Mountains, both in New Mexico, USA (Stiner, unpublished data).

How humans manage to generate transverse breaks on large bone elements is not difficult to understand—ridged anvils and heavy tool edges abound in traditional technologies. And humans easily make these breaks on smaller bones with their teeth or tools. The reasons humans make large numbers of transverse fractures in ungulate assemblages is not altogether settled. Transverse fractures on the bones of large game animals result at least partly from carcass sectioning in preparation for transport or sharing (Stiner 1994). Dividing limbs by sectioning bones rather than disarticulating strong joints makes energetic sense because it may be easier than joint disarticulation for many areas of the ungulate anatomy. A classic example in Paleolithic Mediterranean faunas is sectioning the femoral neck rather than detaching the femoral head from the pelvic socket (acetabulum).

In the case of small game, some authors (e.g., Hockett 1994; Jones 1984) have proposed that transverse breaks on bone tubes such as lagomorph long bones are the products of marrow removal by humans. One or both ends were snapped off with the teeth and the marrow was sucked from the bone tube, or, in the case of birds, the soft ends of the bones were chewed off for the nutrients they contained. These feeding practices are supported ethnographically and are not in dispute here. However, transverse breaks are as common on tortoise limb bones as they are on bird and hare bones in the Wadi Meged series (table 5.11). Tortoise bones do not contain marrow pockets, and the ends of their bones offer little food value. It may be that the prevalence of transverse breaks in human-modified small faunas relates foremost to the mechanics by which

Table 5.7
Incidence of burning damage by taxon in Mousterian Layer E of Hayonim Cave

Taxon (Observed NISP)	% Code 0	% Code 1	% Code 2	% Code 3	% Code 4	% Code 5	% Code 6	Total % Burned
Ungulates								
Gazella gazella (1,912)	84	4	5	6	<	<	<	16
Capreolus capreolus (11)	82	0	0	18	0	0	0	18
Small ungulate (1,996)	86	3	5	6	<	<	0	14
Sus scrofa (176)	87	2	3	6	0	2	0	13
Capra aegagrus (8)	87	0	0	12	0	0	0	13
Equus cf. hemionus (9)	*	0	0	0	0	0	0	*
Dama mesopotamica (1,178)	86	2	5	6	<	<	0	14
Medium ungulate (2,681)	86	2	4	7	<	<	0	14
Large cervid (321)	85	2	4	8	1	0	0	15
Cervus elaphus (155)	84	4	2	10	1	0	0	16
Bos primigenius (281)	84	4	4	7	0	0	0	16
Equus caballus (10)	80	0	10	10	0	0	0	20
Large ungulate (460)	85	2	5	7	1	<	0	15
Dicerorhinus hemitoechus (2)	*	0	0	0	0	0	0	*
Reptiles								
Testudo graeca (5,212)	78	7	9	5	<	<	<	22
Coluber sp. (325)	86	2	5	6	0	0	0	14
Ophisaurus apodus (227)	77	2	10	10	0	0	0	23
Indet. lizard (4)	50	0	0	50	0	0	0	50
Aves								
Small-medium bird (98)	93	1	3	2	1	0	0	7
Large bird (predator) (45)	91	0	2	6	0	0	0	9
Struthio camelus eggshell (91)	40	4	21	26	9	0	0	60
Small mammals								
Lepus capensis (5)	60	20	0	0	20	0	0	40
Sciurus anomalous (12)	92	0	0	8	0	0	0	8
Indet. small mammal (94)	88	3	2	6	0	0	0	12
Erinaceus sp. (2)	*	0	0	0	0	0	0	*
Carnivores								
Hyaenidae (2)	*	0	0	0	0	0	0	0
Vulpes vulpes (18)	83	6	11	0	0	0	0	17
Canis/Lycaon spp. (9)	*	0	0	0	0	0	0	0
Panthera pardus (9)	*	0	0	0	0	0	0	0
Ursus arctos (21)	100	0	0	0	0	0	0	0
Felis spp. (9)	*	0	0	0	0	0	0	0
Martes foina (3)	*	0	0	0	0	0	0	0
Indet. carnivore (23)	96	4	0	0	0	0	0	4
Total (17,097)	83	4	6	6	<	<	<	17

NOTE: Burning codes range from unburned (0) through fully carbonized (3) to fully calcined (6), as defined in chapter 3. "Observed NISP" refers to the total number of observed specimens, usually equivalent to the total sample for the assemblage except in the case of tortoises and snakes.

KEY: (*) Unburned (code 0) value is 100%, but the calculation is suspect due to very small sample size; (<) burning occurred at well under 1%.

Table 5.8
Burning frequencies for common prey groups in the Kebaran and pre-Kebaran assemblages from the Wadi Meged

Burning Code	Tortoise		Hare[a]		Birds		Small Ungulates		Medium Ungulates		Large Ungulates	
	NISP	%	NISP	%	NISP	%	NISP	%	NISP	%	NISP	%
Hayonim Cave, Kebaran, Layer C												
0	224	80	9	100	69	91	1,229	84	571	85	111	91
1	25	9	0	0	2	3	74	5	19	3	5	4
2	16	6	0	0	2	3	68	5	37	5	2	2
3	16	5	0	0	1	1	59	4	35	5	4	3
4	0	0	0	0	2	3	16	1	6	1	0	0
5	1	<	0	0	0	0	9	1	1	<	0	0
6	0	0	0	0	0	0	3	<	0	0	0	0
Total	281	—	9	—	76	—	1,458	—	669	—	122	—
Meged Rockshelter, Early Kebaran (0–199 cm bd)												
0	365	79	75	83	149	87	743	85	252	89	—	—
1	25	5	7	8	3	1	35	4	9	3	—	—
2	32	7	3	3	5	3	26	3	10	3	—	—
3	14	3	2	2	8	5	17	2	3	1	—	—
4	15	3	2	2	6	3	36	4	6	2	—	—
5	3	1	0	0	0	0	12	1	3	1	—	—
6	8	2	1	1	0	0	9	1	1	<1	—	—
Total	462	—	90	—	171	—	878	—	284	—	—	—
Meged Rockshelter, pre-Kebaran (200 cm bd to bedrock)												
0	82	66	7	78	23	88	228	78	95	85	—	—
1	12	10	1	11	2	8	23	8	5	4	—	—
2	9	7	0	0	0	0	15	5	5	4	—	—
3	9	7	1	11	0	0	10	3	5	4	—	—
4	4	3	0	0	0	0	9	3	2	2	—	—
5	4	3	0	0	0	0	3	1	0	0	—	—
6	4	3	0	0	1	4	3	1	0	0	—	—
Total	124	—	9	—	26	—	291	—	112	—	—	—

NOTE: Burning codes range from unburned (0) through fully carbonized (3) to fully calcined (6), as defined in chapter 3. "Observed NISP" refers to the total number of observed specimens, usually equivalent to the total sample for the assemblage, except in the case of tortoises and snakes.

KEY: (<) Burning occurs at well under 1%.

[a] Includes some squirrel and hedgehog remains.

humans processed the bodies of small animals (such as tearing away limbs before or after cooking) and only secondarily to the extraction of bone-locked nutrients from certain kinds of small prey. The eating of cartilage that normally is concentrated in the ends of bird and perhaps also reptile bones could explain some of this pattern, however.

Cone (hertzian) fractures and other kinds of impact damage, such as dents and concentrated surface crushing, are associated mainly with the extraction of bone-locked nutrients from mammal skeletons and of tortoise meat from bony shells. Cone fractures indicate the application of a highly concentrated force,

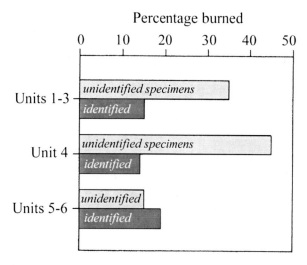

Figure 5.10
Percentages of burned specimens in the unidentified and identified fractions of the Mousterian faunal assemblages from Hayonim Cave.

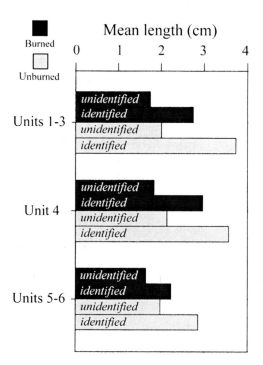

Figure 5.11
Mean fragment lengths for burned and unburned specimens in the unidentified and identified Mousterian faunal fractions.

Table 5.9
Burning frequencies for unidentified and identified faunal specimens from the Mousterian of Hayonim Cave

Depth Range (cm bd)	Unidentified Specimens			Identified Specimens		
	NUSP	Unburned (%)	Burned (%)	NISP	Unburned (%)	Burned (%)
Units 1–3 (300–419)	2,205	65	35	1,783	85	15
Unit 4 (420–465)	251	55	45	8,977	86	14
Units 5–6 (466–539)	1,293	85	15	4,650	81	19
Total	3,749	71	29	15,410	84	16

NOTE: NUSP is the number of unidentified specimens.

such as that from the jaws of a large carnivore or, more commonly, from striking a hard stone hammer against dense bone (e.g., Binford 1978; Blumenschine and Selvaggio 1991; Brain 1981: 141–142; Martin 1907–1910; Potts 1982; Stiner 1994). Because of significant differences in the mechanics of force applied, carnivores seem to produce cone fractures much less often (generally less than 3 percent of NISP in modern and Pleistocene cases; Stiner 1994) than do humans armed with stone hammers.

These damage phenomena are described by prey group in the following sections.

DAMAGE TO REPTILE REMAINS

The remains of tortoises, the most common reptiles (chelonians) in the Wadi Meged assemblages, display clear evidence of human modification, not only in the prevalence of burning damage to them but also in the abundant impact fractures such as cones, dents, and depressions on carapace and plastron parts (table 5.12). Green (spiral) breaks on the shell and other bones and transverse breaks on limb members are also common. Such a damage profile contraindicates the possibility that the tortoises were intrusive to the cul-

Table 5.10
Mean fragment lengths for unidentified (NUSP) and identified (NISP) faunal specimens from the Mousterian of Hayonim Cave

Depth Range (cm bd)	Unidentified Specimens			Identified Specimens		
	No. Observed	Mean Length (cm)	SD	No. Observed	Mean Length (cm)	SD
Unburned						
Units 1–3 (300–419)	1,427	2.01	0.95	1,160	3.74	2.17
Unit 4 (420–465)	137	2.14	0.88	5,681	3.57	2.07
Units 5–6 (466–539 cm)	1,100	1.98	0.95	3,204	2.86	1.75
Burned						
Units 1–3 (300–419 cm)	778	1.75	0.68	250	2.75	1.36
Unit 4 (420–465 cm)	114	1.84	0.67	1,170	2.97	1.67
Units 5–6 (466–539 cm)	193	1.64	0.71	797	2.23	1.18

NOTE: "No. Observed" is the number of specimens for which length measurements were taken.

Table 5.11
Frequencies of transverse fractures on major limb bones of small prey animals in the Wadi Meged assemblages

Assemblage	Birds		Hares		Tortoises	
	Obs. NISP	% TR Fractured	Obs. NISP	% TR Fractured	Obs. NISP	% TR Fractured
Hayonim Kebaran	33	73	5	80	47	70
Meged early Kebaran (UP)	44	70	14	79	94	88
Hayonim Mousterian	81	43	3	~100	321	74

NOTE: Transverse (TR) fractures are old, clean breaks oriented perpendicularly to the main axis of the element. "Obs. NISP" refers to the total number of observed specimens, usually equivalent to the total sample for the assemblage, except in the case of tortoises.

tural deposits (i.e., hibernating individuals that died in situ) or that they were accumulated by large predatory birds at cliff roosts above the two sites (see Paz 1987:64–65 on golden eagles).

Sampson's (2000) comparison of modern Bushmen-collected and raptor-collected tortoise assemblages in and around shelter sites in the Karoo area of South Africa revealed sharp differences in body part representation and bone condition. The nonhuman collectors aggregated unburned tortoise elements, accompanied by small percentages of carapace and plastron parts, and most of these were from relatively small individuals. The human-generated food waste instead contained high frequencies of carapace and plastron parts, accompanied by extensive green-bone breaks and charring (as high as 30–40 percent of NISP). The Wadi Meged tortoise assemblages closely resemble Samp-

son's human-modified examples and are very different from the raptor-collected examples.

The remains of tortoises that die in hibernation also have a distinct appearance. Body part representation under these circumstances can be remarkably complete in undisturbed sediments, often with the body endocast preserved. If the tortoise skeleton were later disassembled by any factor, pieces of the shell would tend to separate along suture lines rather than display spiral or green-bone breaks across bony plates (Stiner 1994:174–176). The Wadi Meged tortoise assemblages, with their high frequencies of spiral fractures on carapace and plastron fragments, cannot represent intrusive deaths or raptor-collected material.

The tortoises consumed at Hayonim Cave and Meged Rockshelter were roasted on open coals and apparently cracked open with stone hammers. High

Table 5.12

Frequencies of impact fractures, spiral fractures, and burning damage on tortoise remains in the Wadi Meged assemblages

Assemblage	MNI	NISP	Cone Fractures (NISP)	Impact Depressions (NISP)	Dented/ Crushed (NISP)	% All Types Impact Damage	% Spiral Fractures	% (All) Burned	% Calcined
Hayonim Kebaran	8	343	2	5	2	3	38	20	<1
Meged E. Kebaran	6	462	3	4	3	2	56	21	6
Meged pre-Keb	2	124	0	0	0	0?	65	34	10
Hayonim MP1	1	14	0	0	0	0[a]	50[a]	21[a]	0[a]
Hayonim MP2	2	48	0	0	0	0	52	21	2
Hayonim MP3	7	426	5	16	2	5	51	20	<1
Hayonim MP4	67	3,155	11	69	22	3	48	20	<1
Hayonim MP5	16	1,417	2	36	14	4	73	18	<1
Hayonim MP6	9	1,184	1	22	1	2	61	30	1
Hayonim MP7	2	113	1	2	1	3	66	40	0

NOTE: "Tortoise remains" refers to fragments of the bony carapace and plastron. All damage counts are relative to NISP. "All Types Impact Damage" includes cone fractures, impact depressions and dents, and localized crushing. Calcined bone includes burning categories 4–6 only. A question mark (?) indicates that calcite concretions on specimens may be responsible for the zero value in this case.

[a] Very small sample makes percentage calculation suspect.

rates of impact damage on marginal (edge) plates of the carapace (14 percent of shell edge NISP) indicate that tortoise shells were often set on edge and struck on the opposite rim (figs. 5.12 and 5.13). The number of cone fractures, impact depressions, dents, and localized crushing on tortoise shell parts is in general agreement with tortoise MNI (table 5.12), assuming that one to three scars representing impact points and opposing anvil damage should exist per animal; some strikes to the shell will have generated halved impact scars, due to splitting of the bone. Many of the major limb bones of the tortoises were ripped or torn away from the body; this is especially apparent on the proximal ends of the humeri (13 percent of specimens in Mousterian Layer E; fig. 5.14).

Tortoise remains in the Mousterian layers of Hayonim Cave were frequently deposited in hearth ash, sometimes as large pieces (fig. 5.15). However, few tortoise bones were burned to the point of calcination, suggesting either that their bones seldom were discarded in live fire beds or, less likely, that trampling subsequently destroyed most bones that were calcined by direct exposure to red heat (see chapter 4).

Speth and Tchernov (2002) attributed differential burning on tortoise shell parts to the method of roasting, with the highest frequencies of burning damage occurring on the carapace. I addressed this possibility for the Wadi Meged assemblages by comparing rates of

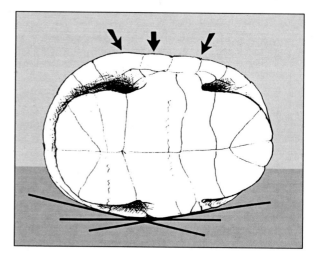

Figure 5.12

Inferred method of splitting tortoise shells by setting them on edge on a hard surface or anvil (represented by base lines) and striking the opposing margin (arrows) with a rock hammer. Percussion fractures are present on shell fragments, frequently but not always on edges. (Reproduced from Stiner and Tchernov 1998.)

burning on shell surfaces, shell margins (edges), and nonshell parts. No significant differences were found among the major body regions (table 5.13), except for slightly lower frequencies of burning on nonshell

Figure 5.13
Impact damage on tortoise shell edges from the Mousterian of Hayonim Cave.

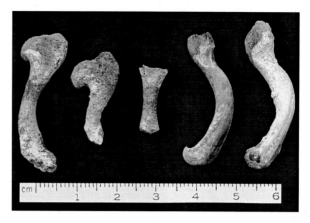

Figure 5.14
Damaged tortoise humeri, some with torn epiphyses and others with transverse fractures.

Figure 5.15
A crushed, burned tortoise carapace in a cemented ash lens from Mousterian Layer E.

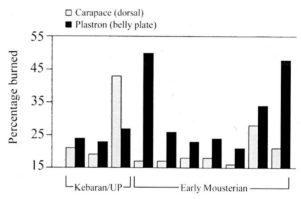

Figure 5.16
Percentages of burning damage on tortoise carapace and plastron fragments in time-ordered Kebaran, pre-Kebaran (UP), and Mousterian assemblages from Hayonim Cave and Meged Rockshelter.

parts. However, the incidence of burning damage on tortoise carapace (dorsal shell) and plastron (belly plate) fragments (table 5.14) differed consistently, with only one exception, the pre-Kebaran assemblage from Meged Rockshelter (fig. 5.16). Generally speaking, plastron fragments were burned more often than carapace fragments, suggesting that the belly plates of tortoises were placed in contact with heat more often or for longer periods than the dorsal parts of the shell. Another possible explanation for this is that the belly plate simply burns more evenly on a flat bed of live coals than does the domed carapace. Regardless, this bias seems to represent more than an idiosyncratic roasting style, since it is repeated across seven Mousterian units and the Kebaran horizons in the Wadi Meged faunal series.

Legless lizards *(Ophisaurus apodus)* and a few large agamid lizards (mainly *Agama stellio*) also appear to have been the prey of humans at Hayonim Cave and

Table 5.13

Frequencies of burning on tortoise shell (carapace and plastron), shell margins, and nonshell skeletal parts from Kebaran and Mousterian layers of the Wadi Meged sites

Assemblage	Shell NISP	% Burned	Shell Edge NISP	% Burned	Nonshell NISP	% Burned
Hayonim Kebaran	234	22	—	—	50	10
Meged E. Kebaran	359	21	—	—	103	21
Meged pre-Kebaran	113	34	—	—	11	27[a]
Hayonim MP1	14	21[a]	0	0	0	0
Hayonim MP2	43	21	9	22[a]	5	20[a]
Hayonim MP3	295	20	39	23	41	15
Hayonim MP4	1,978	20	276	25	172	16
Hayonim MP5	1,350	18	246	18	62	10
Hayonim MP6	1,131	31	219	31	48	23
Hayonim MP7	101	34	9	44[a]	11	18[a]
All MP	4,912	22	798	24	339	16

NOTE: Damage counts are relative to total observed NISP.
KEY: (—) No data.
[a] Very small sample makes percentage calculation suspect.

Table 5.14

Frequencies of burning on tortoise carapace (dorsal shell) and plastron (belly plate) specimens from Kebaran and Mousterian Layers of the Wadi Meged sites

	Carapace		Plastron	
Assemblage	% Burned	NISP	% Burned	NISP
Hayonim Kebaran	21	147	24	87
Meg E. Kebaran	19	161	23	198
Meg pre-Kebaran (UP)	43	51	27	62
Hayonim MP1	17	12	50[a]	2
Hayonim MP2	17	24	26	19
Hayonim MP3	18	180	23	115
Hayonim MP4	18	1,228	24	750
Hayonim MP5	16	875	21	475
Hayonim MP6	28	595	34	536
Hayonim MP7	21	53	48	48
All MP	19	2,968	26	1,945

NOTE: Damage counts are relative to total observed NISP.
[a] Very small sample makes percentage calculation suspect.

Meged Rockshelter. Burning damage occurred on 23 percent of legless lizard specimens from the Mousterian layers of Hayonim Cave (see table 5.7), on a par with the rates of burning observed on tortoise remains. Burning was common on the small sample of agamid lizard specimens as well (table 5.7). Mandibles of legless lizards were few in number (chapter 7), but most of them appeared to have been ripped away from the skull with considerable force (fig. 5.17). Given the large sizes of the mandibles in the Wadi Meged assemblages, as well as the total lengths that these thick-bodied animals are known to attain in the wild (chapter 6), it is not surprising that they were attractive prey to Paleolithic humans. Legless lizards were

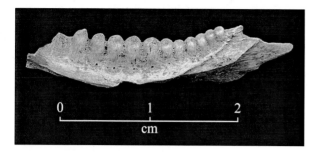

Figure 5.17
A typically ripped mandible of a legless lizard from the Mousterian of Hayonim Cave.

Figure 5.18
Articulated calubrid snake vertebrae from the Mousterian of Hayonim Cave.

never common in the assemblages, however, and they could not have been a staple source of food.

Snake vertebrae occurred in the deposits of Hayonim Cave and to a lesser degree in Meged Rockshelter. These were from nonvenomous calubrid snakes (chapter 6), probably racers of the genus *Coluber*. Their presence in the two sites is more difficult to explain than that of the other reptiles, mainly because snake skeletons by their very nature express few kinds of predator-induced damage, apart from burning. Yet 24 percent of all snake bones from the Mousterian layers were burned, a rate comparable to that for tortoise bones. The problem in the case of snakes is that the species present during the Paleolithic appear to have been the same as those that visit the cave today in search of rodents. Small sets of articulated vertebrae were fairly common in the archaeological layers (fig. 5.18), but these could have derived either from human discard or from deaths in situ. Large, dark, calubrid snakes slip easily beneath any soft duff or matted plant material, and had they died in place, their bones could have been burned by fires built subsequently on the ground surface. On the other hand, snakes are a valuable source of food in many recent human economies, and the individual snakes represented in the Hayonim archaeofaunas were quite large. The high incidence of burning damage on snake bones is in good agreement with the condition of other animal remains that, on the basis of other evidence, clearly represent the prey of humans, making it difficult to exclude snakes from the Paleolithic diet. In recognition of the ambiguity surrounding the snake remains, these animals were not included in the dietary analyses described in the next chapters. Fortunately, removing snakes from the sample did not substantially affect the outcomes of the dietary analyses in any case, because snakes were uncommon in the assemblages.

DAMAGE TO BIRD REMAINS

Many challenges accompany the interpretation of bird remains in Mediterranean shelter sites. This fact stems from the ecological importance of cliffs and high crags as nesting, roosting, and feeding sites for many bird species (Blondel and Aronson 1999). So integral are cliff and cave features to Mediterranean landscapes that they have played important roles in promoting species diversity, especially but not exclusively among birds (chapter 6). Some birds gather into large temporary colonies on ledges; others utilize the many solitary cavities in the rock. The larger birds that inhabit cliffs and caves are carnivorous and may carry parts of their prey to roosts as undigested material or disgorge boluses from their crops, creating microvertebrate series that paleontologists use to study Pleistocene environments. The large raptor and owl species most attracted to cliffs or caves in the Mediterranean area include golden eagles *(Aquila chrysaetos)*, griffon vultures *(Gyps fulvus)*, bearded vultures *(Gypaetus barbatus)*, eagle owls *(Bubo bubo)*, barn owls *(Tyto alba)*, Egyptian vultures *(Neophron percnopterus)*, ravens *(Corvus corax)*, peregrine falcons *(Falco peregrinus)*, and kestrels *(Falco tinnunculus)* (Blondel and Aronson 1999:122). Swifts *(Apus* sp.) and stock doves *(Columba oenas)* also congregate along cliffs. We are certain that owls inhabited the small solution cavities in the ceiling of Hayonim Cave for many years, if not millennia. Most of the microfaunal component of the Hayonim Cave fauna is clearly from regurgitated pellets of larger owls (chapter 6) such as the barn owl and possibly also the little owl *(Athene noctua)* and eagle owl (Tchernov 1981, 1989, 1994, 1996).

Bird remains, large or small, were rare in the Mousterian layers of Hayonim Cave. They were a good deal more common in the Upper Paleolithic and Epipale-

Table 5.15

Frequencies of transverse fractures, tool marks, and burning damage on bird and small mammal remains in the Wadi Meged assemblages

Assemblage	Taxon	Total Observed (NISP)	TR Fractured (NISP)	Tool-Marked (NISP)	Burned (NISP)
Hayonim Kebaran	Hare	7	5	0	0
	Squirrel	2	0	0	0
	Indet. small mammal	23	6	0	2
	Small-medium birds	49	25	0	4
	Large birds	27	9	0	3
		(108)	(42%)	(0%)	(8%)
Meged Early Kebaran (UP)	Hare	25	14	0	4
	Squirrel	1	0	0	0
	Indet. small mammal	74	40	0	13
	Small-medium birds	162	76	0	21
	Large birds	35	13	0	4
		(297)	(48%)	(0%)	(14%)
Hayonim Mousterian (all)	Hare	5	3	0	2
	Squirrel	12	6	0	1
	Indet. small mammal	94	47	0	11
	Hedgehog	2	2	0	0
	Small-medium birds	98	41	0	7
	Large birds	45	10	0	4
		(109)	(43%)	(0%)	(10%)

olithic horizons of the Wadi Meged series (chapter 7) and other Galilee sites (e.g., Munro 2001; Pichon 1983, 1987; Simmons and Nadel 1998; Tchernov 1993a). In the Mousterian assemblages of Hayonim Cave, transverse breaks on bird long bones were present but somewhat less common than expected in comparison with the assemblages of later periods (table 5.11). Only 7 percent of the bones of smaller birds and 9 percent of the bones of larger birds in the Mousterian assemblages were burned. Although charring was especially uncommon for this vertebrate group in the Mousterian period, it seems that burning damage was relatively less common on birds than on other species in later periods as well: 8 percent on smaller birds and 11 percent on larger birds from the Kebaran layer of Hayonim Cave. No bird bone is cut-marked (table 5.15), but transverse fractures are common on the long bones of small and medium-sized birds from the Kebaran and pre-Kebaran assemblages (70–73 percent; table 5.11), though they are considerably fewer for the Mousterian (43 percent). It is impossible to exclude birds altogether from the human diet of this period—they are simply rare in the assemblages (chapter 7) and therefore were presum-

ably of limited economic importance, if they were eaten by Mousterian people at all.

Ground birds (Phasianidae), especially chukar partridges *(Alectoris chukar)*, and doves (Columbiformes) predominate in the medium-sized avian fraction of the Upper Paleolithic and Epipaleolithic assemblages of Hayonim Cave (chapter 7). At Meged Rockshelter, humans exploited raptors almost exclusively; most striking is the remarkable overrepresentation of pedal phalanges, especially the terminal phalanges, or talons, of *Aquila* sp., *Buteo buteo,* and other falconiform birds (table 5.16; Munro in Kuhn et al. 2004). Raptor foot bones obviously held cultural significance for these people and likely served as ornaments, talismans, or piercing tools. Although the raptors from Meged Rockshelter are not considered "game birds," the requirements of capture might not have been very different from those for other large birds, and their edibility in conjunction with raw material acquisition cannot be refuted. It is for this reason that predatory birds were grouped with food species for the economic analyses to be described later, provided that a human connection was established from taphonomic evidence.

Table 5.16
Avian pedal phalanx representation for the Early Kebaran and pre-Kebaran levels of Meged Rockshelter

Body Size	Total Aves NISP	1st Phalanx (NISP)	2nd Phalanx (NISP)	3rd Phalanx (NISP)	% 1st–3rd Phalanges	% 3rd Phalanges
Tiny	14	1	2	1	29	7
Small	32	0	2	10	37	31
Medium	109	10	12	19	38	17
Large	31	3	2	12	55	39
Huge	2	0	0	2	100	100

SOURCE: Munro, in Kuhn et al. 2004

NOTE: Percentages are of total Aves NISP, nearly equivalent to MNE. Most identified bird remains are from raptors.

Ostrich eggshell was uncommon but widespread in the Mousterian and Kebaran layers of Hayonim Cave (and probably also in the Aurignacian layer). None was found in the Natufian of Hayonim Cave or in the early Kebaran and pre-Kebaran assemblages from Meged Rockshelter. It is unclear whether the contents of these huge eggs were eaten as food, whether the shells were conserved as water containers in the manner of Holocene desert cultures of Africa and Asia (e.g., Lee 1979), or both. The ostrich egg specimens from Hayonim Cave were invariably broken (fig. 5.19), and burning damage occurred at a rate of 60 percent in the early Mousterian layers and 43 percent in the Kebaran layer. Such high rates of burning, roughly double that seen on other faunal materials, indicate a consistent spatial association between ostrich eggs and hearth areas in Hayonim Cave in all the periods in which eggs were used.

Figure 5.19
Ostrich eggshell fragments (burned and unburned) from the Mousterian of Hayonim Cave.

DAMAGE TO SMALL MAMMAL REMAINS

Few small mammal prey are present in the Mousterian assemblages of Hayonim Cave, but their relative numbers increase greatly for the Upper Paleolithic and Epipaleolithic (chapters 7–9). All of these remains appear to represent the prey of humans. No tool marks are apparent on them, but transverse fractures are common on the bones of hares from all periods and on those of squirrels from the Mousterian (table 5.15). Bone specimens attributed to the indeterminate small mammal category were burned at a rate of 12 percent in the Mousterian, and the smaller samples of hares and Persian squirrels were burned at rates of 40 percent and 8 percent, respectively (see also tables 5.7 and 5.8). Although small mammals are comparatively rare in the Mousterian assemblages, there is good reason to include hares, squirrels, and perhaps hedgehogs in the list of taxa exploited by humans.

DAMAGE TO CARNIVORE REMAINS

Few traces of gnawing damage from carnivores, if any, were found in the Wadi Meged assemblages, although carnivore skeletal remains were present in low numbers. What damage could be found on large carnivore bones was confined to burning, a small number of cut marks from human tools, and some transverse fractures on long bones (table 5.17). Impact fractures are not terribly relevant to the interpretation of the carnivore remains, because most of the bones are from small species whose skeletons are unlikely to display this kind of damage.

In the Kebaran layer of Hayonim Cave, two phalanges of large carnivores, one a leopard and the other a brown bear, were transversely fractured, and 27 per-

Table 5.17

Frequencies of damage on carnivore remains in the Wadi Meged assemblages

Assemblage	Taxon	Total Obs. (NISP)	TR Fractured (NISP)	Tool-Marked (NISP)	Burned (NISP)
Hayonim Kebaran	Indet. Carnivore	2	1	0	0
	Hyena	1	1	0	1
	Leopard	1	1	0	1
	Wild cat	6	0	0	1
	Jackal	2	0	0	0
	Fox	8	2	0	0
	Brown bear	2	1	0	0
		(22)	(27%)	(0%)[a]	(14%)
Meged Early Kebaran (UP)	Indet. carnivore	1	0	0	0
	Wild cat	3	0	0	1
	Fox	8	3	0	1
	Stone martin	4	1	0	0
		(16)	(25%)	(0%)	(12%)
Hayonim Mousterian	Indet. carnivore	23	9	0	1
	Hyena	2	0	0	0
	Leopard	9	3	0	0
	Wild cat	9	2	0	0
	Wolf/Cape hunting dog	9	0	0	0
	Fox	18	7	1	3
	Brown bear	21	5	1	0
	Stone martin	3	2	0	0
		(94)	(30%)	(2%)	(4%)

[a] A few leopard bones from other areas of the Kebaran excavation, not included in the study sample, are cut-marked.

cent of all large and small carnivore bones were transversely fractured. No tool marks were observed, although two leopard phalanges recovered elsewhere in the Kebaran layer (i.e., not in the study sample) were cut-marked (fig. 5.20). Burning damage occurred on 14 percent of the carnivore bones, including those of wildcats and leopards. In the small samples from the early Kebaran and pre-Kebaran layers of Meged Rockshelter, transverse fractures (25 percent) and burning damage (12 percent) displayed frequencies similar to those in the Hayonim Cave samples. Few archaeologists familiar with Upper Paleolithic and Epipaleolithic cultures would be surprised by these observations, because exploitation of carnivores, especially small "fur bearers," is widely acknowledged.

It is interesting, therefore, that 30 percent of the carnivore remains of all body sizes from the early Mousterian of Hayonim Cave also display transverse

fractures (e.g., fig. 5.21), and 2 percent of all carnivore remains are cut-marked—one a fox scapula and another the first phalanx of a brown bear—consistent with the frequencies of tool marks on ungulate remains in the Mousterian and later cultural layers, to be discussed shortly. Burning damage is less common on the carnivore remains overall (4 percent) than on other taxa and in fact is confined to fox (17 percent) and indeterminate carnivore bones.

One juvenile hyena mandible was found during the 1965–1979 excavations of the Mousterian layer in Hayonim Cave, which might or might not suggest the presence of a den somewhere in the stratigraphic series. However, all of the diagnostic modification evidence on carnivore bones points overwhelmingly to humans. If one accepts the evidence for human collection and modification of carnivore remains in the Kebaran and pre-Kebaran horizons, then the evidence for human-

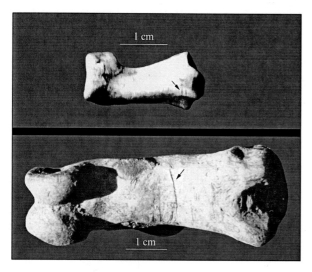

Figure 5.20
Cut-marked first and second phalanges of leopard from
Kebaran Layer C of Hayonim Cave.

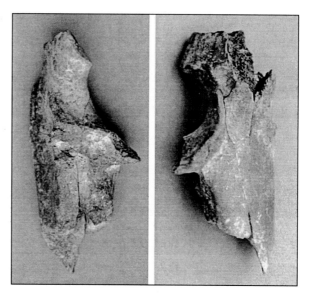

Figure 5.21
Two views of a human-modified brown bear ulna with
transverse fractures and localized crushing, from Mousterian
Layer E of Hayonim Cave.

collected and -modified carnivore bones in the early
Mousterian is equally compelling. Carnivore remains
are fragmentary and rare in all of the assemblages
(table 5.17 and chapter 7), and gnawing damage from
carnivores is virtually absent.

DAMAGE TO UNGULATE REMAINS

Ungulate bones constitute the bulk of the Wadi Meged
assemblages, and because the bodies and bones of
these species are quite large, the ungulate remains display some of the clearest evidence of human modification. The frequencies of tool marks, fractures, and
burning damage are summarized in table 5.18 for
small, medium, and large ungulates. The damage
counts are based on NISP, and total MNE values are
also presented for comparison. Consistent with other
Mediterranean faunas, such as those in Italy, Portugal, and Turkey, the incidence of cut and chop marks
is low in the Wadi Meged faunal series, ranging
between 1 and 4 percent in the larger samples, and no
temporal trends are apparent for burning damage or
tool marks (fig. 5.22). Cut marks testify to human
activities and occur in association with major muscle
masses on bone and, in a few cases, in places where
one would expect to find skinning marks on the distal
extremities (*sensu* Binford 1978, 1981, among others).

Burning damage to ungulate bone ranges between
12 and 22 percent in most of the assemblages, irrespective of ungulate body size. Impact fractures and
tool marks (e.g., fig. 5.23) are universal in these assem-

blages. Impact damage is dominated by cone fractures,
dents, and localized crushing, along with large, semi-
detached cracks. Cone fractures are too abundant to
have been produced by carnivores (Stiner 1994:106,
153–156); the collective frequencies of impact damage
associated with marrow processing and carcass dismemberment range from 3 to 22 percent. Impact damage is most common in the Kebaran assemblages and
the earliest assemblages of the Mousterian sequence.

Clean-edged splits, aligned with the main axis of the
element, and transverse fractures, perpendicular to the
main axis, occur on the bones of ungulates of all body
sizes (table 5.18). However, transverse fractures (and
to a lesser extent splits) are more common in the
smaller species, especially mountain gazelles (fig.
5.24), regardless of culture period (fig. 5.22). This difference may be a product of size-dependent variation
in the way humans went about sectioning carcasses.
The locations of transverse fractures on limb bone elements are generally similar for gazelles and fallow deer
(figs. 5.25 and 5.26). They center on the shoulder and
pelvic girdles, the middle leg areas, and the metapodials just above the feet. The prevalence in the assemblages of these kinds of fractures, which lack any
subsequent modification of break edges by carnivores,
is also taken to indicate human modification alone.

Figure 5.27 displays the "thoroughness" with which
Mousterian humans at Hayonim Cave opened the
major medullae—the central marrow cavities—of

Fracture forms

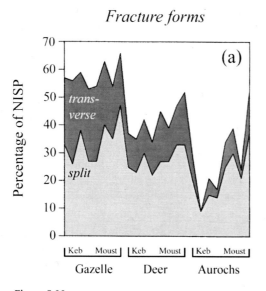

Burning & tool damage

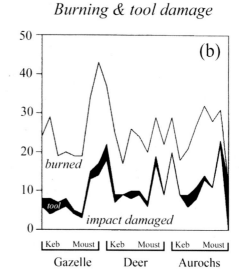

Figure 5.22
Frequencies of (a) transverse and split fractures and (b) burning and tool damage on gazelle, deer, and aurochs remains across the Kebaran and Mousterian periods in Hayonim Cave.

ungulate limb bone elements (data in table 5.19). The term *cold marrow processing,* used here and later, refers to simple methods that may employ reflective heat but do not require boiling or extensive pulverization. Elements are rank-ordered for each ungulate body size group according to medullary cavity size, largest to smallest. The percentage of unopened bone medullae is calculated as the number of complete or nearly complete elements relative to the estimate of total MNE (fragmented or otherwise) for that element, excluding the effects of recent breakage. All Mousterian units are combined, since there were no apparent differences in carcass treatment among units.

The percentage of unopened elements varies most clearly with the size of the ungulate but also with the size of the element in question. There was little waste, and marrow evidently was nearly always in a condition worth eating during the occupations. Cold marrow extraction stopped short, however, of humans' cracking open most of the second and third phalanges and calcanei of gazelles and the third phalanges of fallow deer, all of which contain small marrow reserves. Every element medulla of aurochs bones was opened, so those data are not plotted in figure 5.27. Marrow processing was equally thorough in the Kebaran period (not shown) and followed the same basic patterns seen in the Mousterian. Marrow extraction appears to have been more intensive during the Natufian period,

involving pulverization of some spongy bones of gazelles, possibly indicating more intensive grease rendering of some kind (see Munro 2001).

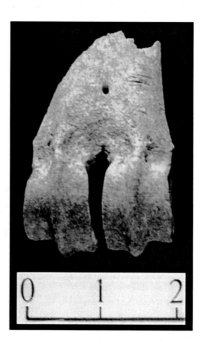

Figure 5.23
Example of cut marks and transverse fracture on a distal metapodial of gazelle from Mousterian Layer E of Hayonim Cave.

Table 5.18
Frequencies of impact fractures, spiral fractures, and burning damage on ungulate remains in the Wadi Meged assemblages

Assemblage	Total Bone MNE	NISP	Cone Fractures (NISP)	Dented/ Crushed (NISP)	% All Types Impact Damage	% Tool-Marked (NISP)	% NISP Split	% NISP Transverse	% NISP Burned
Small Ungulates									
Hayonim Kebaran	518	1,458	85	5	6	2	33	24	16
Meged E. Kebaran	268	878	22	6	3	—	41	23	15
Meged pre-Keb (UP)	130	291	11	1	4	—	39	31	22
Hayonim MP1	35	76	3	0	4	4	26	30	21
Hayonim MP2	49	130	6	0	5	2	38	21	12
Hayonim MP3	135	512	25	5	6	2	27	26	12
Hayonim MP4	450	1,798	49	19	4	1	27	27	14
Hayonim MP5	239	778	18	6	3	1	40	23	15
Hayonim MP6	150	557	68	4	13	2	35	19	19
Hayonim MP7	23	57	8	0	14	3	47	19	26
Medium ungulates									
Hayonim Kebaran	135	669	115	6	18	4	25	12	15
Meged E. Kebaran	78	284	26	3	10	—	27	22	11
Meged pre-Keb (UP)	47	112	6	0	5	—	42	24	15
Hayonim MP1	13	43	1	2	7	2	23	12	16
Hayonim MP2	16	64	6	0	9	0	30	12	8[a]
Hayonim MP3	98	566	39	4	8	2	22	12	16
Hayonim MP4	474	2,709	210	41	9	1	27	18	14
Hayonim MP5	109	572	32	4	6	1	27	12	13
Hayonim MP6	69	322	51	5	17	2	33	14	10
Hayonim MP7	26	70	6	0	9	0	33	19	13
Large ungulates									
Hayonim Kebaran	28	122	22	2	20	0	20	9	9
Hayonim MP1	2	11	0	1	9	0	9	0	9[a]
Hayonim MP2	7	34	2	0	6	3	15	6	12[a]
Hayonim MP3	17	98	6	2	8	3	14	3	16
Hayonim MP4	61	319	28	14	13	1	25	9	18
Hayonim MP5	33	158	16	2	11	0	30	9	17
Hayonim MP6	20	113	25	0	22	1	21	3	8
Hayonim MP7	1	8	0	0	0	12[a]	37[a]	25[a]	0[a]

NOTE: Bone MNE excludes all dental elements.
KEY: (—) no tool mark data available for Meged Rockshelter assemblages due to light calcite encrustations.
[a] Very small sample makes percentage calculation suspect.

SUMMARY OF ASSEMBLAGE FORMATION HISTORIES

The taphonomic analyses described in this and the previous chapter were designed to address the basic questions of how the skeletal accumulations in the Wadi Meged shelters were formed, which agents were responsible for producing them, and what the probable extent was of bone disturbance and in situ loss following primary deposition. From the information presented it is clear that the list of human prey species is long and diverse and that the patterns of body part

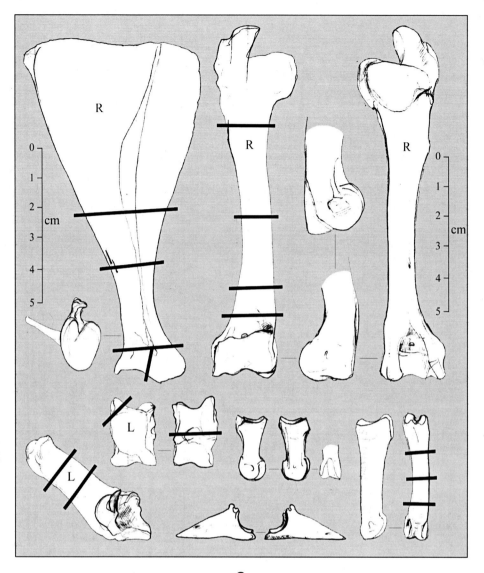

Figure 5.24
Anatomical illustrations and common placements of clean transverse and split fractures on gazelle appendicular bones: (left) scapula, humerus, astragalus, calcaneum, and phalanges; (center) radius and femur; (right) tibia, ulna, and metapodials.

a

representation reflect foremost the transport and processing behaviors of human beings. The Paleolithic inhabitants of the Wadi Meged sites consumed large herbivores (mountain gazelles, fallow deer, and aurochs, among other ungulates), along with a variety of small animals. They occasionally exploited carnivores as well. Macroscopic damage to the bones can be attributed to hominid activities almost without exception. Few traces, if any, of gnawing by large carnivores were observed, although large carnivore bones occurred in the assemblages in small numbers. What damage could be found on the remains of large carnivores was confined to burning and cut marks from human tools. One of the more striking outcomes of

the taphonomic analyses was the demonstration of the importance of small game animals, mainly tortoises, in Middle Paleolithic diets of the Wadi Meged.

The quality of bone preservation was quite good in the Mousterian assemblages overall but somewhat less so in samples from the base of the Mousterian and in the later Kebaran assemblages. Attrition, whatever its cause or causes, did not severely distort the original patterns of prey body part representation (see also chapter 10) or the ungulate mortality patterns (chapter 11). It is unclear whether greater attrition in the Kebaran and pre-Kebaran assemblages was the result of intensified bone processing or of noncultural factors, although the latter seems more likely.

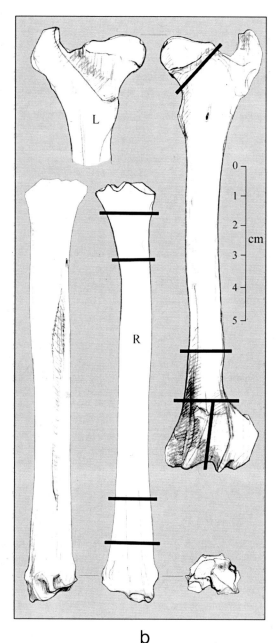

b

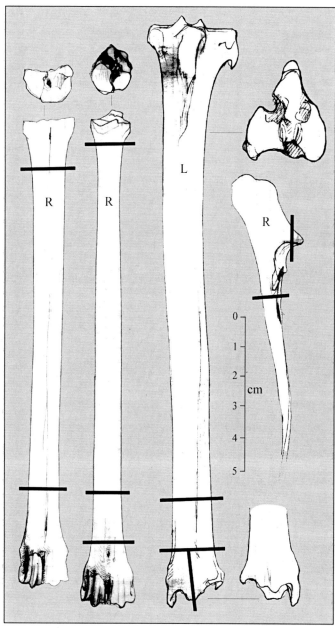

c

BUTCHERING AND MARROW PROCESSING

Generally speaking, there are two broad families of techniques among human cultures for extracting nutrients from bone (mostly medullary marrow and bone grease) that first emerged in Paleolithic times. The first of these was in fact the only technique in use for the bulk of prehistory: "cold" marrow processing, which, as defined earlier, is focused almost exclusively on removing consolidated marrow from limb medul-

lae and the mandible, along with the brain from the cranial vault. This purely mechanical procedure is highly portable and requires only a few simple tools. It can be facilitated by gentle application of heat but is easily done without it. The peripheral elements of the body—legs and heads—are generally the main targets of the cold marrow extraction technique. The second kind of bone-processing technique relies on heated liquid to mobilize the grease locked in small bone pores. This is usually accomplished by greatly fragmenting

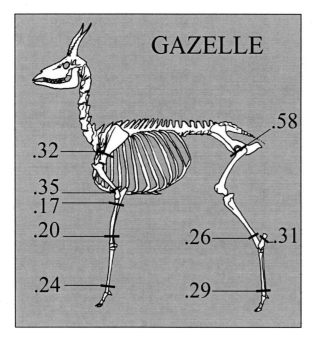

Figure 5.25
Frequencies of transverse fractures on limb elements of
gazelles in Mousterian Layer E.

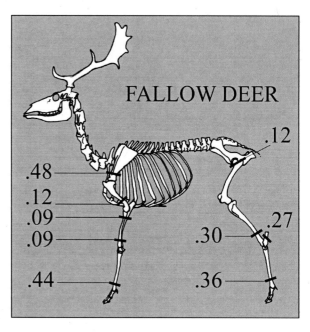

Figure 5.26
Frequencies of transverse fractures on limb elements of fallow
deer in Mousterian Layer E.

or pulverizing spongy bone parts (e.g., the vertebrae, certain limb ends) and then boiling them in water with the aid of preheated stones (Binford 1978; Brain 1981; Brink 1997; Gifford 1977; Lupo 1995; Lupo and Schmitt 1997; O'Connell et al. 1988a; Yellen 1977, 1991a). The fatty components eventually float to the top of the mixture and can be skimmed off, purified, and stored. The goal of the second technique is to obtain grease from the bone—unconsolidated nutrients that cannot be lifted out manually.

Heat-in-liquid grease processing is both more elaborate and more labor intensive than cold marrow techniques, but it allows human consumers to squeeze more food value out of any carcass they obtain. In this sense, heat-based grease rendering can be taken as an indication of economic intensification, wherein people get greater food value per procurement event without increasing the prey population's productivity. Heat-in-liquid rendering of bone-locked nutrients leaves characteristic debris, not only the exceptional comminution of bones (broken while fresh and collagen intact) but also a prevalence of heat-fractured and heat-scarred stone if stone boiling is involved, along with stone anvils (or heavy mortars) for opening and smashing the bones. The containers needed for this procedure can be anything from simple hide-lined pits to leak-proof hide or fiber containers and ceramic ves-

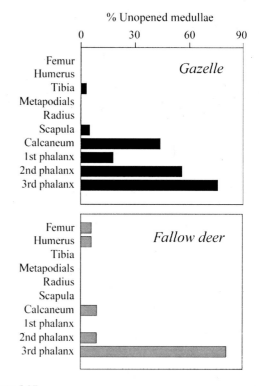

Figure 5.27
The "thoroughness" with which the major limb element medullae of gazelles and fallow deer were opened by Mousterian humans at Hayonim Cave. Elements are rank-ordered according to medullary cavity size, largest to smallest. All aurochs elements were opened (not shown).

Table 5.19
Percentages of ungulate limb bones from Mousterian Layer E in which medullary cavities were not opened prior to discard

Element	Small Ungulates		Medium Ungulates		Large Ungulates	
	Total MNE	% Unopened	Total MNE	% Unopened	Total MNE	% Unopened
Femur	26	0	17	6	2	0
Humerus	34	0	17	6	8	0
Tibia	38	3	27	0	9	0
Metapodials	67	0	40	0	5	0
Radius	30	0	21	0	6	0
Scapula	40	5	25	0	7	0
Calcaneum	36	44	22	9	2	0
1st phalanx	76	18	58	0	12	0
2nd phalanx	89	56	44	9	5	0
3rd phalanx	83	76	59	81	4	0

NOTE: Calculations are based on total MNE for the designated ungulate body size group.

sels, the first of these being the most widely available to foragers. Pit features may also be found in archaeological sites where liquid grease rendering was practiced, though pits are less likely to be preserved than are characteristic fire-cracked rock and heavily comminuted bone. Heat-in-liquid methods are bound to have gained importance in forager systems as human population densities increased and encounter rates for large mammal prey and mobility options for foragers declined.

I find no evidence for heat-in-liquid techniques in the Mousterian of the Levant or elsewhere on the Mediterranean Rim (Stiner 1994). Limb ends were fragmented, but large volumes of spongy bone remained intact. A comparison of ungulate vertebrae condition across the Mousterian and pre-Kebaran periods for the Wadi Meged sites indicates no directional changes in the intensity of fragmentation, calculated as the ratio of MNE to NISP and expressed as a percentage (table 5.20). This is important in that there is little reason for foragers to break up vertebrae beyond limited splitting, as seen in figure 5.28, unless grease in spongy bone pores is the target of their processing efforts. On the other hand, cold marrow extraction was pushed nearly to its maximum potential in these Paleolithic periods, a pattern typical of later arid-land cultures (reviewed by Stiner 1994:225–230). There simply is no additional technological and time investment indicated by the cases examined here.

More intensive processing of ungulate carcasses was an important feature of later Paleolithic game use in several parts of Eurasia. This behavior certainly is in evidence in some Epipaleolithic cases, such as the Magdalenian of France and Germany (Audouze 1987; Weniger 1987) and even the early Gravettian in some European regions (Portugal, Stiner 2003). There also is reasonably good evidence for pulverization of gazelle bones in the Natufian layer of Hayonim Cave (Munro 2001, 2004).

Figure 5.28
Example of a top-split gazelle vertebra from the Mousterian of Hayonim Cave. This form of splitting was common on the axial elements of gazelles and deer.

Table 5.20

Degree of fragmentation of ungulate vertebrae (index = MNE/NISP) in the larger assemblages from the Wadi Meged sites

	Small Ungulates		Medium Ungulates	
Assemblage	MNE/NISP	Index value	MNE/NISP	Index value
Hayonim Kebaran	30/142	.21	10/52	.19
Meged Early Kebaran	12/64	.19	7/20	.35
Meged pre-Kebaran (UP)	9/26	.35	5/15	.33
Hayonim MP3	19/70	.27	11/40	.27
Hayonim MP4	64/272	.24	94/249	.38
Hayonim MP5	25/121	.21	18/57	.32
Hayonim MP6	25/113	.22	17/48	.35

NOTE: Data exclude the atlas and axis vertebrae. Large ungulate remains were too few to be considered.

NUMBERS OF BONES VERSUS STONE ARTIFACTS IN MOUSTERIAN UNITS

Figure 5.29 provides one last look at bone preservation and bone frequencies throughout the Mousterian (MP 3–7) units of Hayonim Cave. This is a plot of the density of occurrence of identifiable bones (NISP) and lithic artifacts (large blanks, retouched pieces, etc.) per cubic meter of sediment. The calculations for bones were based only on areas for which sediment chemistry indicated a favorable preservation environment. Lithic calculations were based on larger areas of the excavation, since stone tools were unaffected by diagenesis and their distributions were usually quite even (chapter 4). Although the scale of frequency variation is far greater for bones than for lithic artifacts, changes in each vary in the same direction (up or down) from Mousterian units 3 through 6. A different relation is apparent for unit 7, where lithic artifact numbers increase as numbers of bones decrease. The counts of bones in MP unit 7 may be problematic, consistent with the results indicating a higher degree of density-mediated attrition. Although ratios of bones to stones can vary for economic reasons (Stiner and Kuhn 1992), unit 7 in Hayonim Cave also represents a significant departure from the preservation conditions as indicated by sediment chemistry and diagenetic history— and below this point, almost no bones were preserved. Systematic comparisons of lithic and bone occurrences in the younger layers were not possible, because much of the upper deposits was removed during an earlier excavation campaign at Hayonim Cave, and lithic and faunal studies of the older collections used different counting criteria.

In all, the taphonomic results indicate that the prospects for behavioral research on the Wadi Meged faunal series are bright. The presentation may now proceed to questions about dietary composition, dietary breadth, and other aspects of human predator-prey relationships.

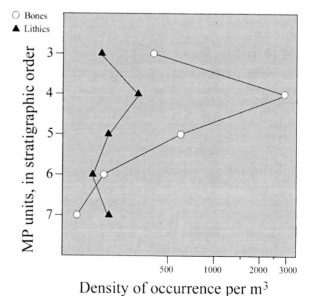

Figure 5.29

The density of occurrence of identified bones (NISP) and lithic artifacts per cubic meter of sediment in the Mousterian sequence (Layers E and F) in the central trench of Hayonim Cave.

■

CHAPTER SIX

Mediterranean Ecology and Species Diversity

■

Nature in the Mediterranean Basin is beautiful not so much for its grandeur but for the richness of its textures and its seemingly infinite variety. Mediterranean forests, for better or for worse, generally do not make humans feel small or inconsequential. These forests form hierarchical patchworks with smaller woody and herbaceous plants, and no glade looks just like another. The structure of Mediterranean biota has been extensively modified by human disturbance, both recently and in the remote past. Many scientists have remarked on the resemblance of Mediterranean biotic communities to a medieval garden tapestry—at whose center, I would add, rests an ungulate in a collar. Mediterranean ecosystems are among the most complex and human-tamed landscapes on earth. Yet many qualities of Mediterranean ecosystems were established long before humans became a part of them. In this chapter I review some essential aspects of Mediterranean ecology and species turnover during the Pleistocene and the characteristics of the species represented in the Wadi Meged faunas.

The Mediterranean Basin is remarkably rich in endemic species—indigenous or native taxa of restricted distribution (e.g., Gomez-Campo 1985; Myers 1990). Yet populations of any one species tend to be comparatively small. In the case of marine life, this is readily apparent relative to populations in the world's oceans, mainly because the Mediterranean Sea is comparatively small and partly isolated. On land, the complex geography, topography, and interdigitation between land and sea have wrought similar effects on animal and plant diversity. The uniqueness of Mediterranean ecosystems is relevant to understanding human behavioral evolution, predator-prey systems, and land use strategies in the region. In the following paragraphs, I review the relevant ecological history and community properties of the Mediterranean Basin; much of this information comes from Blondel and Aronson 1999, Cheylan 1991, di Castri 1991, Groves 1991, Tchernov 1984b and 1992b, and Trabaud 1991.

VEGETATION

Mediterranean habitats would not be what they are if it were not for cool wet winters and hot dry summers. Recurrent fires are also important to the basin's ecology and have been so for a very long time (Trabaud 1991). Not surprisingly, the highest rates of speciation in Mediterranean plants have been manifested in short-lived taxa, not in the long-lived types, apparently because of rapid cycles of stress from drought, fire, and, more recently, human disturbance from livestock grazing and woodcutting (Naveh and Whittaker 1979; Trabaud 1991). Yet another factor determining the character and extent of Mediterranean vegetation is soils, which are generally poor in phosphorus and nitrogen (Groves 1991).

The diverse structure of Mediterranean forests greatly influences the food supplies available to generalist foragers such as humans, bears, and pigs and to browsers such as aurochs and deer, among others. Up to 40 tree species may be common in any given corner of the basin, many of them fruit- or nut-bearing types, and more than 50 additional species occur sporadically (Blondel and Aronson 1999:112–113). Quézel (1976b) noted that Mediterranean forests may contain nearly 100 tree species, in contrast to roughly 37 in the vast central and northern European forests.

Interspersed with forest in the Mediterranean area are scrubwood plant communities of a type known variously as *matorral* (the term preferred here), *garrigue, macchia, monte bajo,* or *xerovumi.* Through time, matorrals have varied in size and degree of spatial isolation (Blondel and Aronson 1999), but pollen sequences indicate the presence of scrubland formations in many areas throughout the Pleistocene (Groves 1991; Trabaud 1991). Matorral plant communities are remarkably stable in the face of recurring fires and other sources of disruption, regenerating rapidly via rhizomes, corms, bulbs, and suckers springing from old stumps (Trabaud 1991). Most other plants that successfully invade matorrals are short-lived types that germinate from seeds already present in the soil; these are regionally endemic annuals, and they usually are eliminated by the aggressive growth of long-lived plants within a few years after burning or clear-cutting. That plants of all sorts are adapted to frequent burning testifies to the deep history of this phenomenon in the Mediterranean Basin (Trabaud 1991:184–187). No single factor, however, can be said to control the structure of Mediterranean plant communities. Rather, many factors appear to work together to maintain the character of biota of the region. Climate and soil characteristics are easily as significant as fire, drought, chance historical events, and, more recently, human cultural effects (di Castri 1991).

A great many ecological niches potentially exist in matorrals, and the number of plant species that can be found is high. Di Castri (1981) defined matorrals as typically dominated by evergreen shrubs with tough (sclerophyllus) small leaves, often with a complex understory of smaller trees, diverse annuals, and herbaceous perennials. Today matorral vegetation has replaced many forests, a shift provoked by human land use practices, usually some combination of overgrazing, woodcutting, cultivation, terracing, and range burning. But recent humans have merely expanded the distributions of matorrals and distorted their contents and textures (Groves 1991; Trabaud 1991). Although matorral vegetation is particularly quick to recover from disturbance, the return to true forest is slow (Blondel and Aronson 1999; Trabaud 1991). Thus, the disturbances brought about by humans, natural fires, and climate changes can substantially and often permanently alter vegetation patchiness and the proportions of open land and forest. The environmental history of the Levant is one case in point: observed or hypothesized environmental changes center on the transformation of forests to semiopen and open vegetation from the Late Pleistocene through the Holocene.

LIFE ZONES, ECOLOGICAL QUADRANTS, AND THE WADI MEGED

The Mediterranean Basin is said to contain a minimum of eight life zones, whose distributions are governed by multiple mountain ranges with abrupt elevation gradients and a wide range of latitudes (Blondel and Aronson 1999:90; Ozenda 1975; Quézel 1985). The eight life zones include the infra- and thermo-Mediterranean zones of low-altitude North Africa and the Near East, the latter dominated by olive (*Olea europaea oleaster*), carob (*Ceratonia siliqua*), lentisk (*Pistacia lentiscus*), and laurel (*Laurus nobilis*) trees, among other woody evergreen trees and shrubs. More widely recognized and in some sense more classic are the meso- and supra-Mediterranean life zones, the former typified by evergreen oaks (especially holm-oak, *Quercus ilex*) and Calabrian pine (*Pinus brutea*) in the west-central basin and *Q. calliprinos* and Aleppo pine (*P. halepensis*) in the eastern end of the basin. The Wadi Meged is associated most closely with these four life zones, either directly or because, in Paleolithic times, the transition zones to them were not far away.

The supra-Mediterranean life zone is dominated by deciduous oak forests at elevations between about 500 and 1,000 m, and some species in these forests are cold sensitive. Higher up, successively, are the montane-, oro-, and alti-Mediterranean life zones, typified, respectively, by beech (*Fagus*) or a mix of conifer forests, then pine forests, and finally a more open combination of dwarf junipers and mixed perennial grasses of the genera *Bromus, Festuca, Poa,* and *Phleum,* among

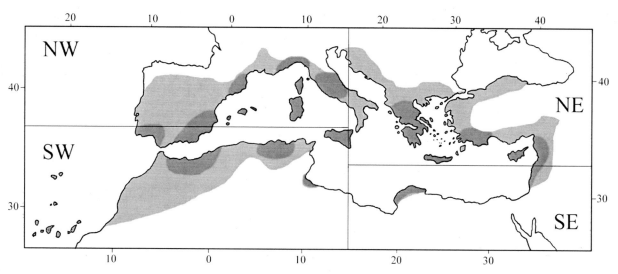

Figure 6.1
Biogeographical quadrants of the Mediterranean Basin, with modern Mediterranean-type biotic distributions indicated by light shading. Dark shading indicates major refugia during glacial maxima (adapted from Blondel and Aronson 1999:8, 28). The southern Levant, including the Galilee, where the Wadi Meged empties into the Mediterranean Sea, lies at the eastern end of the basin, near the top of the southeast quadrant.

others. The highest life zone of all, the cryo-Mediterranean zone, is poorly vegetated except for small alpine plants (e.g., *Saxafraga*) that take root in crags and pockets of gravel scree.

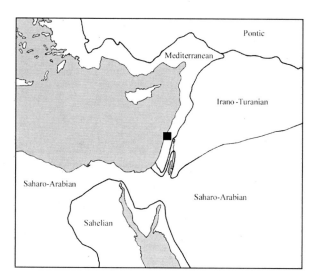

Figure 6.2
Biogeographical provinces of the eastern Mediterranean area, based principally on phytogeographic distributions but also representing important differences in the distributions of animal (especially mammal) species (after Blondel and Aronson 1999:38). The filled square denotes the location of the Wadi Meged in the Galilee.

Discussions of Mediterranean ecology generally divide the basin into four large quadrants, with Mediterranean-type ecosystems generally hugging the sea rim (fig. 6.1). This zooarchaeological study is situated principally in the southeastern quadrant, which broadly includes the southern Levant of Israel, the southern edge of Lebanon, and part of Jordan. The southeastern quadrant is a permanent contact or "suture" zone between Eurasia and Africa. There one encounters remarkable variation in precipitation and elevation over short distances, such as from the coast of Israel to 400 m below sea level at the edge of the Dead Sea in the Jordan Rift and on to the heights of Mount Edam in central Jordan (Blondel and Aronson 1999:101–105).

Prior to extensive human development, the warm (thermo-Mediterranean) coastal dunes of what is now Israel were rich in endemic plant species, some of them of Saharo-Arabian origin, and many reptile species (*T. graeca* being common). As the land rises to elevations of 400 to 500 m inland, evergreen oaks and other warmth-loving plants give way to mixed deciduous-coniferous woodlands of the meso-Mediterranean life zone. Hayonim Cave and Meged Rockshelter lie only 20 km from the modern seacoast and may have straddled at least two life zones throughout much of prehistory. Eastward of the low coastal mountains lies an abrupt rain shadow, because the heights block the flow of moist air from the Mediterranean Sea. The starkness of this rain shadow coincides with remarkable biotic transitions between the Mediterranean province, at the edge of the Euro-Siberian region, the cold central Asian deserts of the Irano-Turanian province, and the hot deserts of the Saharo-Arabian province (fig. 6.2). Species richness in the Mediterranean Basin reaches its highest levels here (Blondel and Aronson 1999).

CAUSES OF DIVERSITY IN THE MEDITERRANEAN BASIN

Before the Miocene, floras of the Mediterranean area were a mixture of tropical forests and savanna grasslands, very different from the region's vegetation today. A "Mediterranean-type" climate, driven principally by winter rainfall and dry summers, was established in the Oligocene and especially from the Pliocene (3.2 million years ago [MYA]) onward. The modern biota is largely the consequence of multiple invasions and local isolation since the Miocene epoch (di Castri 1991; Groves 1991). Indeed, the Mediterranean has been a hub for the intricate interpenetration of species, owing to its unique position at the intersection of three continents.

Although the geographic sources of colonizing species for the Mediterranean Basin are many, Irano-Turanian and Euro-Siberian sources have predominated (e.g., Casevitz-Weulersse 1992; Cheylan 1991). Southern European mammals are basically Euro-Siberian in origin; 70 to 80 percent of species come from there. Although the imprint of Euro-Siberian elements in Mediterranean communities is most obvious, the contribution of Irano-Turanian elements has also been important, especially in the northeastern and eastern areas of the basin; these are species generally adapted to extremely hot summers and very cold, dry winters. Zohary (1969) has argued that adaptations to such harsh conditions made possible the largely unidirectional, aggressive incursions of Irano-Turanian plants (e.g., *Artemesia, Ephedra, Pistacia, Salsola*) into mesic communities of the Mediterranean, with few Mediterranean taxa expanding in the opposite direction.

During glacial maxima, biomes shifted south as much as 10 to 20 degrees in latitude (Blondel and Aronson 1999) and at times fragmented into refugia (fig. 6.1). Mediterranean-type vegetation was almost completely eradicated from Europe during the Würm glaciation in particular (especially after 30 KYA, or oxygen isotope stages [OIS] 3 to 2), with just a few refugia remaining in Andalucia, Sicily, Crete, the Peloponnese, and certain other areas (Cheylan 1991). Mainland refugia generally were surrounded by broad-leaved deciduous forests and, beyond, steppe or tundra. Today the Mediterranean vegetation belt is largely unbroken east to west (fig. 6.1), except by the industrial development and agriculture of the historic period.

Incursions by tropical African taxa have been rare throughout the ecological history of the Mediterranean Basin. This rarity is explained by the great desert barrier formed much of the time by the Saharo-Arabian province and by east-west trending mountain chains and seas that have made north-south biological exchanges difficult since the Miocene-Pliocene boundary some 5–6 MYA (Blondel and Aronson 1999; Cheylan 1991). Even in the Near East, Euro-Siberian mammals dominate, although species from other geographic sources are somewhat better represented there than in Europe; the closer relations between mammalian faunas of North Africa and the Near East are a consequence of the desert belt that links the southern parts of the two regions. Species contributions from the Saharo-Arabian province are limited principally to frost-free areas of the eastern Mediterranean.

Overall, the major groups of Mediterranean biota that exist today have a distinctly Holarctic flavor, with only minor and disparate influences from the vast tropical areas to the south. It is for these reasons that simple out-of-Africa models for large suites of animal taxa during the Middle and Late Pleistocene are neither satisfactory nor credible. This observation does not deny the possibility that a few species slipped through from time to time. One must ask how and why it happened when it did with certain species, among them humans, spotted hyenas, and the rare cape hunting dog.

Species turnover (successional change) is not the only process that determines species diversity in ecosystems. Considerable sorting has also occurred in the Mediterranean Basin—a process in which species possessing certain life-history traits are preferentially favored under distinctly regional patterns of selective pressure (e.g., Blondel and Aronson 1999; Diamond 1994). Close-knit variations in elevation, climate, and geopedologic complexity have played major roles in fostering and maintaining species diversity, such that regional and local endemics abound. Mediterranean flora is among the richest in species in the Old World (Greuter 1991, 1994; Quézel 1985), and much of this diversity evolved in mountainous and island habitats (Cheylan 1991). The great contrast in species richness between the Mediterranean Basin and large areas to the north is probably explained by periodic waves of extinctions in the latter, caused in large part by the cycles of the ice ages (Blondel and Aronson 1999; Cheylan 1991; Latham and Ricklefs 1993; Tchernov 1984b, 1992c, 1994). There have been considerably fewer losses of species within the basin since the Miocene, presumably because conditions remained more stable in the aspects most critical to species persistence, perhaps especially in refugia.

Interestingly, few Mediterranean species became differentiated during the Pleistocene; most phylogenetic differentiation of this period was confined to the subspecies level, even though it occurred to an impressive extent in some lineages (Blondel and Aronson 1999:52; Hewitt 2000; Taberlet et al. 1998). These subspecies radiations are often attributed to population changes that took place while species were in glacial refugia, followed by range expansion during periods of climate amelioration. Cheylan (1991), among others, has argued that speciation rates in Mediterranean refugia might have been slowed by species saturation. Theoretically, species diversity is negatively correlated with niche overlap among taxa, and high species diversity in conjunction with a high incidence of endemic species implies relatively efficient packing of the maximum available resource niches in the ecosystem (reviewed by Pianka 1978).

The great biodiversity of the Mediterranean area is a matter of the exceptional number of endemic species, which by definition occur in spatially limited areas rather than being spread homogeneously throughout the basin. For this reason, species diversity at the most local spatial scale often is not tremendous. It is very high, however, at the scale of Mediterranean subregions and groups of neighboring habitats (Blondel and Aronson 1999:107–109). Variation at this larger scale (such as in association with so-called ecotones) might be what attracted human foragers to some sites repeatedly (in the Levant, Jelinek et al. 1973; Rabinovich and Tchernov 1995; Saxon 1974).

As in terrestrial situations, species diversity in the Mediterranean Sea is extremely high among plants, invertebrates, fishes, and especially mammals, and half of the marine species known are endemic (Tortonese 1985). Even prior to the influx of Indo-Pacific species into the eastern Mediterranean with the opening of the Suez Canal in 1869, it was observed that marine species of the Mediterranean Sea originated in many geographic areas. Mass extinctions resulted from the Messinian Salinity Crisis, which desiccated most of the sea, and it was not until the connection with the Atlantic Ocean was renewed some 5 MYA that this body of water was repopulated, mainly but not exclusively by Atlantic biota (Blondel and Aronson 1999:82–84). Climatic and hydrologic conditions differ in the western and eastern halves of the Mediterranean Sea, also contributing to high biological diversity overall. The western end is dominated by essentially temperate-zone biota, whereas the southeast has many more subtropical species.

The coast lining the western Galilee, with its vast stretches of low-energy, sandy beaches and its limited river discharge, generally makes for poor foraging at the littoral margin. Some impoverishment may be a recent consequence of the damming of the Nile, but this is not a full explanation. Shellfish exploitation must have occurred at some southern Levantine localities during the Paleolithic, but to the best of current knowledge, shellfish were seldom if ever a major source of meat for foragers of that area. Small game exploitation, however, is a central theme of this book, and we know that humans exploited marine resources elsewhere on the Mediterranean Rim from the early Middle Paleolithic onward (Italy, Palma di Cesnola 1965, 1969; Stiner 1994, 1999; Gibraltar, Barton et al. 1999; Cyrenaica, Klein and Scott 1986). Some of these data figure in the interregional comparisons presented in chapters 7 and 9.

A final point about perceptions of species richness in the Mediterranean Basin concerns the practice of classification. Much of the reported diversity in plant and animal species is justified, but counts of species have been amplified for some organism groups by the sheer enthusiasm of taxonomists and by the great length of time over which animals and plants have been studied in this part of the world (Blondel and Aronson 1999; Cheylan 1991). Perhaps nowhere is this amplification more apparent than for mollusks, but it also applies to other life forms such as insects and certain vertebrates such as tortoises (Stubbs 1989). The "lumper-splitter" dichotomy among the researchers who dedicate themselves to systematics cannot be ignored by any serious consumer of the literature on evolutionary ecology in the Mediterranean Basin. Population variation at the subspecies level is important to some kinds of biological research, but it is often trivial for studies of human foraging ecology, except to the extent that it affects expectations and tests for changes in body size, an issue taken up for spur-thighed tortoises in chapter 7.

DIVERSITY AND HISTORY OF THE MAJOR ANIMAL GROUPS

REPTILES AND AMPHIBIANS

Strongly opposed patterns of species richness and distribution exist for reptiles and amphibians in the Mediterranean Basin (Blondel and Aronson 1999: 71–73; Cheylan and Poitevin 1998; Delaugerre and Cheylan 1992). Reptiles generally are the more abundant and richer group, with species diversity increasing markedly from north to south and from west to east in conjunction with greater warmth and aridity. Indeed, most Mediterranean reptiles originated in western Asia and North Africa (Meliadou and Troumbis 1997). Almost half the reptile species in North Africa are endemic types, as are a third in the Near East but only a quarter in Iberia. In comparison with other Mediterranean animal groups, reptiles are somewhat richer in tropical relicts (e.g., crocodiles, chameleons) that became established in the basin long ago. Reptilian faunas have undergone much selection and extinction in situ, and they appear to have been quite stable geographically relative to mammals, whose distributions underwent changes during the Pleistocene. Successful invasions of Mediterranean-type habitats by reptiles since the Miocene or Pliocene have been extremely rare (di Castri 1991).

BIRDS

Surveys by Covas and Blondel (1998) indicate that as many as 366 bird species exist today in the Mediterranean Basin, as opposed to only about 500 for the whole of Europe. Avifaunas were established in the basin mainly during the Plio-Pleistocene (Blondel 1991; Blondel and Aronson 1999:75–79), and the constituent taxa originated in as many as nine distinct biogeographical provinces. Species incursions were facilitated primarily by the spread of matorral habitats and invasions from the northern and Saharo-Arabian regions; very few Mediterranean bird species are of tropical origin. Partridges, which were important game animals to humans from the Upper Paleolithic onward, are closely associated with matorral habitats, as are many warbler species, hares, and tortoises. A distinctive feature of Mediterranean avifaunas is that their species distributions are a good deal more homogeneous than is true for other animal groups, and regional variation is least marked in forests (Covas and Blondel 1998). Endemism also is less pronounced in birds, with comparatively few species having evolved

in the Mediterranean Basin proper. A much higher rate of subspecific variation exists among birds there than in the Palaearctic, a difference thought to have resulted from the high geographical diversity within the basin and from barriers to dispersal (Blondel 1991; Blondel and Aronson 1999).

MAMMALS

Nearly 200 mammal species occur in the Mediterranean Basin, roughly a quarter of which are endemic (Cheylan 1991). Nonflying mammal assemblages have been strongly influenced by the proximity of the three continental landmasses, by repeated disturbances and turnovers provoked by the climate variations of the Pliocene and Pleistocene, and, ultimately, by the activities of prehistoric humans, especially from the Neolithic onward. Mammalian faunas of the quadrants of the Mediterranean Basin (fig. 6.1) differ conspicuously. All four quadrants harbor a rich assortment of mammal species, but the diversity is greatest in the eastern end, with 117 species; as many as 23 of Asian origin are unique to this part of the basin. North Africa is the second richest in species (91), followed closely by the Aegean-Balkans area (82 species); the numbers decline somewhat to the west (71 in Italy, 74 in Iberia) (Cheylan 1991: 233).

Recent mammals of the northern Mediterranean Rim are basically Euro-Siberian in origin; the Villafranchian, or Lower Pleistocene, faunas that preceded them were mainly tropical types that disappeared from Europe at the transition to the Middle Pleistocene about 780,000 years ago. Many new species arrived in Europe from Asia with the ice ages, coming mainly from temperate and boreal regions (Cheylan 1991). The "Levantine Corridor" at the eastern end of the Mediterranean Basin was the conduit for many faunal exchanges during the Neogene and Quaternary (Tchernov 1984b, 1992c). Eurasian brown bears, aurochs, fallow deer, red deer, roe deer, wild boars, wild goats, and ibex colonized North Africa by this route during the climate amelioration of the Riss-Würm Interglacial (OIS 5) some 110 to 70 KYA, following the spread of Mediterranean-type habitats along Africa's northern edge. The Würm Glaciation subsequently forced many of these species into southern Mediterranean refugia. Thus, the story of large mammals in the Levant is roughly paralleled by that for North Africa. Species richness gradually increased during the Last Interglacial; the number of large herbivores (ungulates) and carnivores rose from 17–20 species to 29 species beginning in the late Riss and slowing in the Würm (110–14 KYA). Species richness declined rapidly only in the Holocene (Tchernov 1984b). The richness curves for rodent species are very different between the two areas, however, with the number of rodent species increasing steadily during the Holocene (from 10 to 17 genera) in North Africa but remaining fairly constant (about 15 genera) from the Last Interglacial to the present in the Levant (Cheylan 1991).

As the Sahara Desert reexpanded northward with recent global warming, some Palaearctic mammals were replaced by arid-adapted species, many of them browsers, whose spread into the Levant was facilitated by a separate desert corridor that now links the arid Near East and North Africa (Cheylan 1991). Still, the number of tropical mammals that have colonized the Near East from Africa, mainly via the Nile Valley, are few in number: genet *(Genetta)*, mongoose *(Herpestes)*, hyrax *(Procavia)*, hartebeest *(Alcephalus)*, spiny mouse *(Acomys)*, a large frugivorous bat *(Rousettus)*, and ghost bat *(Taphozous)* (Blondel and Aronson 1999:81). Rare visits to the northern rim of Africa by cape hunting dogs *(Lycaon)* and cheetahs *(Acinonyx)* have also been reported.

The number of carnivore species in the Mediterranean Basin in the Pleistocene and early Holocene was quite high and suggests considerable saturation of potential niche hypervolume space by this group *(sensu* Hutchinson 1957; Pianka 1978:239–245). The richness in carnivore species may also explain indirectly the prevalence of carnivores in cave faunas in many Mediterranean habitats (Gamble 1986). Carnivore influences are particularly evident, for example, in Pleistocene and recent cave deposits in Italy (Stiner 1991b, 1994; Tozzi 1974), Spain (Lindly 1988; Straus 1982), southern France (Brugal and Jaubert 1991; Villa and Soressi 2000), Mediterranean Turkey (Stiner et al. 1996), Morocco (Wrinn n.d.), and the Levant (Horwitz 1998; Horwitz and Smith 1988; Speth and Tchernov 1998). Oddly, this is not the case for the Wadi Meged shelter sites in the Levantine study area (see chapter 5).

The foregoing observations about Mediterranean faunal diversity and ecogeographical history imply that one should look to mammals if one wants to track shifts in community composition. Even within the mammals, changes can be relatively subtle, because gains and losses of species in the Late Pleistocene were limited (Stiner 1994; Tchernov 1992b, 1994). Given the high levels of endemism in reptiles and birds of the Mediterranean Basin, one should almost certainly avoid them for the purpose of studying climate-driven shifts in community structure during the Middle and Late Pleistocene. Mammals are the notable if partial

exception, and this is the logic behind Tchernov's (1975, 1984b) long-term focus on rodent community turnover in the eastern Mediterranean Basin. Such arguments are less easily applied to the larger mammals, although some insights can be gained even from them (chapter 7).

SPECIES IN THE WADI MEGED ARCHAEOFAUNAS

The full range of species other than microfauna in the Wadi Meged series is presented in table 6.1, along with common names and summary information on body size ranges. The species in this list generally will not surprise scientists familiar with the Pleistocene Near East. However, the large samples retrieved from Hayonim Cave yielded a few exceptional taxa, including a well-preserved mandible of *Lycaon,* the cape hunting dog, from the early Mousterian. The cultural deposits also proved to be rich in tortoise remains. General observations about the ecological significance of the archaeofaunal taxa are provided in the following sections; quantitative analyses of species abundance are postponed until chapters 7 and 9.

Large Mammals

Carnivora
Species of the order Carnivora were relatively rare in the Mousterian through Natufian deposits in Hayonim Cave and Meged Rockshelter. In most cases, the carnivore bones appear to have been modified by humans, either as food or for other uses (chapter 5). Exceptions may be rare specimens of *Lycaon* sp. and hyenas. The types and abundances of carnivore families represented in the faunas are potentially important for understanding site formation history, and so they are presented in some detail here.

Most hyena (Hyaenidae) remains were too fragmentary to diagnose to species, and two specimens from MP unit 7 were coprolites. The candidate species are *Hyaena striata* (a.k.a. *H. hyaena*), which continues to inhabit western Asia today, and *Crocuta crocuta,* which occurs in other Pleistocene cave sites of the Galilee but no longer exists in the region (e.g., Bate 1937a, 1937b; Saxon 1974). It is significant that only two coprolites were found during the extensive recent excavations of the Mousterian layer in Hayonim Cave, indicating very limited hyena activity there, in contrast to some of the late Mousterian and early Upper Paleolithic units in Kebara Cave on Mount Carmel (Speth and Tchernov 1998, 2001).

Table 6.1

Macrovertebrate taxa identified or suspected to exist in the Wadi Meged faunal series

Scientific Name	Common Name	Weight Range (kg)
Ungulates		
Gazella gazella	Mountain gazelle	17–23
Capreolus capreolus	Roe deer	10–17
Sus scrofa	Wild pig	40–300
Capra aegagrus	Wild goat	40–90
Equus cf. hemionus	Wild ass/onager	160–240
Dama mesopotamica	Mesopotamian/Persian fallow deer	50–120 (+)
Alcelaphus buselaphus	Hartebeest	120–160
Cervus elaphus	Red deer	90–250 (+)
Bos primigenius	Wild cattle/aurochs	500–1,000
Equus caballus	Wild horse	250–400 (+)
Dicerorhinus hemitoechus	Rhinoceros	800–1,000
Reptiles		
Testudo graeca	Spur-thighed tortoise	1–2 (+)
Coluber sp.	Racer snake (nonvenomous)	0.5–1.0 (+)
Ophisaurus apodus	Legless lizard	1–2 (+)
Agama stellio	Agamid lizard	0.2–0.4
Aves		
Small birds	Songbirds	<0.1
Medium birds	Mainly galliformes, columbiformes	
Alectoris chukar	Chukar partridge	0.535–0.620
Columba sp.	Doves	0.28–0.37
Large birds—Strigidae	Owls	0.2–1.3
Large birds—Falconiformes	Smaller raptors	0.2–2.0
Huge birds—Falconiformes	Large raptors, including vultures	5–8
Struthio camelus	Ostrich*	1–2*
Small mammals		
Lepus capensis	Cape hare	1.6–4.0
Sciurus anomalous	Persian squirrel	0.2–0.3
Erinaceus sp.	Hedgehog	0.6–0.7
Carnivores		
Vulpes vulpes	Red fox	2.2–5.7
Canis aureus	Golden jackal	6.5–12.0
Lycaon sp.	Cape hunting dog (archaic form)	~20–50
Canis lupus	Wolf (low-latitude form)	25–38
Panthera pardus	Leopard	20–50
Ursus arctos	Brown bear (small form)	95–240
Hyaena hyaena	Striped hyena	28–55
Crocuta crocuta	Spotted hyena	47–80
Felis sylvestris	Wild cat	2–5
Felis cf. chaus	Large hunting cat	3–8
Martes foina	Stone martin	1
Vormela peregusna	Marbled polecat	0.6

Continued on the next page

Table 6.1, continued

Scientific Name	Common Name	Weight Range (kg)
General animal categories		
Indet. lizard	—	0.1–0.4
Small ungulate	—	10–23
Medium ungulate	—	40–250
Large cervid	—	90–250
Large ungulate	—	250–1,000
Indet. carnivore	—	2.2–80
Indet. small mammal	—	0.2–4.0
Indet. medium-large mammal	—	10–250

SOURCES: Dunning 1993 for body mass estimates for birds; Silva and Downing 1995 for body mass estimates for mammals. Weight data are based on populations for diverse regions but emphasizing western Asia to the extent possible.

KEY: (+) Animal can weigh considerably more than this, depending on the population.

* Eggs of ostriches only; no ostrich bones were found.

Several canids (Canidae) are represented in the Wadi Meged faunal series, notably the red fox *(Vulpes vulpes)* and species of *Canis* that include the golden jackal *(C. aureus)* and probably also the wolf *(C. lupus)*. Foxes were exploited for their fur and canines (as ornaments) during the Levantine Aurignacian (Belfer-Cohen and Bar-Yosef 1981), Kebaran, and Natufian periods (Munro 2001). The significance of fox remains in the Mousterian is less clear, because they are never common.

Most remarkable among the canid remains is the nearly complete right mandible of an archaic form of *Lycaon* (fig. 6.3). This specimen was found in MP unit 5 of the main trench in Hayonim Cave and is described elsewhere (Stiner et al. 2001b). As unusual as this fossil is, there is no question about its proper generic attribution, because three permanent molars are preserved (fig. 6.4), as well as critical features of the horizontal ramus and condyle. The morphology of the mandible is quite primitive, retaining ancestral features in the dentition and an exceptionally robust horizontal ramus. The tiny entoconid of the first lower molar (M_1) is pleisiomorphic, judging from comparisons with fossil canids by Martínez-Navarro (in Stiner et al. 2001b). A larger version of this feature occurs in *Canis (X.) falconeri*, but it is entirely absent in modern *Lycaon*. Pleisiomorphy is also evidenced for the M_2 of the Hayonim specimen. The general dentition, as well as the thickness and outward cant of the anterior-dorsal horizontal ramus, closely resembles that of modern *Lycaon pictus*. Although the sizes of the M_1 and M_2 of the Hayonim specimen are generally consistent with those of

Lycaon pictus, these teeth have somewhat different proportions. Also, and more striking, the horizontal ramus of the Hayonim mandible is much higher dorso-ventrally relative to thickness than is observed for any modern *Lycaon* population, and the M_3 is relatively large. These last two features bear a strong resemblance to those of the ancestral species, *Canis (X.) falconeri*. Dating to early in OIS 6 or perhaps late in OIS 7 of the later Middle Pleistocene, the Hayonim specimen is the only unequivocal example of *Lycaon* recorded outside of the African continent for any prehistoric period.

One may wonder what the presence of *Lycaon* in Hayonim Cave means for transcontinental biotic exchanges. The mandible is associated with a relatively dry phase of the late Middle Pleistocene and with a mix of Eurasian and Afro-Arabian micromammal community elements. This may have been a time of greater linkage between the vast grasslands of northern Africa and western Asia. *Lycaon pictus* was widespread in sub-Saharan Africa until very recently, avoiding only thick rainforests and the driest deserts (Girman et al. 1993). The dispersal capabilities of recent *Lycaon pictus* populations in Africa are great, owing to the large litters supported by each pack (Ewer 1973; Fanshawe 1989; Maddock and Mills 1994), vast hunting territories (Fuller and Kat 1990), and an ability to traverse long distances in short periods (McNutt 1996). Thus it is not surprising that these dogs have made brief appearances in North Africa in the historic period and that such appearances could have occurred in the Levant during the Pleistocene.

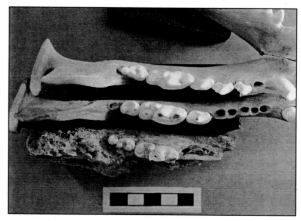

Figure 6.4
The Hayonim *Lycaon* mandible (darkest specimen at bottom of trio) compared with modern *Lycaon pictus* (top) and *Canis lupus* (middle). Note the shorter distance between the maximum posterior extension of the mandibular condyle and the posterior edge of the M_3 alvaeolus in modern and middle Pleistocene *Lycaon* relative to that of modern *C. lupus*. (Reproduced from Stiner et al. 2001b.)

Figure 6.3
Partial right hemimandible of *Lycaon* sp. (cape hunting dog), found at 489 cm below datum in square K21c (H-K21-547, Layer E) in the central excavation area of Hayonim Cave, cleaned and stabilized. Top: lingual view; bottom: lateral, or buccal, view. Note that the breaks on the posterior portion of the mandible (ascending ramus) occurred while the bone was relatively fresh. The breaks on the anterior portion and extending through the lower first molar (M_1) occurred during excavation from brecciated sediments. Scratches on the mandible surface also are from excavation damage. (Reproduced from Stiner et al. 2001b.)

These facts about *Lycaon* do raise the question of why the genus did not spread easily into Eurasia on the whole. Other large predators such as felids, hyenids, and hominids dispersed from Africa during the Pleistocene (Martínez-Navarro and Palmqvist 1995, 1996; Turner 1984, 1986), although these events were few relative to those that contributed the greater numbers of species from Eurasia. The explanation may be the existence in Eurasia of *Canis lupus* and *Cuon alpinus*, two other social hunting canids that came to occupy vast areas of the northern hemisphere. Their predecessors might have limited expansions of *Lycaon* populations during much of the Pleistocene and Holocene. *Cuon* and *Lycaon* are especially similar in their ecology (Kurtén 1968:114), and *Cuon* is well documented in some areas of western Asia and Europe from the Middle Pleistocene onward (Baryshnikov 1996; Crégut-

Bonnoure 1996; Kurtén 1968). Today, *Lycaon pictus* is the more tropically adapted of the pair (Girman et al. 1993:458), which may also explain its apparently brief stratigraphic appearance in western Asia and its complete absence from Europe (cf. Kurtén 1968).

The most common felid (Felidae) in the Wadi Meged series is a small cat that is almost certainly *Felis sylvestris*. At least one other species of *Felis* is represented; it is larger than *F. sylvestris* and probably is *F. chaus*. Leopard *(Panthera pardus)* remains occurred in very low frequencies in Mousterian units 4 and 6 and in the Kebaran layer of Hayonim Cave. A small-bodied variant of brown bear *(Ursus arctos,* Ursidae) occurred in low frequencies throughout the Mousterian and Kebaran layers. To the extent that the condition of any of these rare specimens reveals its taphonomic history, some of the cats and bears appear to have been the prey of humans (chapter 5). Not all bones show signs of modification by humans, however. Remains of Mustelidae are rare except in the Natufian layer (Munro 2001), and only the stone martin *(Martes foina)* is unequivocally represented. Mustelids may have been exploited for their fur in the Kebaran and later culture periods, but their significance in the Mousterian is unclear, because of the very small samples represented.

Ungulates and Megafauna

Browsing species tend to be well represented in Mediterranean animal communities (Blondel and Aronson 1999; Cheylan 1991). Of these, deer (Cervi-

Table 6.2

Pooled osteometric statistics for gazelles and fallow deer in the Kebaran/Pre-Kebaran and Early Mousterian periods in the Wadi Meged faunal series

Element	Type of Measurement	No. Observed	Mean (mm)	SD	Minimum (mm)	Maximum (mm)
Kebaran/pre-Kebaran						
Gazella gazella						
Astragalus	Distal breadth	34	16.9	0.93	15.4	19.0
	Length	28	27.7	1.35	25.9	30.0
Mandible	Condylar breadth	3	11.6	1.17	10.7	12.9
Tibia	Distal breadth	3	21.7	1.36	20.8	23.3
	Distal anterior-posterior	4	18.9	0.33	18.5	19.2
Humerus	Distal breadth	13	24.4	1.23	22.6	26.2
	Distal anterior-posterior	12	19.0	2.43	16.3	24.1
Dama mesopotamica						
Astragalus	Length	3	44.3	3.34	41.3	47.9
Mousterian						
Gazella gazella						
Astragalus	Distal breadth	40	17.1	0.86	15.4	19.2
	Length	38	28.6	1.50	25.0	32.3
Mandible	Condylar breadth	23	14.0	1.39	11.3	19.0
Tibia	Distal breadth	17	24.0	3.07	18.1	34.0
	Distal anterior-posterior	11	19.5	1.25	17.7	21.9
Humerus	Distal breadth	11	26.2	3.89	21.9	35.0
	Distal anterior-posterior	5	17.9	0.52	17.2	18.6
Dama mesopotamica						
Astragalus	Distal breadth	10	30.3	6.63	24.0	43.3
	Length	7	44.0	4.18	37.0	48.0
Mandible	Condylar breadth	9	20.1	2.79	16.0	24.0
Tibia	Distal breadth	3	39.4	0.49	39.1	40.0
	Distal anterior-posterior	3	33.6	0.67	33.0	34.3

NOTE: "No. observed" is the count of measurable specimens. Data are only for those elements and anatomical positions for which three or more specimens could be measured. Right and left sides are combined.

dae) are prominent in the Wadi Meged faunal series, although only the Mesopotamian fallow deer *(Dama mesopotamica)* is common; red deer *(Cervus elaphus)* and roe deer *(Capreolus capreolus)* occur in small numbers, the latter being very rare. The Mesopotamian fallow deer is relatively large bodied in comparison with *D. dama* of Europe (limited osteometric data are presented in table 6.2).

The Bovidae are a richly represented ungulate group in the Wadi Meged faunal series, including mountain gazelle *(Gazella gazella)*, aurochs or wild cattle *(Bos primigenius)*, wild goat *(Capra aegagrus)*, and rarely hartebeest *(Alcelaphus buselaphus)*. Of these, only mountain gazelles are common. The mean body size of the gazelles from the early Mousterian slightly exceeds

that for the pre-Kebaran/Kebaran specimens, judging from linear osteometric comparisons (table 6.2 and chapter 11). Wild boar *(Sus scrofa,* Suidae) is present throughout the Wadi Meged series as well but is never common.

Members of the Equidae are rare in the Wadi Meged series, but two species are represented, the caballine horse, *Equus caballus,* and a smaller asine type that is either *E. hemionus* or *E. hydruntinus* (see Tchernov 1994; Uerpmann 1981). Rhinoceros (Rhinocerontidae) is rare in Paleolithic sites of the Levant, and only a few fragmentary remains occurred in the Kebaran, Aurignacian, and early Mousterian layers of Hayonim Cave. The species could not be determined from these fragments but is most likely *Dicerorhinus hemitoechus.*

SMALL ANIMALS

Reptiles

Tortoises (Testudinae, Chelonia) were one of the most important prey animals in the Wadi Meged series, from the early Middle Paleolithic through the Natufian period. This comes as a surprise in light of earlier studies of Paleolithic faunas of western Asia, in which tortoise remains and their condition were seldom reported in discussions of human subsistence. Tortoise bones were plainly abundant in the faunas we excavated and, upon examination of the older collections from Hayonim Cave and Kebara Cave, were found to be abundant there as well. Until 1992 there had been no systematic examination of the taphonomic question of human attribution for the tortoise remains (Speth and Tchernov 2002; Stiner and Tchernov 1998).

The taxonomic affinity of the Wadi Meged tortoises appears always to be the Mediterranean spur-thighed tortoise, *Testudo graeca terrestris*. This is not, however, a cut-and-dried matter in the minds of some (Speth and Tchernov 2002; Y. Werner, Hebrew University, personal communication, 1996). Because species and subspecies classifications are important to interpretations of body size variation in tortoises over time, the tortoises of Wadi Meged receive additional consideration in chapter 7.

By way of introduction, the genus *Testudo* is an aggregate of many species; it is the largest and most widespread genus of nonmarine (terrestrial) turtles in the Old World (Cogger and Zweifel 1998; Pope 1956). *Testudo graeca* is particularly widely distributed, from southern Spain in the west across the north coast of Africa and eastward to Israel and Syria, and from northern Greece, Romania, and Bulgaria through Turkey to the Transcaucasus, Russia, and Iran (Ernst and Barbour 1989:266–270). Such a distribution, spanning nearly every modern country lining the Mediterranean Sea (Cogger and Zweifel 1998:121), testifies to the great adaptability of the species. Other species of the genus have much more limited distributions, with *T. marginata* found only in Greece and Sardinia, and *T. kleinmanni* in Egypt.

T. graeca typically is found throughout semiarid hill scrub and matorrals, as well as in coastal dune habitats and open pine and oak forests (Hailey 1988; Stubbs 1989). Its altitudinal distribution ranges from sea level to about 3,000 m, but it is most commonly found below 2,000 m. As latitude increases, altitudinal distribution seems to decrease, suggesting an adaptation to higher temperatures and more arid conditions than is true for *T. hermanni*. *T. graeca* lives on the ground

surface and does not excavate true burrows, although tortoises may bury themselves shallowly during dormancy (Cogger and Zweifel 1998). Mating generally takes place from April to July, when courting males chase, ram, and push at females, sometimes nipping at their limbs and head (Ernst and Barbour 1989). Most nesting (egg laying) occurs in May and June, but it may continue into July. The tortoise's diet consists almost exclusively of plants, but occasional consumption of worms, snails, and insects is reported (Cogger and Zweifel 1998).

In the northern extent of its range, *T. graeca* hibernates in winter, but in hot southern areas it may estivate in summer instead. Seasonal activity in *T. graeca* varies considerably among study areas. In Spain, tortoises are active from February to June, with a short estivation period in the height of summer, before activity is renewed in September through November (Andreu 1988, cited in Stubbs 1989). In the modern Wadi Meged, we frequently found tortoises on the move during our excavation season from June through early August. Dry plant stalks line the wadi in the summer months, and tortoises were discovered mainly from the noise they made as they moved through the vegetation. Most of the tortoises that our team spotted were adults and subadults, consistent with the experiences of biologists who work on the species. Some of the bias against very young tortoises might have been explained by our excavation schedule—the Hayonim excavation season closed each year before hatchlings normally emerged from their nests—but censuses of wild populations invariably indicate that infants are much more difficult to find than adults, irrespective of season (Hailey et al. 1988; Lambert 1982, 1984). The availability of tortoises to Paleolithic foragers must have been somewhat seasonal, but the months of availability are difficult to infer because the activity schedules of the species vary so much today (Cogger and Zweifel 1998; Ernst and Barbour 1989).

The legless lizard (Anguidae) in the Wadi Meged archaeofaunal series is *Ophisaurus apodus*. The largest of all anguids, this species possesses a thick body up to 1.4 m in length and a very long, fragile tail that breaks away easily (Cogger and Zweifel 1998; Pope 1956). The genus *Ophisaurus* is widespread, and *O. apodus* occurs in southwestern Asia and southeastern Europe. It feeds on small mammals, snails, insects, and reptiles. It lives on the ground surface, in loose organic matter and under rocks, and it reproduces by laying eggs, which the female remains with but does not actively protect. *O. apodus* is severely threatened in northern Israel today and is seldom seen, although it continues to be found in remote areas. In addition to its distinctive dentition (chapter 5, fig. 5.17), another important skeletal characteristic of *O. apodus* is its bony osteoderm plates or scales, which preserve well in sediments. Legless lizards must have been more common in the Wadi Meged in prehistory, because their remains occurred among other archaeofaunal debris in several excavated Paleolithic layers. The individuals represented in the archaeofaunas were always medium to large in size and thus would have represented substantial amounts of meat.

One other large, diurnal, terrestrial lizard, *Agama stellio* (Agamidae), also occurred in the Wadi Meged faunas. The condition of the bones (chapter 5) suggests that at least some of them fell prey to humans. This large, vegetarian lizard, characterized by unsocketed acrodont teeth on the rims of its jaws and a nonshedding tail, is common in the Wadi Meged and attains a length of up to about 50 cm. Preferring rocky terrain, it lives mainly on the ground surface, not in burrows (Cogger and Zweifel 1998:134–136).

Nonvenomous calubrid snakes (Calubridae, Serpentes) are prevalent in western Asia and southern Europe and include the genera *Natrix* (water snakes), *Elaphe* (rat and chicken snakes), *Coluber* (racers and whip snakes), and *Contia* (Pope 1956). Snake remains occurred in appreciable frequencies in many of the Mousterian and later Paleolithic assemblages of Wadi Meged. The vertebrae best resemble those of *Coluber* sp. The association of snake bones with cultural remains was most consistent in Hayonim Cave, and many of the bones were also burned (chapter 5). Despite the general spatial association of snake bones with artifactual material, it is unclear whether snakes were the prey of humans or were repeatedly attracted to deposits disturbed by humans. Large snakes recolonized our excavations in Hayonim Cave each fall and winter as annual plants reseeded in the entrance of the cave. The snakes tended to burrow just below the surface of loose vegetable matter. Thus, it is possible that some individuals died in burrows or surface litter in the past and that the remains were subsequently burned by fires built by humans. Unlike the other reptile remains, the snake vertebrae preserved no signs of human modification other than burning, owing mainly to their simplified skeleton—leaving open the question of whether snakes were human prey (see chapter 5).

Small Mammals

Leporidae (Lagomorpha) are represented in the Wadi Meged faunal series by the cape hare *(Lepus capensis)*, which occurs in substantial numbers in the assem-

blages from the later Paleolithic periods. One member of Sciuridae (Rodentia), the Persian squirrel *(Sciurus anomalous)*, is present in very low frequencies, as are hedgehogs *(Erinaceus* sp., Erinicidae, Insectivora). Hares clearly represent human prey in the Wadi Meged faunal series (chapter 5), and squirrels probably were as well. The hedgehog remains are more difficult to evaluate, and although there is little reason to exclude them as possible prey on the basis of bone damage data, they were never a significant source of meat.

Birds

Ostrich *(Struthio camelus)* eggshell fragments were present throughout the Paleolithic sequence of Hayonim Cave, except during the Natufian period, when ostriches seem to have disappeared from the area. There is no evidence for bead-making from this material in any period, although marine shells were used as ornaments from the Upper Paleolithic onward (D. Bar-Yosef 1989; Hovers et al. 1988; Kuhn et al. 2004). Many of the ostrich eggshell fragments are burned (chapter 5). Conjoinable pieces were sought but not found, but in any case the fragments would have amounted to just one or a few eggs per cultural layer. In addition to food, the eggs could have been used to carry or store water during any of the periods in which they occurred, including the early Mousterian. No bones of ostriches were found.

Partridges, especially chukars *(Alectorus chukar)*, were a major food source in the Galilee from the Upper Paleolithic onward, but they occur only rarely in the early Mousterian assemblages. Other birds became important in archaeofaunas at the same time as partridges, including doves *(Columba)* and raptors (Falconiformes). Although the presence of raptors and owls in Mediterranean shelter assemblages can be associated with pellets or middens accumulated by roosting birds, it is clear from bone damage patterns (chapter 5) that most of the large carnivorous birds (including large vultures) in the Wadi Meged Upper and Epipaleolithic assemblages were obtained and modified by humans (see also Munro 2001; Pichon 1983, 1984; Tchernov 1993a). Owls, on the other hand, appear to have been responsible for most of the rich microfaunal accumulations during the Mousterian and later Paleolithic occupations.

Micromammals

Some environmental changes are indicated by the microfauna in the Paleolithic series of Hayonim Cave (Tchernov 1981, 1994, and personal communication, 2001). Sixteen rodent species were identified, five of which occurred only in the lower part (earliest Mousterian) of the sedimentary sequence. Two of these five species (*Arvicanthis ectos* and *Praomys [Mastomys] batei*) belong to rat genera (Muridae) that originated in Africa long ago but are now endemic to the southern Levant and are most typical of dry savanna habitats. The third species, *Gervillus dasyurus* (Gerbillidae), is of Arabian origin and is an obligate dweller of barren, rocky slopes. *Ellobius fuscocallillus* (Arvicollidae) is an Asiatic subterranean vole; it appeared for the first time in the southern Levant in the early Mousterian period, soon dispersing into North Africa, although its occupation of these regions was brief. The fifth species, *Mesocricetus auratus* (Cricetidae), is another typically Asiatic element that appeared intermittently in the southern Levant. All of the other rodent species are either Palaearctic forms or restricted to the eastern Mediterranean region (e.g., *Myomimus judaicus, Spalax ehrenbergi, S.* cf. *kebarensis*).

According to Tchernov (1992c, 1994), invasion of the Mediterranean Levant by Afro-Arabian micromammal species and their apparent mixing with Asian, arid-adapted species may correlate with a late dry phase of OIS 7. These taxa disappeared during subsequent cold phases of OIS 6 and then reappeared late in OIS 5 at the site of Qafzeh. Although this scenario may seem at odds with some of the radiometric dates for the early Mousterian of Hayonim Cave, it should be noted that the ESR dates are averaged values that center on 170,000 years ago but represent a sediment series more than 1 m thick (see chapters 1 and 2). Thus, the ESR results do not exclude the possibility of greater age for the lowermost Mousterian faunas. By the Late Pleistocene, rodent communities had stabilized, and they changed little thereafter (Tchernov 1984b).

SUMMARY

The last 200,000 years in the Galilee have seen many changes in the conditions of life but surprisingly few in the basic composition and diversity of macrofaunal and plant communities. The high incidence of endemic taxa in the Mediterranean Basin is testimony to considerable community stability, particularly among the reptiles, birds, and large mammals. Variation in the relative proportions of these taxa in the archaeofaunas is another story, some of which certainly is explained by climate change, but some of which is more closely tied to changes in human diet and ecology.

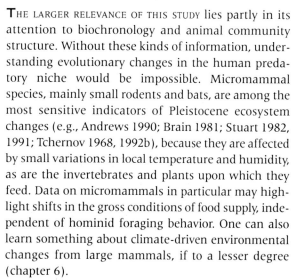

CHAPTER SEVEN

Paleocommunities and Trends in Species as Prey

THE LARGER RELEVANCE OF THIS STUDY lies partly in its attention to biochronology and animal community structure. Without these kinds of information, understanding evolutionary changes in the human predatory niche would be impossible. Micromammal species, mainly small rodents and bats, are among the most sensitive indicators of Pleistocene ecosystem changes (e.g., Andrews 1990; Brain 1981; Stuart 1982, 1991; Tchernov 1968, 1992b), because they are affected by small variations in local temperature and humidity, as are the invertebrates and plants upon which they feed. Data on micromammals in particular may highlight shifts in the gross conditions of food supply, independent of hominid foraging behavior. One can also learn something about climate-driven environmental changes from large mammals, if to a lesser degree (chapter 6).

The Mediterranean region is characterized by exceptionally high diversity in animal and plant species on land and in the sea (Blondel and Aronson 1999). As explained in chapter 6, however, mammals are the only major animal group whose species distributions differ markedly among the four quadrants of the Mediterranean Basin (Cheylan 1991). Mammal species comparisons across space and time began early in the Levant with the work of Dorothea Bate (1937a, 1937b), who explored, among other questions, the causes of frequency shifts in deer and mountain gazelles during the Pleistocene.

In this chapter I am concerned with variation in species representation and related trends in the Wadi Meged faunal series from 200 to 10 KYA. At least two sorts of questions may be asked of the data on taxonomic abundance. The first has to do with trends in the relative occurrences of key prey species, particularly but not exclusively mountain gazelles, fallow deer, and tortoises, as well as the overall proportions of small to large game. The second kind of question is focused more on animal community composition and what it might reflect about patterns in vegetation and precipitation.

The animals in the Wadi Meged macrofaunal series represent prey of one kind of predator—Paleolithic humans—and this behavioral filter has biased the composition of the faunal assemblages. The filter, however, is a consistent one, and thus relative differences in prey composition are informative about the communities from which the prey animals derived. Gazelles, as the conventional wisdom goes, favor arid open lands (as do goats, wild asses, and horses). Fallow deer are more versatile, tending to rely on browsing in mixed open

woodlands and perhaps requiring higher precipitation. The relative proportions of these two ungulates varied through time in the Paleolithic Levant (Bate 1937a, 1937b, 1940; Davis 1982; Saxon 1974).

Spur-thighed tortoises are endemic to the region and appear to flourish under a great range of environmental conditions. Taxonomic distinctions among tortoises are unlikely in principle to be diagnostic of climate-induced changes, but tortoises' growth rates are sensitive to the rate of energy flow in ecosystems and are controlled especially by the length of the growing season and the abundance of green plants, their main source of food (Cogger and Zweifel 1998; Ernst and Barbour 1989). Information on mean body size variation in tortoises can be compared with independent data on climate change, as well as with changes in the contents of the archaeofaunas.

Learning about the boundaries of the hominid diet requires that the full suite of faunal remains, large and small, be examined without prejudice toward what modern folk regard as edible, scientifically interesting, or "important." Much of zooarchaeological research on the Paleolithic has proceeded with just such biases, and as is demonstrated later, perceptions of early human diets have in some regions leaned too heavily toward questions about large game exploitation.

To understand how species variation in the Wadi Meged faunal assemblages might or might not relate to changes in environmental conditions and animal community structure, the debate about the ratio of mountain gazelles to fallow deer is rekindled first. Trends within the small game fractions are also examined, and tortoise frequencies and body sizes are considered against the relative abundances of the two ungulate species. The results are synthesized further through trend analyses of macrovertebrate species representation. The prevalence of tortoises in the Wadi Meged series, along with some questions raised previously about the implications of tortoise size reduction (Speth and Tchernov 2002), demands a closer look at the nature and taxonomic identity of these animals. In addition to examining the tortoise data for the timing and intensity of body size diminution, in the final sections of this chapter I address variation in the maximum sizes attained, possible evidence of age structure distortion, sex biases, and the morphology of the tortoises.

ZOOARCHAEOLOGICAL METHODS: IDENTIFICATION AND QUANTIFICATION

Classifications of the Wadi Meged macrofaunal specimens at the level of genus or species were based on comparative material in the vertebrate collections of the Zoology Laboratory in the Department of Evolution, Systematics, and Ecology at the Hebrew University of Jerusalem. In this presentation I generally avoid the complex taxonomic notation favored by paleontologists, because much of the descriptive work has already been done for macrovertebrates in the Levant. Rather, I comment on systematics only as it is relevant to this study of hominid subsistence and paleocommunity in the Galilee.

The piece-plotted and screen-recovered fractions of the faunas from Hayonim Cave were joined so that the full array of game animals in Paleolithic diets could be evaluated (chapter 1). The counting unit for most of the analyses of taxonomic abundance was the number of identified faunal specimens (NISP, see appendix 1). NISP is the closest approximation to a continuous variable among the counting units available to zooarchaeologists and is the least subject to aggregation error in archaeological contexts (Grayson 1984). Here it is used for *relative* comparisons of taxonomic abundance. A second approach uses the minimum number of individuals (MNI) based on bony features as a way to correct the species abundance data in terms of prey biomass. Dental remains were not used to estimate MNI in this analysis, for the sake of comparability, because many of the prey taxa in the Wadi Meged series did not possess teeth.

Although NISP has several advantages as an irreducible quantitative unit, the possibility of differential fragmentation among assemblages can affect comparisons of species abundance (Grayson 1984). Ranking the abundances of common prey animals by body size on an ordinal scale (fig. 7.1) reveals similar gradients (or slopes) in the mean fragment sizes in the Natufian, Kebaran, and Mousterian samples from Hayonim Cave (table 7.1, appendix 10). Mean fragment sizes in the Mousterian sample tend to be somewhat greater overall (fig. 7.2), but differences between the Mousterian and the two later periods are minor. The mixes of prey change with time, but the ratio of small to large game NISP is trendless, and there is no reason to suppose that taxonomic variation introduced consistent (systematic) size-related distortions in prey type abundance in the comparison. The only exception to this

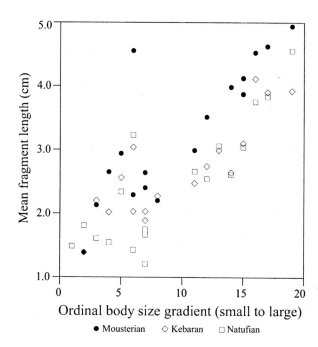

Figure 7.1
Mean skeletal fragment lengths by prey body size on an ordinal scale for Natufian, Kebaran, and Mousterian samples from Hayonim Cave. Smallest body size = 1, *Agama stellio;* largest = 20, large ungulate.

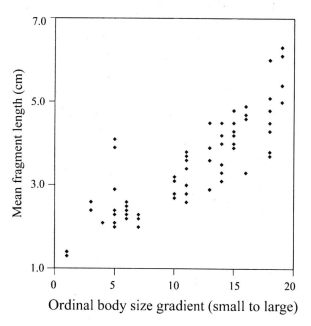

Figure 7.2
Mean skeletal fragment lengths by prey body size for each of the seven Mousterian units in Layer E of Hayonim Cave.

conclusion is the greater presence of large-bodied ungulates in the Mousterian, which I discuss later.

The deposits of Meged Rockshelter were considerably more concreted than those in Hayonim Cave, resulting in greater breakage of bones during excavation and making fragment size comparisons with the Hayonim samples difficult. Less formal observations nonetheless indicate that predepositional fragmentation of bone in Meged Rockshelter was on a par with that of the other assemblages in the Wadi Meged series, particularly with the Kebaran sample from Layer C of Hayonim Cave.

TRENDS IN SPECIES ABUNDANCE

Because many aspects of community assembly and faunal turnover in the Pleistocene remain poorly understood, little can be taken for granted about the environmental affinities of the prey species. Simple questions about the robustness of species associations through time are addressed here first. Specifically, which taxa "adhere to" or "repel" one another most consistently through time? Seeking trends in frequency data is complicated by the fact that only a few taxa dominate the Wadi Meged faunal assemblages,

so that a decline in one usually means an increase in another. This important criticism of Bate's (1937a, 1937b) gazelle–fallow deer index was raised by Davis more than two decades ago (1982). The gazelle–fallow deer index is calculated as

gazelle NISP/(gazelle NISP + fallow deer NISP).

Negative correlations should be expected of the percentage calculations a priori. In recognition of the problem, only positive correlations and statistical independence between species pairs are considered particularly informative. The trend analyses to follow therefore represent a process of weeding out spurious associations and isolating those most likely to reflect historical (climate-induced or human-induced) processes.

Temporal relations among animal taxa in the Wadi Meged faunal series are examined in three ways. The first analysis compares the relative frequencies of key taxa on an ordinal time scale using first NISP and then biomass counting units; it then examines variation in the proportions of key taxa via correlation statistics. The second approach employs detrended, interspecific comparisons of the amount of change between consecutive taxon pairs. The third analysis focuses on tortoise diminution and age structure over time and across oxygen isotope stages (following Martinson et al. 1987). Other aspects of species abundance, productivity, and diversity are taken up in chapters 8 and 9.

Table 7.1

Spearman's correlation statistics for the relation between mean fragment length (cm) of identified faunal specimens (NISP) and animal body size (ordinal scale) for Hayonim Cave

Assemblage	No. Body Size Groups Observed	Spearman's rho	p
By culture period			
Natufian (Layer B)	19	0.782	.001
Kebaran (Layer C)	18	0.834	.001
Mousterian (Layer E)	19	0.838	.001
Within the Mousterian layer			
MP unit 1	5	0.821	.083[a]
MP unit 2	8	0.994	.001[a]
MP unit 3	14	0.819	.001
MP unit 4a	13	0.946	.001
MP unit 4b	14	0.846	.001
MP unit 5	12	0.984	.001
MP unit 6	10	0.952	.001
MP unit 7	6	0.886	.020[a]

NOTE: Rare taxa were removed from consideration in all assemblages.

[a] The very small sample size from this Mousterian unit may be problematical.

SPECIES TRENDS BASED ON PERCENTAGES OF NISP

Two ungulate species—mountain gazelle and Mesopotamian fallow deer—are very common in the Paleolithic faunas from the Wadi Meged sites, along with spur-thighed tortoises. The full spectrum of prey species is much more diverse (see appendixes 11 and 12; table 6.1), consistent with the natural content of Mediterranean ecosystems. Figure 7.3 summarizes some gross patterns in prey representation from the Middle Paleolithic through the Epipaleolithic period. It is immediately apparent that carnivores are rare and ungulates generally dominate the assemblages. Small animal representation varies through time, but their remains are always present (tables 7.2 and 7.3). No trend is apparent for the relative proportions (NISP) of ungulates and small game (tables 7.4 and 7.5) based on the small–large game index:

small game NISP/(small game NISP + ungulate NISP).

There is a marked increase in small game exploitation for the early Natufian, as noted previously by Davis et al. (1988), Munro (2001, 2004), and Tchernov (1992d). However, the next highest incidence of small game occurs in the early Mousterian period.

Great changes took place in the *types* of small game emphasized by human foragers over 200,000 years,

due mainly to the significant addition of birds and hares beginning in the Upper Paleolithic. Tortoises are the most consistently represented type of small game throughout the culture sequence ("slow type" in fig. 7.3); the use of tortoises merely gave way to greater use of birds and small mammals. It is striking that tortoises were the only major small game animal during the Middle Paleolithic, since paleontological evidence indicates that warm-blooded prey were also present in these ecosystems throughout the time frame considered, and animal species composition was particularly stable in the Late Pleistocene communities (Tchernov 1992b, 1994, 1998c). The few bird bones found in the Hayonim Middle Paleolithic assemblages are from large predatory species, mostly owls that roosted in the cave. A very different suite of avian species appears in the Upper Paleolithic assemblages, one rich in ground birds (especially chukar partridges) and certain other medium-large birds. Raptors (vultures, eagles, and falcons) are also common in some of the assemblages. Taphonomic evidence for human exploitation of the latter in various Galilee sites is apparent from cut marks (e.g., Munro 2001; Pichon 1983, 1987) and the great overrepresentation of large raptor claws or talons discussed in chapter 5. Predatory birds were widely exploited by Levantine Upper Paleolithic and Epipaleolithic peoples, apparently for their skins, fine bone

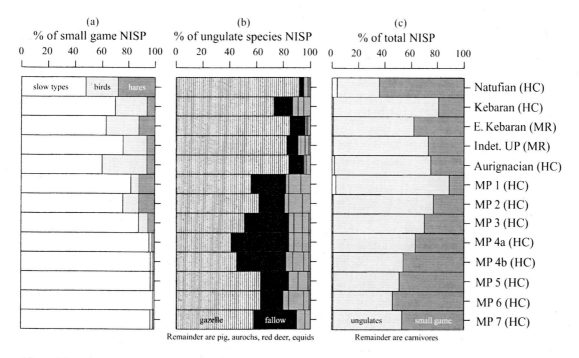

Figure 7.3
Patterns in prey representation in the Wadi Meged faunal series. (a) Slow prey, quick flying prey, and quick running prey as percentages of small game NISP; (b) gazelles, fallow deer, and other ungulate species as percentages of ungulate NISP; (c) carnivores, ungulates, and small game as percentages of total NISP in each assemblage.

Table 7.2
Percentages of carnivores, ungulates, and small game animals in the Wadi Meged faunal series

Assemblage (Period, Site, Layer, Unit)	Period	Total NISP	% Carnivores	% Ungulates	% Small Game
EP, HC, B	Natufian	15,395	4	31	64
EP, HC, C	Kebaran	3,680	1	80	19
EP, MR, <200 cm	Kebaran	1,978	1	61	38
UP, MR, >199 cm	Pre-Kebaran (UP)	597	1	72	27
UP, HC, D	Aurignacian	10,463	2	73	25
MP, HC, E, unit 1	Mousterian	152	3	86	11
MP, HC, E, unit 2	Mousterian	306	1	76	23
MP, HC, E, unit 3	Mousterian	1,733	1	70	29
MP, HC, E, unit 4a	Mousterian	4,536	<	63	37
MP, HC, E, unit 4b	Mousterian	3,903	<	53	46
MP, HC, E, unit 5	Mousterian	3,083	<	51	49
MP, HC, E, unit 6	Mousterian	2,254	1	45	54
MP, HC, F, unit 7	Mousterian	261	1	52	46

SOURCES: Hayonim Natufian data are from Munro 2001. Hayonim Aurignacian data are based on Stiner's NISP counts of small game taxa and Rabinovich's (1998) counts of carnivores and ungulates.

NOTE: Tortoise/lizard category is dominated by tortoises (> 95%). Small mammals are mainly hares, along with low frequencies of squirrels. Birds include mainly the orders Galliformes and Colombiformes. Snakes were omitted from consideration.

KEY: (EP) Epipaleolithic; (UP) Upper Paleolithic; (MP) Mousterian/Middle Paleolithic; (HC) Hayonim Cave; (MR) Meged Rockshelter. (<) Percentage value is greater than zero but much less than 1. (~) Percentage calculation is questionable due to small sample size.

Table 7.3

Percentages of taxa in the small game fraction of the Wadi Meged faunal series

Assemblage (Period, Site, Layer, Unit)	Period	Total Small Game NISP	% Tortoise & Lizard	% Ostrich Eggshell	% Bird (All Types)	% Hare and Other Small Mammal
EP, HC, B	Natufian	10,488	48	0	24	28
EP, HC, C	Kebaran	713	68	1	23	7
EP, MR, <200 cm	Kebaran	751	62	0	25	12
UP, MR, >199 cm	Pre-Kebaran (UP)	163	76	0	18	6
UP, HC, D	Aurignacian	2,579	60	0	34	6
MP, HC, E, Unit 1	Mousterian	17	~82	~0	~6	~12
MP, HC, E, Unit 2	Mousterian	68	73	3	12	12
MP, HC, E, Unit 3	Mousterian	510	88	<	7	5
MP, HC, E, Unit 4a	Mousterian	1,669	95	1	2	2
MP, HC, E, Unit 4b	Mousterian	1,808	96	1	2	1
MP, HC, E, Unit 5	Mousterian	1,510	96	2	2	1
MP, HC, E, Unit 6	Mousterian	1,224	98	1	1	<
MP, HC, F, Unit 7	Mousterian	121	96	1	2	1

SOURCES: Hayonim Natufian data are from Munro 2001. Hayonim Aurignacian data are based on Stiner's NISP counts of small game taxa.

NOTE: Tortoise/lizard category is dominated by tortoises (>95%). Small mammals are mainly hares, along with low frequencies of squirrels. Birds include mainly the orders Galliformes and Colombiformes. Snakes were omitted from consideration.

KEY: (EP) Epipaleolithic; (UP) Upper Paleolithic; (MP) Mousterian/Middle Paleolithic; (HC) Hayonim Cave; (MR) Meged Rockshelter. (<) Percentage value is greater than zero but much less than 1. (~) Percentage calculation is questionable due to small sample size.

tubes, phalanges, and feathers. Whether or not the flesh of raptors was eaten is a matter of debate. More important from an ecological point of view is that the basic procurement techniques that allowed humans to add more birds to their diets developed in tandem with the integration of raw materials from raptors into their material culture.

Small mammals are also rare in the Middle Paleolithic assemblages. Hares increase rather suddenly in the Upper Paleolithic and are common in some Epipaleolithic assemblages (appendixes 11 and 12). Their numbers seem to explode with the early Natufian, a period of community aggregation in the Mediterranean hills (Bar-El and Tchernov 2001; Munro 2001). The proportion of hares declines in the late Natufian to levels more typical of the Kebaran period (Munro 2001).

Among the ungulates, the proportions of mountain gazelles, aurochs, deer, and wild pigs varied the most over time (table 7.6). Gazelles increased slowly but steadily from the early Middle Paleolithic through the Natufian period (fig. 7.4), at the expense of fallow deer, aurochs (wild cattle), and pigs. The proportions of deer and aurochs also fluctuated relative to another,

presumably in response to vegetation changes (*sensu* Bate 1937a, 1937b). The earlier faunas consequently contained more large-bodied prey (see also Speth and Tchernov n.d. on Kebara Cave). The potential explanations for the decline in the sizes of the dominant ungulate species are many. It is important to note here that body size in ungulates generally correlates inversely with the rate of reproductive output, generation turnover rate, and prey population resilience. This is because small ungulate species, such as gazelles, mature in a much shorter time (one year or less; reviewed in Munro 2001:217–235) than larger deer (about two years for fallow deer, two to three years for red deer; reviewed in Stiner 1994:333–335), and all of these species mature faster than large ungulates such as aurochs (two to four years).

TRENDS IN PREY IMPORTANCE AS A FUNCTION OF BIOMASS

Because prey animals have drastically different body weights, the amount of food obtained from one or two individuals of a large species may greatly outweigh the

Table 7.4

Trends in game indexes and tortoise mean size in the Wadi Meged faunal series

Assemblage (Period, Site, Layer, Unit)	Period	Gazelle–Fallow Deer Index	Tortoise Mean Size (mm)	Tortoise as % of Small Game NISP	Small-Large Game Index	Inferred Oxygen Isotope Stage	Ca. KYA
EP, HC, B	Natufian	.97	3.2	49	.67	1	11–13
EP, HC, C	Kebaran	.84	3.2	63	.19	2	14–17
EP, MR, <200 cm	Kebaran	.89	3.2	61	.38	2	18–19
UP, MR, >199 cm	Pre-Kebaran (UP)	.91	(^)	76	.27	2	19–22
UP, HC, D	Aurignacian	.89	3.3	43	.25	3	26–28
MP, HC, E, unit 1	Mousterian	.68	3.6	82	.11	5	70–90
MP, HC, E, unit 2	Mousterian	.78	(^)	70	.23	5	70–90
MP, HC, E, unit 3	Mousterian	.61	4.3	83	.30	6	~150
MP, HC, E, unit 4a	Mousterian	.49	4.3	88	.37	6	~170
MP, HC, E, unit 4b	Mousterian	.55	3.9	93	.46	6	~170
MP, HC, E, unit 5	Mousterian	.75	3.9	94	.49	6–7	~170
MP, HC, E, unit 6	Mousterian	.79	(^)	97	.55	7	~200
MP, HC, F, unit 7	Mousterian	.64	(^)	93	.47	7	~200

SOURCES: Natufian data are from Munro 2001. Aurignacian data are based on Stiner's NISP counts of small game taxa and Rabinovich's (1998) counts of carnivores and ungulates. Oxygen isotope stages (OIS) follow Martinson et al. 1987.

NOTE: Gazelle–fallow deer index is species-specific gazelle NISP divided by the sum of species-specific gazelle and fallow deer NISP. Small-large game index is calculated as small game NISP divided by the sum of small game and ungulate NISP, each category including species-specific and more general identifications. Snakes were omitted from consideration.

KEY: (EP) Epipaleolithic; (UP) Upper Paleolithic; (MP) Middle Paleolithic; (HC) Hayonim Cave; (MR) Meged Rockshelter. (^) Mean value represents unit combined with one or more of those above. (~) Approximate age.

total food obtained from many individuals of a smaller species. Another means for comparing patterns of game use employs estimates of the yields of meat plus bone—biomass—by time period. Here, biomass is simply MNI multiplied by the estimated average carcass weight (following White 1953) of each taxon (table 7.7, appendixes 13 and 14). The biomass percentage for each taxon is then calculated relative to total prey weight for each period. The seven Mousterian units are combined in this comparison, because taxonomic variation was fairly limited within the period, and only the more common prey types are considered. The Aurignacian assemblage is omitted because the information needed to generate accurate MNI data is incomplete.

About 97 percent of total prey biomass came from ungulates in every period except the early Natufian, when it declined to 83 percent. In the Mousterian— somewhat counterintuitively with respect to the foregoing NISP-based comparisons—wild cattle were the major source of meat, with fallow deer a distant second. Gazelle biomass appears to have been as minor as

Table 7.5

Percentages (of NISP) of major prey classes for faunal assemblages from Meged Rockshelter

Taxon	<200 cm bd	200–214 cm bd	≥ 215 cm bd
Ungulates	62%	68%	74%
Reptiles	24%	23%	20%
Birds	10%	8%	4%
Small mammals	5%	1%	2%
NISP	1,966	154	439

NOTE: Terrestrial carnivores were excluded, but avian predators (raptors) were included.

that obtained from red deer and wild pigs and even from tortoises. By the pre-Kebaran, at roughly 20 KYA, ungulate prey biomass was primarily from fallow deer, followed closely by gazelles. Gazelle biomass overtook that of deer in dietary importance during the Kebaran and expanded further in the Natufian.

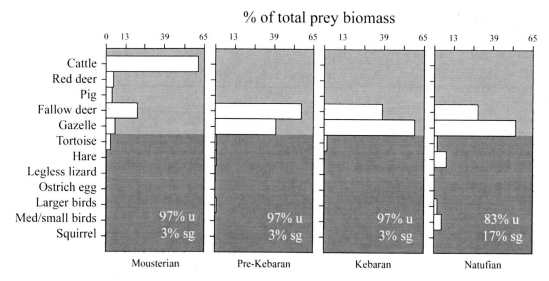

Figure 7.4
Percentages of total prey biomass consumed from size-ordered prey species in the Wadi Meged faunal series. Only relatively common taxa are considered in order to avoid aggregation errors for those represented by few remains. Key: (u) total ungulate biomass percentage; (sg) total small game biomass percentage. No data are available for the Aurignacian.

Table 7.6
Ungulate taxa as percentages of species-specific NISP in the Wadi Meged faunal series

Assemblage (Period, Site, Layer, Unit)	Period	Total Species-Specific Ungulate NISP	% Gazelle	% Pig	% Aurochs	% Fallow Deer	% Red Deer	% All Equids
EP, HC, B	Natufian	2,839	92	3	1	3	1	<
EP, HC, C	Kebaran	1,428	73	4	4	14	5	<
EP, MR, <200 cm	Kebaran	530	85	1	<	11	2	<
UP, MR, >199 cm	Pre-Kebaran (UP)	192	83	2	3	8	3	<
UP, HC, D	Aurignacian	7,419	84	1	1	11	3	<
MP, HC, E, unit 1	Mousterian	54	55	2	9	26	7	0
MP, HC, E, unit 2	Mousterian	114	62	3	10	17	6	0
MP, HC, E, unit 3	Mousterian	478	51	4	6	33	5	1
MP, HC, E, unit 4a	Mousterian	1,028	41	4	6	43	5	1
MP, HC, E, unit 4b	Mousterian	816	45	5	8	37	4	<
MP, HC, E, unit 5	Mousterian	746	63	7	7	21	1	1
MP, HC, E, unit 6	Mousterian	438	63	4	11	17	4	<
MP, HC, F, unit 7	Mousterian	50	58	0	6	32	4	0

SOURCES: Natufian data are from Munro 2001. Aurignacian data are based on Rabinovich's (1998) counts of ungulates.

NOTE: Total ungulate NISP represents only the sum of the species named. *Capra*, *Capreolus*, and *Alcelaphus* also occur in some of the assemblages but are rare if present.

KEY: (EP) Epipaleolithic; (UP) Upper Paleolithic; (MP) Middle Paleolithic; (HC) Hayonim Cave; (MR) Meged Rockshelter. (<) Percentage value is greater than zero but much less than 1.

Table 7.7
Carcass biomass (% total weight) for common prey species in the Wadi Meged faunal series

Taxon	Multiplier (kg)	Natufian		Kebaran		E. Keb/UP		Mousterian	
		MNI	%	MNI	%	MNI	%	MNI	%
Comparison by prey taxon									
Ungulates									
Gazella gazella	20.0	17	54	14	59	16	40	23	6
Sus scrofa	150.0	~	0	~	0	~	0	2	4
Dama mesopotamica	90.0	2	29	2	38	5	57	19	21
Cervus elaphus	200.0	~	0	~	0	~	0	2	5
Bos primigenius	700.0	~	0	~	0	~	0	7	61
Reptiles									
Testudo graeca	1.0–2.0[a]	14	2	8	2	8	1	104	3
Ophisaurus apodus	1.5	~	0	1	<	1	<	7	<
Aves									
Small-medium birds	0.5	65	5	3	<	~	0	6	<
Large birds	1.0	10	2	1	<	8	1	2	<
Struthio eggs	1.0	0	0	1	<	0	0	4	<
Small mammals									
Lepus capensis	2.0	26	8	1	<	3	1	1	<
Sciurus anomalous	0.25	~	0	~	0	0	0	3	<
Erinaceus sp.	0.6	~	0	~	0	0	0	1	<
(Total food weight in kg)		(628.5)		(475.0)		(793.5)		(8,000.8)	
Comparison by general game categories									
Large game (ungulates)			83		97		97		97
Small slow game			2		2		1		3
Small quick game			15		1		2		<

NOTE: The multiplier is the estimated average carcass weight for each taxon. Total food weight estimates are intermediate values based on ranges for populations from environments closest to or within western Asia.

KEY: (~) Individual represented by a very small number of parts and thus assigned a value of zero. (<) Percentage value is greater than zero but much less than 1.

[a] The greater weight estimate is used for the Mousterian tortoises, and the lower one for Upper Paleolithic and later tortoises, to account for diminution effects on forager returns.

Biomass-corrected counts across the entire prey body size spectrum indicate greater use over time of faster-reproducing species, including smaller ungulates. Although the proportional contribution of small game to Paleolithic diets was constant at 3 percent until the Natufian (late Epipaleolithic) period, there was a continuous downward shift in prey size overall, and correspondingly greater emphasis was placed on the more biologically productive gazelles. This phenomenon is also apparent in the small game fractions of the Wadi Meged faunal series (chapter 8).

FAUNAL CORRELATIONS AND INDEXES

How closely do ungulate species track one another in the Wadi Meged faunal series? And what is their relation to changes in the prevalence and mean body sizes of tortoises in the same assemblages? Although no significant correlations exist between time and the relative frequencies of carnivore, ungulate, and small game remains, trends are suggested in figure 7.5 for the relation between the gazelle–fallow deer ratio and tortoise

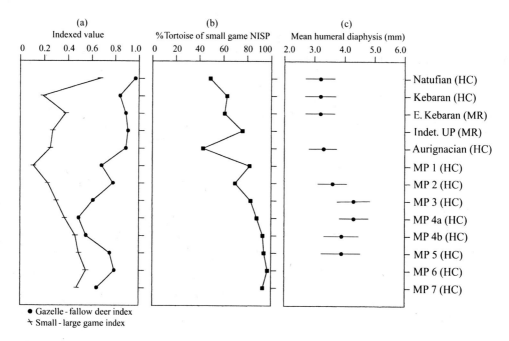

Figure 7.5
Trends in the Wadi Meged series for (a) the gazelle–fallow deer index and small-large game index; (b) tortoises as a percentage of small game NISP; and (c) changes (with standard deviations) in mean breadth of the tortoise humeral diaphysis.

abundance, as well as for the relation between those measures and reductions in tortoise mean body size.

Species-specific ungulate NISP counts for gazelles, fallow deer, and aurochs are mildly correlated with time (tables 7.8 and 7.9). Only fallow deer and gazelles are strongly correlated with each other, negatively in this case, with gazelles increasing in the later periods. Aurochs and red deer frequencies loosely follow those of fallow deer. Pigs and equids may be mildly correlated with each other, although their relations with other ungulate species are unclear. Within the small game fraction, three major small prey categories—slow animals, quick-flying birds, and quick-running small mammals—are very strongly correlated with time (table 7.8) as well as with one another (see chapter 8).

The next question is how the large game and small game fractions, previously examined independently, relate to each other. The gazelle–fallow deer index displays a mildly negative correlation with the percentage of tortoises in the small game fraction (table 7.10), but both of these indexes are more strongly correlated with time. There is no relation between the small–large game index and the relative frequency of tortoises in small game fractions of the assemblages overall. In other words, one does not predict or explain the other.

The implications of such data for climate-induced shifts versus human effects on prey populations are

Table 7.8
Spearman's correlation statistics for taxonomic abundance and time for the Wadi Meged faunal series

Variable (% of Total NISP)	Time (Ordinal)
Carnivores	-0.464
Ungulates	-0.368
Small game	0.368
Tortoises and lizards	**0.932**
Birds (all types)	**-0.882**
Small mammals	**-0.865**

NOTE: Number of assemblages = 13, df = 11. Bold, underscored Spearman's *rho* values are significant at the .001 level of probability. Following archaeological convention, time is counted backward from the present, which is why negative values correspond to an increase in the designated taxa in more recent time periods.

notoriously difficult to sort out. Both factors could easily have been involved in shaping the Wadi Meged faunal series. Could the rising importance of gazelles over time have been connected to the rise in birds and hares in local environments, such as might occur with the expansion of open land habitats? Bate (1937a, 1937b, 1940) explicitly called upon climate change to explain the decline of fallow deer in animal communities of

Table 7.9

Spearman's correlation statistics for taxonomic abundance and time for ungulate categories in the Wadi Meged faunal series

Variable (% of Total NISP)	Time (Ordinal)	% Gazelle	% Fallow Deer	% Pig	% Aurochs	% Red Deer
Gazelle	**-0.652**	—	—	—	—	—
Fallow deer	**0.704**	**-0.967**	—	—	—	—
Wild pig	0.283	-0.402	0.406	—	—	—
Aurochs (wild cow)	**0.678**	-0.645	0.573[a]	0.450	—	—
Red deer	0.153	-0.671	0.551[a]	0.058	0.561[a]	—
Equids (large and small)	0.495	-0.454	0.413	0.672[a]	0.334	-0.084

NOTE: Number of assemblages = 13, df = 11. Bold, underscored Spearman's rho values are significant at the .001 level of probability; those in boldface only are significant at the .01–.05 level of probability. Following archaeological convention, time is counted backward from the present, which is why negative values correspond to an increase in the designated taxa in more recent time periods.

[a] For correlations among ungulate taxa pairs, only those with positive values are considered informative beyond what the correlations with time already indicate, because negative relations are expected a priori in these percentage-based comparisons.

Table 7.10

Spearman's correlation statistics for two taxonomic indexes, mean tortoise size, and time in the Wadi Meged faunal series

Variable	Time (Ordinal)	Tortoise Mean Size Index	Gazelle– Fallow Deer Index	Small–Large Game Index
Tortoise mean size index	**0.838**	—	—	—
Gazelle–fallow deer index	**-0.828**	**-0.893**	—	—
Small–large game index	0.217	0.051	0.042	—
% Tortoises of small game NISP	**0.883**	0.735	-0.854	0.250

NOTE: The number of assemblages is reduced to 9 (of the original 13), because some were aggregated to increase sample size for the calculations of tortoise humeral diaphysis means. Bold, underscored Spearman's *rho* values are significant at the .001 level of probability; those in boldface only are significant at the .01–.05 level of probability. Following archaeological convention, time is counted backward from the present, which is why negative values correspond to an increase in the designated taxa in more recent time periods.

the Near East toward the end of the Late Pleistocene and the corresponding increase in gazelles. This idea was challenged to some extent by S. Binford (1968), Ducos (1968), Hooijer (1961), Saxon (1974), and Jelinek et al. (1973), who variously suggested that shifts in ecotone positions and possibly changes in humans' hunting and foraging strategies could account for the temporal variation in large mammal composition at any one site. These objections are nearly as convincing as Bate's argument for large-scale environmental change but are equally difficult to test with available evidence. It remains to be established

just how strongly and consistently gazelles, fallow deer, and tortoises "attract" or "repel" one another in the Wadi Meged faunal series.

DETRENDED COMPARISONS OF GAZELLE, FALLOW DEER, AND TORTOISE ABUNDANCE

Perhaps the greatest obstacle to understanding quantitative and ecological links between the gazelle, fallow deer, and tortoise frequencies in the Wadi Meged faunal series is their overall dominance throughout, which almost guarantees significant negative statistical correlations among them. These animals are the most common prey animals in the Wadi Meged assemblages, almost without exception, and at no point did any of them disappear from the menu. Simple correlation analysis, though useful, does not test all of the potential ecological links among prey types. So far it is clear that there are no trends in the relative representation of carnivores, small game, and large game. The exceptionally strong correlations of each of the three small prey types with time (table 7.8) are also informative.

By detrending the NISP data, which essentially removes time as a variable, one can focus on within-taxon differences—positive and negative—for each of the three taxa (gazelles, fallow deer and tortoises) between consecutive assemblage pairs. One can also examine the difference values for correlations between gazelle, fallow deer, and tortoise abundances. The approach distinguishes long-term trends from fine-scale variations. Running the data in this way reveals that fallow deer frequencies are not controlling the appearance of trends in the Wadi Meged series (fig. 7.6, table 7.11). The variation in fallow deer frequencies is of a decidedly short-term nature. What remains is a strong negative relation between tortoises and gazelles. The importance of gazelles in controlling the

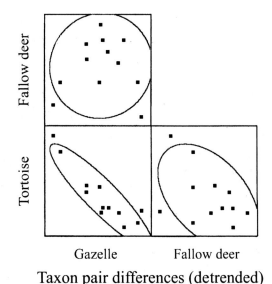

Taxon pair differences (detrended)

Figure 7.6
Detrended frequency relations between major taxon pairs: gazelle–fallow deer, $r = 0.033$; gazelle-tortoise, $r = -0.906$; tortoise–fallow deer, $r = -0.452$. Only the relation between gazelle and tortoise is significant ($p < .001$, no. observations = 12, df = 10).

Table 7.11
Differences between consecutive assemblage pairs in percentages of gazelles, fallow deer, and tortoises in the Wadi Meged faunal series[a]

Assemblage (Period, Site, Layer, Unit)	Period	Gazelle %	Gazelle Diff.	Fallow Deer %	Fallow Deer Diff.	Tortoise %	Tortoise Diff.	Summed NISP
EP, HC, B	Natufian	35	−26	1	−11	64	37	7,475
EP, HC, C	Kebaran	61	14	12	6	27	−20	1,690
EP, MR, <200 cm	Kebaran	47	−7	6	1	47	6	973
UP, MR, >199 cm	Pre-Kebaran (UP)	54	−22	5	−5	41	27	300
UP, HC, D	Aurignacian	76	24	10	−14	14	−10	8,169
MP, HC, E, unit 1	Mousterian	52	1	24	10	24	−11	58
MP, HC, E, unit 2	Mousterian	51	22	14	−5	35	−17	139
MP, HC, E, unit 3	Mousterian	29	11	19	0	52	−11	826
MP, HC, E, unit 4a	Mousterian	18	2	19	6	63	−8	2,329
MP, HC, E, unit 4b	Mousterian	16	−7	13	5	71	2	2,356
MP, HC, E, unit 5	Mousterian	23	5	8	3	69	−8	2,044
MP, HC, E, unit 6	Mousterian	18	0	5	−5	77	5	1,537
MP, HC, F, unit 7	Mousterian	18	—	10	—	72	—	158

NOTE: Pearson's correlations values are as follows: gazelle–fallow deer, $r = 0.033$; gazelle-tortoise, $r = -0.906$; tortoise–fallow deer, $r = -0.452$. Only the relation between gazelle-tortoise is significant (very significant in this case at $p < .001$; no. observations = 12, df = 10).

[a] Percentage values are based only on the sum of species-specific counts of gazelles, fallow deer, and tortoise remains.

long-term trend in large mammals over 200,000 years in the Wadi Meged is in agreement with Bate's original observation (cf. Davis 1982).

One might conclude from the detrended results that a greater proportion of gazelles signals greater environmental aridity. In addition, one might surmise that this occurred at the expense of tortoise populations. However, biogeographical studies indicate that reptiles generally are favored in Mediterranean habitats as aridity increases, and reptile communities are most diverse and richest in endemic species in the eastern end of the Mediterranean Basin, particularly in coastal plains and low valley areas. Both spur-thighed tortoises and chukar partridges thrive in matorral habitats as well (chapter 6; Alkon 1983; Blondel and Aronson 1999; Hailey 1988; Potts 1986; Stubbs 1989). Thus the general decline of tortoises in later Paleolithic diets is unlikely to have been the result of habitat loss suffered by tortoises, and the expectation is that tortoise and partridge abundances should be complementary rather than opposing. Further, Munro (2001) reported considerable variation in the use of tortoises relative to other small game and to gazelles *within* the Natufian period, for which fine-scale temporal comparisons were feasible. Gazelles were the overwhelming type of ungulate prey throughout the Natufian (Munro 2001; see also Bate 1940; Cope 1991; Davis 1978, 1980a, 1982), but exploitation of tortoises was relatively low during climate amelioration in the early Natufian, 12–13 KYA (interpreted as being due to overexploitation by humans), and relatively high during the late Natufian, coinciding with the Younger Dryas (ca. 11–10 KYA), an abrupt cooling and drying event during which Natufian populations dispersed (Munro 2001).

The frequency relations and ecological dynamics of gazelles and tortoises are correlated in time but not with respect to cause. These results do not offer good support for blanket interpretations about the effects of aridity on the natural abundances of ungulates versus reptiles (*contra* Speth and Tchernov 2002). The two animals appear to have responded differently (i.e., rather independently) to whatever environmental changes took place over the Middle and Late Pleistocene. More gazelles in archaeofaunas may well signal removal of a geographic barrier to the south, an expansion of open land habitats, a fragmentation of Mediterranean forests, and perhaps intensified cycles of fire disruption, but the abundance of gazelles in Pleistocene environments is weakly connected to that of tortoises. An alternative way of examining the causes of frequency changes in tortoises is via body size diminution and related changes in tortoise population structure.

TORTOISE DIMINUTION, POPULATION STRUCTURE, AND MORPHOLOGY

The mean body size of a tortoise population can shift as a result of two distinct factors—changes in forage quality and predator pressure. Body size in animals that grow throughout their lives is a nonlinear proxy for individual age. The growth rate of tortoises is intimately connected to precipitation, temperature, and climate. However, age-specific mortality can also affect a population's age and size structure. Because tortoise diminution correlates in a general way with gazelle frequencies in the Wadi Meged series, one question is whether any consistent relation can be found between size reduction in the tortoises and independent data on climate cycles (OIS chronology). To address the separate possibility of species replacement—that is, the possibility of genotypic differences that might have affected the maximum sizes attained by hypothetically distinct subspecies, one of which replaced the other—we also need to know whether there were differences in the maximum adult sizes reached by the tortoises of the Natufian, Epipaleolithic, Upper Paleolithic, and Mousterian periods. If only the average size of the tortoises changed, and if distortion of the population structure is evidenced, then it is likely that predators were causing diminution in tortoise populations through overexploitation, partly or wholly refuting the inference of climatic effects. I address these matters in the following sections, along with the possibility of population replacement, which I examine through morphological comparisons of recent and Pleistocene specimens. Finally, sex ratios are analyzed for evidence of distortion from predators' preference for larger individuals (which were more likely to be females of reproductive age) or from proposed seasonal biases in exploitation of the two sexes (*sensu* Speth and Tchernov 2002).

DIMINUTION TREND

Tortoise humeri reveal a clear trend toward reduction in size (table 7.12), judging from a measurement of the medio-lateral breadth of the shaft at its narrowest point (measurement 4 in fig. 7.7). The dimensions of the humerus, a major weight-bearing member in tortoises, respond directly to increases in body mass (Castanet and Cheylan 1979; Walker 1973), and habitual loading during locomotion translates in a predictable way to bone shaft diameter and cross-sectional area (Wainwright et al. 1976:7). Holding

Table 7.12

Measurements of tortoise humeral diaphysis breadth for the Wadi Meged faunal series

Assemblage (in Chronostratigraphic Order)	No. Humeri Measured (MNE)	Mean Breadth (mm)	SD
Natufian (Layer B, HC)	109	3.2	0.5
Kebaran (Layer C, HC)	63	3.2	0.6
Early Kebaran (MR, all cuts)	58	3.2	0.5
Aurignacian (Layer D, HC)	74	3.3	0.5
Mousterian Units 1–2 (HC)	15	3.6	0.5
Mousterian Unit 3 (HC)	25	4.3	0.7
Mousterian Unit 4a (HC)	39	4.3	0.6
Mousterian Unit 4b (HC)	29	3.9	0.7
Mousterian Units 5–7 (HC)	20	3.9	0.8

NOTE: Some stratigraphic units were collapsed to increase sample size. A hiatus separates the Aurignacian and Middle Paleolithic in the Wadi Meged series.

KEY: (HC) Hayonim Cave; (MR) Meged Rockshelter.

geographical locality constant is essential for this morphometric comparison, because differences in energy flow (food supply and quality) among habitats may lead to significant differences in average adult sizes (Blasco et al. 1986–1987; Lambert 1982). Eliminating geographical variation from the equation—and ignoring for the moment the taxonomic question—reduces the potential explanations for tortoise diminution in the Wadi Meged to predator effects and climate-induced changes in tortoises' food supply or quality.

The tortoises collected by Mousterian foragers in the Wadi Meged were larger on average than those collected by later humans in the same valley (fig. 7.8).

Rather than displaying a gradual pattern of diminution, the means in this series form two groups. An analysis of variance shows the size differences at the shift point to be nonrandomly distributed in the time-ordered samples (F-ratio = 24.15, p < 0.001, df = [8, 414]). The early Mousterian tortoise means vary, but none except that for the youngest of the Mousterian assemblages approaches the low values of the Upper Paleolithic and Epipaleolithic phases.

A hiatus in human occupations and sediment accumulation separates the Mousterian and the Aurignacian in the Wadi Meged faunal series. Because this hiatus lasted some 70,000 years, the Wadi Meged data

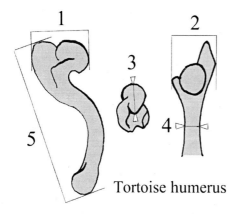

Figure 7.7
Placement and orientations of measurements taken on tortoise humeri. The length of the proximal surface articulation (3) and the lateral section of the humeral diaphysis at its narrowest point (4) are emphasized in the comparisons.

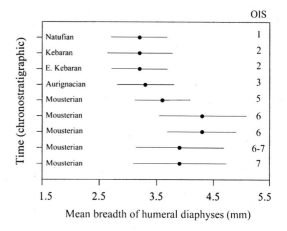

Figure 7.8
Size reduction trend in Mediterranean spur-thighed tortoises in the time-ordered Wadi Meged assemblages, based on mean values and standard deviations for humeral diaphysis breadth. OIS denotes oxygen isotope stage.

Anterior-posterior (ap)

Medio-lateral (ml)

Figure 7.9
Placement and orientations of measurements taken on the nuccal bony plate of the tortoise carapace. The shape index used for sex ratio analysis is the anterior-posterior measurement (ap) divided by the lateral measurement (ml). Adult males display significantly lower index values than do adult females, with the division between the sexes occurring at around 7.3 millimeters.

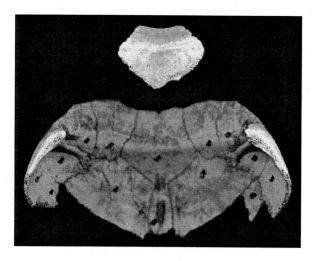

Figure 7.10
The nuccal bony plate of a *Testudo graeca* carapace in articulation and (above) broken free along suture lines.

do not specify the exact timing of tortoise diminution. The late Mousterian and early Upper Paleolithic (Ahmarian) levels from Kebara Cave on the western face of Mount Carmel help to fill this gap. Kebara Cave today is situated in a somewhat richer vegetation zone than that of the Wadi Meged, and for this reason tortoises there might have been larger overall. Nonetheless, a significant decline in the mean size of tortoises occurred in Kebara Cave between the late Middle Paleolithic, dated to 70–55 KYA (4.5 cm, $n = 169$, s.d. = 0.6), and the earliest Upper Paleolithic (Bar-Yosef et al. 1996), dated to 44 KYA (4.0 cm, $n = 31$, s.d. = 0.8), a result recently confirmed by Speth and Tchernov (2002) with larger samples. Size suppression in tortoises in the Galilee therefore began at least 44,000 years before the present.

Diminution is also evidenced by measurements of the nuccal plate of the carapace; measurement orientations are shown in figure 7.9 (data in table 7.13). The nuccal plate tends to retain its contours in fragmented assemblages, often separating neatly from the rest of the carapace at well-defined sutures (fig. 7.10). Figure 7.11 demonstrates that the maximum sizes attained by adult tortoises were about the same for all periods at Hayonim Cave. Only the mean sizes of the tortoises changed significantly with time. Plots of the proximal articulation surface length of the humerus and the diaphysis (measurements 3 and 4, fig. 7.7) for late Mousterian and early Upper Paleolithic tortoises from Kebara Cave (fig. 7.12) similarly indicate that although mean sizes declined with time, the maximum sizes attained by long-lived individuals were about the same for both

periods (nuccal plate measurements are not available for these assemblages). Could such an abrupt decline in size be the result of tortoise population replacement? Given the strong tendency for endemic species in the Mediterranean Basin (chapter 6), the only possibility would be a geographic shift in subspecies distribution following a climate-driven shift in effective latitude.

Climatic variation does not account for the major size decline in tortoises after the late Mousterian or for the fact that diminution was sustained throughout the Upper Paleolithic and Epipaleolithic periods. Numerous shifts in local climate took place over the last 200,000 years (see Bar-Yosef 1981, 1995, 1996; Tchernov 1992b, 1992c). The early Mousterian faunas in Hayonim Cave formed from the late Middle Pleistocene (OIS 7) through the Last Interglacial (OIS 5 and into 4; see chapters 1 and 2). The early Kebaran of Meged Rockshelter dates to the Last Glacial Maximum (Kuhn et al. 2004), and the later Kebaran to early phases of global warming. The early Natufian dates to just before the Pleistocene-Holocene transition, when the pace of global warming had already accelerated (Bar-Matthews et al. 1999; Baruch and Bottema 1991; Martinson et al. 1987). Taken together, the Wadi Meged Paleolithic assemblages span five or more oxygen isotope stages (OIS 7–1). Mean tortoise size shifted suddenly downward only in the middle of OIS 3 (fig. 7.13), and tortoise sizes remained low for the rest of the Pleistocene sequence. Munro's (2001) fine-grained analyses of tortoise dynamics in the early and late phases of the Natufian period also contradict climate-based explanations. The early phase corresponded to a period of climate

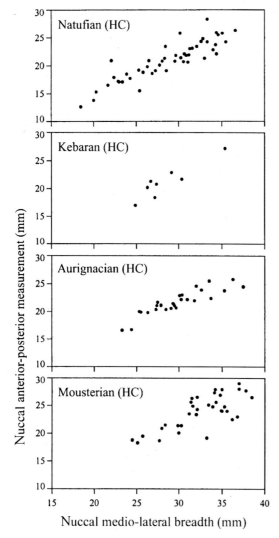

Figure 7.11
Size distributions of tortoise nuccal plates (carapace) in four cultural layers of Hayonim Cave. Each point represents one tortoise.

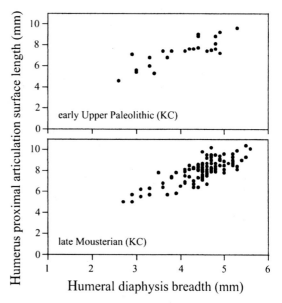

Figure 7.12
Size distributions of tortoise humeri (diaphysis and proximal articulation surface length) in the late Mousterian and early Upper Paleolithic layers of Kebara Cave. Each point represents one humerus.

amelioration, and the late phase to the Younger Dryas, yet the average sizes of tortoises were small across both phases.

AGE STRUCTURE DISTORTION (SKEWING)

There also is evidence of distortion in the size (age) structures of the tortoises during the later Paleolithic in the Wadi Meged and at Kebara Cave. The significance of this phenomenon is rooted in the life history strategy of tortoises: infant mortality normally is very high, and adult mortality very low (Hailey 1988; Shine and Iverson 1995; Wilbur and Morin 1988; see chapter 8). Healthy tortoise populations tend to pack habi-

tats with long-lived adults as a rule, and one of the greater challenges to young tortoises under normal conditions is finding enough life space in which to develop. The high densities typical of healthy, unmolested tortoise populations potentially create a resource that can be mined by predators. But tortoise populations possess limited ability for resurgence if they are exploited heavily.

Although spur-thighed tortoises do well in a variety of Mediterranean regions, Hailey et al. (1988) and Lambert (1982, 1984) found that the opening up of habitats by human disturbance of any sort (overgrazing, extensive plowing, increased incidence of range burning) reduced hatchling survival and intensified competitive interference from adult tortoises. This situation produced an age bias in favor of adults (Hailey et al. 1988; Lambert 1984), skewing the size structure toward the larger (older) end of the continuum. Otherwise, the population displayed a normally distributed, bell-shaped pattern of the linear metric trait. These population states are modeled in terms of size variation in figure 7.14. If the size distribution is measured in terms of "skewness," then negative values and those close to zero reflect natural tortoise populations that have not been subjected to strong depredation by humans, even if human-induced environmental disturbance occurs. Strongly positive values indicate an unstable structure, or distortion, of the tortoise population due to preda-

Table 7.13
Measurements of tortoise carapace nuccal plates from Hayonim Cave

Assemblage	Medio-Lateral Breadth			Anterior-Posterior Length		
	No.	Mean (mm)	SD	No.	Mean (mm)	SD
By culture period						
Natufian (Layer B)	52	28.7	4.5	48	21.0	3.4
Kebaran (Layer C)	9	29.0	3.5	8	21.2	3.1
Aurignacian (Layer D)	31	29.7	3.6	28	21.7	2.2
Mousterian (all units of Layer E)	48	34.0	5.0	40	24.5	3.4
By Mousterian (MP) unit						
MP Units 1–3	9	33.7	4.7	9	24.7	2.4
MP Unit 4	24	34.4	4.6	19	24.3	3.5
MP Units 5–7	13	35.1	5.0	11	25.7	3.2

NOTE: Some Mousterian units were collapsed to increase sample size.

tion in particular, either from hunting or, in recent decades, the capturing of individuals for sale as pets (see chapter 8). Apart from human-induced effects, this is not a normal situation for tortoise populations, which have few predators in adulthood because of their larger size and thick protective shell.

Several cases, including the younger Mousterian tortoise assemblages from Hayonim Cave (fig. 7.15), but not those from Kebara Cave (fig. 7.16), are skewed to the left, or the small end (more positive values) of the body size continuum. Thus, healthy tortoise size/age structures are in evidence for much of the Mousterian and the earliest Upper Paleolithic, but less so for the later periods. Such distortion indicates "destabilization" of tortoise populations and overexploitation by humans. The result refutes the hypothesis that the size changes in the Wadi Meged tortoises were the result of genotypic differences in the potential for maximum growth (population replacement) or of the opening up of vegetation in association with increasing aridity. The results for the size distributions of tortoises in the vicinity of Kebara Cave during the late Mousterian and earliest Upper Paleolithic also indicate populations in an undistorted state, meaning that chronic destabilization occurred only after that time.

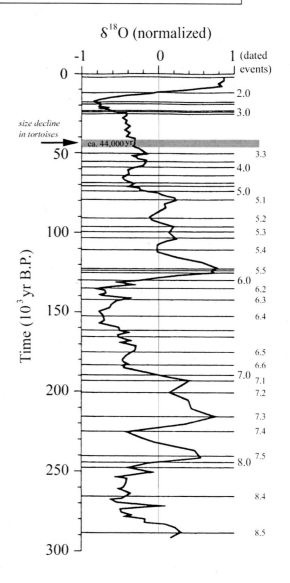

Figure 7.13
Dated oxygen isotope (OIS) chronology spanning the last 300,000 years, showing the timing of tortoise mean size reduction in Galilee sites (OIS 3, ca. 44 KYA). (After Martinson et al. 1987:fig. 18.)

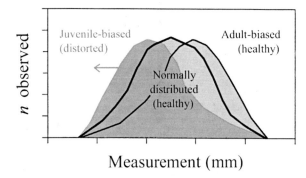

Figure 7.14
Model of size (= age) skewing and its implications for population structure based on a morphometric trait in tortoises and other K-selected, continuously growing animals. Normally distributed (bell-shaped) size/age patterns and those biased toward adults indicate healthy population structures, because adult mortality is low in natural settings and juvenile mortality high. A bias toward juveniles (skewing to the left) reflects abnormal distortion of the population's structure due to abnormally high adult mortality, which is known to result from predation.

TORTOISE SEX RATIOS

Speth and Tchernov (2002) suggested that seasonal variations in site occupation by humans might account for mean size differences in tortoises over time at Kebara Cave and possibly also in the Wadi Meged series. Female tortoises tend to be larger than males of the same age (Blasco et al. 1986–1987; Lambert 1982), and Speth and Tchernov proposed that their seasons of activity and availability could be quite different. A direct test of the idea can be made from the size distributions of sex-dependent features in the Wadi Meged tortoises.

The nuccal bony plate of the tortoise carapace is morphologically unique, and because there is only one per tortoise, it is suitable for counting individuals. What is more, the shape of the nuccal plate of the carapace differs slightly between adult males and females, those of males being shorter in the anterior-posterior dimension relative to the medio-lateral breadth (fig. 7.17). Some of the sex-based differences in plate shape may arise early in carapace development, but most of this shape dimorphism seems to result from the mechanics of tortoise courting habits. Adult males profess their interest by ramming and butting females with the front of their shells, often taking a running start. With age, the anterior-posterior dimension of the nuccal plate in males becomes increasingly squashed relative to that of females. The sex distributions of adult tortoises in archaeofaunal assemblages therefore can be investigated by using a shape index based on

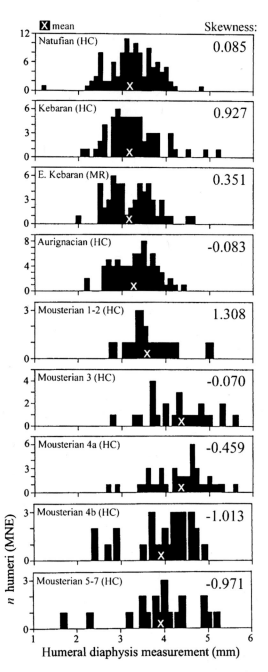

Figure 7.15
Skewing in the size distributions of tortoise humeral diaphyses in the Wadi Meged series. Positive values indicate abnormal age structure distortion; negative values and those near zero are typical of healthy tortoise populations. HC denotes Hayonim Cave, and MR, Meged Rockshelter.

the anterior-posterior (ap) and medio-lateral (ml) breadth measurements of the nuccal bony plate (fig. 7.9), an index calculated as ap/ml. Adult females will have higher values than adult males because their nuccal plates are less compressed front to back.

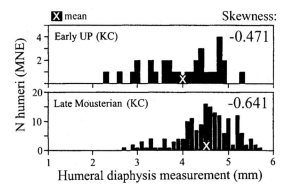

Figure 7.16
Skewing in the size distributions of tortoise humeral diaphyses in the late Mousterian and early Upper Paleolithic layers of Kebara Cave.

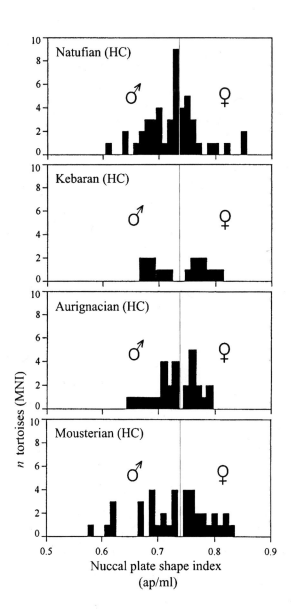

Figure 7.17
Comparison of nuccal plate shapes of female and male tortoises. The female specimen is longer anterior-posteriorly relative to its breadth than the male specimen; the anterior notch of the male specimen is shallower.

Application of the nuccal plate index to the tortoise assemblages from Hayonim Cave reveals bimodal distributions that certainly represent the two sexes (fig. 7.18). A gap between index values 0.71 and 0.74 (avg. 7.3) appears consistently in the assemblages. Females are slightly better represented in some periods, but the overall pattern is one of relatively balanced sex ratios across periods. These data indicate that size dimorphism between adult male and female tortoises did not greatly affect Paleolithic humans' decisions to collect adult spur-thighed tortoises. If the sexes are capable of distinct schedules of seasonal activity, as was suggested by Speth and Tchernov (2002), there is no evidence for such an effect in the Wadi Meged series. (Nuccal measurement data are not yet available for the samples from Kebara Cave.) In addition, there is no evidence of extreme distortion of the reproductive core of these tortoise populations in any of the periods and cases considered.

Figure 7.18
Sex ratios of adult tortoises from four cultural layers of Hayonim Cave, based on the nuccal plate index (ap/ml).

WHAT SORT OF TORTOISE? BIOGEOGRAPHY AND MORPHOLOGY

The question of whether body size diminution in tortoises occurred within a single taxon rather than representing the geographical replacement of an inherently (genotypically) larger-bodied population by a smaller-bodied one can be examined from the perspectives of biogeography and morphology. One might propose, for example, that the tortoises from the Pleistocene deposits in Hayonim Cave could be *T. graeca, T. hermanni,* or even *T. kleinmanni* on grounds that the Pleistocene distributions of these taxa are poorly known. Specifically, the diminution trend in the Wadi Meged tortoises might represent the replacement of *T. hermanni* with *T. graeca;* alternatively, a larger subspecies of *T. graeca* might have been replaced by a distinctly smaller subspecies (Y. Werner, personal communication, 1996). This is not the general pattern in Mediterranean reptiles, judging from the high incidence of endemic taxa in the region, but chance biogeographical events cannot be ruled out in principle.

A simple way to check for population replacement at the level of species is through a comparison of carapace morphology. *T. hermanni* possesses a readily visible division of the anal, or supracaudal, plate, in contrast to *T. graeca.* The shape and degree of flaring along the rear edges of the carapace also help to distinguish among *Testudo* species. *T. hermanni* does not occur in the Levant today. *T. graeca* occurs in much of the Levant, in addition to many other Mediterranean regions. Hermann's tortoise is more tolerant to cold. It tends to be roughly the same size as *T. graeca* where the two species coexist (Hailey et al. 1988), but it is smaller on average than certain other populations of *T. graeca* (Ernst and Barbour 1989). Other candidate species might be suggested to include *T. kleinmanni,* but the lack of pronounced outward curvature of carapace edges in the Wadi Meged specimens excludes this possibility.

Subspecies differences among *T. graeca* are less easily addressed, but shell form is still fairly diagnostic (fig. 7.19). *T. graeca* is subdivided into several subspecies, only four of which are widely recognized (Stubbs 1989). Populations vary in body size across a large geographic range of the species, with eastern races tending to be larger-bodied than those in the west. *T. graeca terrestris* occurs today in southern Asiatic Turkey (Hatay region), northern and western Syria, Lebanon, western Jordan, and northern and central Israel. *T. g. ibera* occurs to the north and west of this range, along the coasts of Greece and southwestern Turkey and in Asia Minor and Anatolian Turkey (Stubbs 1989: 31).

Figure 7.19
Shape correspondences of anal carapace (supracaudal) plates of Pleistocene (Mousterian) tortoises and modern counterparts from the Wadi Meged. Note that the anal plate is undivided in both specimens.

Body size variation in the four subspecies relates most clearly to the degree of environmental aridity (Ernst and Barbour 1989:266–270) and less so to taxonomic distinctions. *T. g. terrestris* occurs throughout the Levant today and is one of the smaller-bodied subspecies (up to 20 cm). Two of its notable skeletal characteristics are the lack of or limited flaring on the posterior marginal plates of the carapace and a relatively smooth shell (fig. 7.20). *T. g. ibera* can be somewhat larger (up to 30 cm) and is found in regions adjacent to the Levant (Anatolian Turkey, the Balkans, and other areas to the east; Cogger and Zweifel 1998; Lambert 1982). The posterior marginal plates of the carapace of *T. g. ibera* are more flared, and the carapace has a lumpier, sculpted appearance (fig. 7.20), with the most pronounced or ridged growth rings developing in the driest of habitats, such as the deserts of Morocco (Lambert 1982:170–174). The tortoises of the Wadi Meged series all appear to be *T. g. terrestris* on the basis of the morphologic criteria just named.

There is also the question of how well the large tortoises of the Mousterian compare to the largest tortoises found in the Wadi Meged today. Examination of corresponding carapace plates indicates comparable body sizes for large adults from the early Mousterian of Hayonim Cave and the largest of the modern tortoises found in the study area (see fig. 7.19). More large adults were present in tortoise populations during the Mousterian period, but the maximum plate

Figure 7.20
Comparison of carapace morphologies of *Testudo graeca terrestris* (top) and *T. graeca ibera* (bottom).

sizes attained then and today are about the same, shell lengths of roughly 15 and 20 cm based on informal projections of individual plate dimensions.

The data presented here contradict the hypothesis that diminution in the Wadi Meged tortoises was due to replacement of one population by another that possessed a significantly different potential for maximum body size. Three observations contribute to this conclusion. First, no climate-based correlation is evidenced in the diminution trend, so available moisture and prevailing temperatures were not regulating species distributions or body size patterns. Second, the maximum sizes attained by old adult tortoises were about the same in the early Mousterian through the Natufian periods and most recently in the Wadi Meged. And third, species or subspecies replacement is not apparent from bony carapace and plastron morphologies, based on anal (supracaudal) plate structure and the degree of flaring of the posterior carapace. A

chance biogeographical shift, unrelated to climate change, is therefore also rejected on the basis of these observations. At the very least, the size trend in tortoises was the result of a combination of human-induced and climatic effects, but major human involvement cannot be denied. Predator pressure was the main cause of the diminution trend and age structure distortion in the Wadi Meged tortoise and Kebara Cave archaeofaunal populations.

CONCLUSION

The Wadi Meged faunal sequence, a composite of assemblages from Hayonim Cave and nearby Meged Rockshelter, presents a rare opportunity to compare humans' reliance on large and small game throughout the Mousterian, Aurignacian, Kebaran, and Natufian periods at a single locality. Prey species ratios in the Wadi Meged faunal series changed significantly with time, but the relative occurrences of the common prey taxa do not trend in a directly opposing manner. The most obvious trends in archaeofaunal species abundance involve a decline in tortoises and deer (and large-bodied ungulates in general) and a corresponding increase in mountain gazelles and quick small game animals, especially hares, partridges, and other birds. The increase in gazelles with time may well suggest reorganization of plant community structure in the area, following Bate's reasoning (1937a, 1937b), although the total range of taxa exploited changed hardly at all. Another, less obvious trend is the increasing use of more productive prey species with time. This is apparent both within the ungulates, coincident with greater use of smaller-bodied species, and within the small game category, in the form of greater use of highly productive lagomorphs and game birds (chapter 8). Tortoises were exploited in significant numbers throughout the sequence, but less so in the later periods. With this there is also evidence of size diminution in tortoises due to increased harvesting pressure on local populations.

The adage that correlation does not necessarily identify cause is relevant to the arguments presented here. The negative relation between the frequencies of gazelle and fallow deer in the Wadi Meged faunal series is deceiving and, when examined in detrended format, proves to be insignificant. Similarly, the negative relation between the frequencies of tortoises and gazelles, which seems so robust in the long view of time, turns out to be little more than random in the short term (as demonstrated within the Natufian). The general decline in tortoises from the early Mousterian through the Natufian was not a simple function of the expansion of grassland habitats in the area, assuming that gazelles generally are favored by such a vegetative change. Wildlife studies show that the expansion of open habitats fosters a greater bias in favor of adult tortoises in the living population, the opposite of that which occurs in the later part of the Wadi Meged archaeofaunal series, when gazelles become more abundant than deer and other browsing ungulates. We can be sure that more than one cause contributed to the trends in prey species frequencies. Climate change is the most commonly cited factor and is not insignificant. However, much of the variation in the small animals procured by Paleolithic foragers is not explained by climate change or by random biogeographical events. The variation in key small game types was largely the product of changes in human predatory behavior. The same could be true for the increasing proportions of smaller, more productive ungulate species with time.

Small game for Mousterian foragers came almost exclusively in the form of slow-moving reptiles, animals that could be gathered. Substantial additions of quick small animals (birds and hares) to Upper Paleolithic and Epipaleolithic repertoires suggest a broadening of human diets in the later periods. A significant worry in interpreting the diminution trend in tortoises has been whether it represents one population's response to climate change or a complete replacement of that population by another whose inherent potential for maximum body size was smaller than the first's. The consistently low mean body sizes of the tortoises of the Upper Paleolithic through Natufian periods, which span many changes in local climate, indicate heavier or more frequent exploitation of these animals by humans. Population replacement should be evidenced by substantial reduction in the sizes of the largest individuals in each assemblage, in addition to a reduction in mean size. The data contradict the second hypothesis because the maximum sizes reached by adult tortoises in the Wadi Meged were about the same in the Mousterian, Upper Paleolithic, and Epipaleolithic culture periods and today. Species or subspecies replacement of the sort that could affect tortoise body size was not responsible for the trend in tortoise diminution.

Although evidence of small game use is widespread for the early Mousterian of the Wadi Meged, this practice seems relatively rare for the Mousterian of Eurasia as a whole. With the exception of a few other Mediterranean cases involving the exploitation of shellfish, reptiles, or both (Cyrenaica, Klein and Scott 1986; Gibraltar, Barton et al. 1999; Italy, Stiner 1994), the Mousterian of Hayonim Cave is difficult to reconcile with much of the literature on archaic human diets in Eurasia. Small game exploitation generally increases as latitude decreases among recent hunter-gatherers (Kelly 1995; Kuhn and Stiner 2001), owing to the nature of animal community assembly and evolution and foragers' responses to opportunity. Even so, small game remains seem too rare in Mousterian faunas according to the coarse rules of effective latitude. The apparent absence of small game in many northern (high-latitude) Middle Paleolithic faunas is probably an accurate reflection of diet, because there must have been fewer small species available to exploit. But this is not expected for areas at lower latitudes, such as the Mediterranean Basin, where the assortment of small animals is inherently richer.

Historical factors relating to the way science is practiced may have contributed to the perceived lack of small game in Mousterian sites situated in warmer environments, a problem that stems from the way materials are recovered and how analyses following excavation are divided among researchers. On the other hand, it is already clear that most small game exploitation during the Mousterian was focused on gatherable animals rather than those that run fast, fly, or swim. Thus there is something very odd (or narrow) about Mousterian small game exploitation from a modern human perspective, something that is not explained simply by incomplete reporting. The remarkably limited use of small game resources in the Middle Paleolithic of Eurasia in general, and the surprisingly narrow use of small animals in the Mediterranean area during the same period, demands an explanation. Chapters 8 and 9 treat issues of small game use through the application of foraging theory and demographic concepts, in an attempt to better appreciate what can be learned from this class of zooarchaeological evidence.

■

CHAPTER EIGHT

Computer Simulation Modeling of Small Game Predator-Prey Dynamics

■

Mary C. Stiner, Natalie D. Munro, and Todd A. Surovell

GIVEN THE GREAT ABUNDANCE OF UNGULATE REMAINS in Paleolithic faunas, it is not surprising that so many archaeological discussions of human predatory behavior have been focused on large game. There exists, however, another dimension of the archaeofaunal record—small game exploitation—that provides unique information about the demographic conditions under which human predator-prey relations evolved. Though well short of dominant in most Paleolithic archaeofaunal collections, small animals were important to human diets in the Mediterranean Basin from at least the early Middle Paleolithic onward, and the types of small game emphasized have changed a great deal over the last 200,000 years (chapter 7). This is most striking if one considers the distinct biological properties of the small animals that commonly appear in Paleolithic human diets in Eurasia: littoral shellfish, tortoises, partridges, rabbits, and hares. These small animals differ greatly in their reproductive potentials, maturation rates, and capacities for population recovery under conditions of heavy exploitation. They also differ in the ease with which they can be caught without special tools. Changes in humans' interest in these animals therefore may testify to shifts in the organization of forager adaptations and potentially also to rising human population densities during the later Pleistocene (Stiner et al. 1999, 2000).

A striking trend in small game use from the early Middle Paleolithic through the Epipaleolithic period in the Galilee of present-day Israel was demonstrated in chapter 7. Briefly, this trend was one of increasing use of quick-fleeing small animals (birds and hares) relative to slow-moving prey (tortoises) with time. Trends of a generally similar nature have been observed elsewhere in the Mediterranean Basin, although the timing varies. The repetition of this phenomenon across disparate environments suggests that circumstances leading into the Epipaleolithic and the Neolithic were already ripe for change. Here we explore the ecological ramifications of the small game trends from the perspective of predator-prey interactions, specifically asking how prey escape behaviors, prey reproductive characteristics, and predator numbers relative to available prey interact to produce patterns at the scale of populations. This is accomplished through predator-prey computer simulation modeling for three common small prey taxa—tortoises, partridges, and hares.

Formal diet breadth models predict that lower-ranked species, defined as those that yield relatively low returns for the foraging effort, will be sought as encounter rates for highly ranked types, which provide

the greatest returns, decline. These models, however, do not specify all of the appropriate criteria for ranking resources, because one kind of organism might interact with its environment differently from the next (Stephens and Krebs 1986; chapter 9). The habits and physical properties of prey animals must affect humans' access to them. Small animals are broadly equivalent with respect to protein content and package size, but they differ tremendously with respect to handling costs and the number of individuals that can feasibly be harvested at once, all of which may affect their rank. Prey items may also differ with respect to the long-term price of heavy exploitation (Belovsky 1988; Botkin 1980; Harpending and Bertram 1975; Winterhalder et al. 1988).

Differences in prey population resilience and the work of capture constrained Paleolithic people's uses of small game in predictable ways. *Resilience* here refers to a prey population's ability to withstand heavy, cyclical predation; it is linked to individual maturation rate. Prey population resilience is especially important in situations where predator densities are high. *Work of capture* concerns the way a prey animal's defense and escape mechanisms affect a predator's ability to obtain it. Work of capture may be the cost of searching for prey, investment in technological aids, or the energetics of a true chase. It may also include processing costs, although in the study areas under consideration people used fire to do much of the processing work for them (chapter 5; Stiner 1994, 1999).

SMALL GAME TRENDS ACROSS THE MEDITERRANEAN RIM

Figure 8.1 compares changes in the contribution of quick small game to total small game taken by Paleolithic foragers from 200 to 9 KYA in the Wadi Meged area of Israel with those for coastal Italy (Stiner et al. 2000; Stiner 1994, 1999). As discussed in chapter 7, tortoises were collected in the Wadi Meged from the Middle Paleolithic through the Kebaran and Natufian periods, but their numbers eventually gave way to greater use of birds (especially but not exclusively galliforms) and small mammals (mainly lagomorphs). Analogous trends involving somewhat different suites of prey have been observed elsewhere on the Mediterranean Rim. The Italian sequence comes from mainly coastal shelters in Liguria and Lazio and spans 110,000 to 9,000 years ago (tables 8.1 and 8.2). It includes the Middle to Upper Paleolithic technological transition and terminates not long before the end of Paleolithic

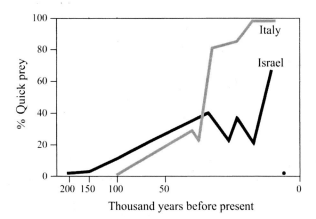

Figure 8.1
Relative abundance of quick-moving prey animals in the small game fractions of the Paleolithic faunal series from the Wadi Meged in Israel (200–11 KYA) and coastal Italy (110–9 KYA). The isolated point represents Layer A of the Italian site of Riparo Mochi, a layer deposited at a time when site function shifted from residential to that of a special use camp.

lifeways in the region. The faunal samples are from Riparo Mochi in Liguria (Blanc 1953; Cardini and Biddittu 1967; de Lumley-Woodyear 1969; Kuhn and Stiner 1992, 1998a; Laplace 1977), which preserves Late Epigravettian, Early Epigravettian, Gravettian, Middle Aurignacian, and Early Aurignacian faunas (36–9 KYA); from the Middle Paleolithic coastal cave sites of Grotta dei Moscerini (110–70 KYA, Kuhn 1995; Stiner 1994) and Grotta Breuil (33–40 KYA, Bietti et al. 1990–1991; Kuhn 1995; Stiner 1994) in coastal Latium; and two inland assemblages from the Epipaleolithic Latium shelters Grotta Polesini (Radmilli 1974) and Grotta Palidoro (Bietti 1976–1977; Cassoli 1976–1977). Counts by taxonomic category are based on the number of identified specimens (NISP) for vertebrates and the minimum number of individuals (MNI) for mollusks, the latter to ensure specimen size comparability to vertebrate remains in the frequency comparisons (see chapter 7).

For the Italian sites, there is no trend in the relative contribution of small game to total animal intake based on the small game–ungulate index (table 8.1). Large game species in the Italian archaeofaunal series are mostly red deer (*Cervus elaphus*), aurochs (*Bos primigenius*), roe deer (*Capreolus capreolus*), and, in some instances, ibex (*Capra ibex*), wild ass (*Equus hydruntinus*), and wild boar (*Sus scrofa*). The proportions of three categories of small animals within the small game fraction shift significantly with time (table 8.2). Relatively "sessile" (immobile or slow-moving) animals, mainly edible shellfish (*Patella* spp., *Mytilus gal-*

Table 8.1

Percentages of carnivores, ungulates, and small game prey animals in the Paleolithic series from Italy

Culture Period, Site, Phase (Layer)	Total Game	% Carnivores	% Ungulates	% Small Game	Small Game–Ungulate Index	OIS	Geographical Setting
EP RM Late Epigravettian (A)	980	1	17	81	.83	1	Coastal
EP GPo Late Epigravettian	44,403	3	94	3	.02	1	Inland
EP GPa Evolved Epigravettian	2,079	2	97	1	.01	2	Inland
EP RM Early Epigravettian (C)	2,971	1	59	40	.40	2	Coastal
UP RM Gravettian (D)	3,653	2	77	21	.21	2	Coastal
UP RM Middle Aurignacian (F)	930	2	53	45	.46	3	Coastal
UP RM Early Aurignacian (G)	1,592	2	53	44	.46	3	Coastal
MP GB Middle Paleolithic	1,571	4	96	<1	.01	3	Near coastal
MP GM Middle Paleolithic	1,422	<1	53	46	.47	4–5	Coastal

SOURCES: Stiner et al. 2000. Data for Grotta Polesini are from Radmilli 1974; for Grotta Palidoro, Cassoli 1976–1977. OIS chronology follows Martinson et al. 1987.

NOTE: Total game counts are NISP for vertebrates but MNI for mollusks, in order to correct for significant differences in mean fragment sizes. Birds are mostly partridges, along with doves, waterfowl, and large passerines. Edible shellfish are mainly mussels, limpets, and turbans (Stiner 1994, 1999; Kuhn and Stiner 1998a).

KEY: (EP) Epipaleolithic; (UP) Upper Paleolithic; (MP) Middle Paleolithic. Site codes are (RM) Riparo Mochi; (GPo) Grotta Polesini; (GPa) Grotta Palidoro; (GB) Grotta Breuil; (GM) Grotta dei Moscerini.

Table 8.2

Percentages of main taxa in the small game prey fraction of the Paleolithic series from Italy

Culture Period, Site, Phase (Layer)	Total Small Game	% Tortoises	% Lagomorphs	% Other Small Mammals	% Birds	% Shellfish
EP RM Late Epigravettian (A)	797	0	<1	0	0	100
EP GPo Late Epigravettian	889	0	41	1	58	0
EP GPa Evolved Epigravettian	30	0	17	0	83	0
EP RM Early Epigravettian (C)	1,191	0	45	6	34	14
UP RM Gravettian (D)	769	0	23	15	43	19
UP RM Middle Aurignacian (F)	420	0	2	9	12	76
UP RM Early Aurignacian (G)	710	0	4	6	18	71
MP GM early Middle Paleolithic	660	6	1	0	0	93

SOURCES: Stiner et al. 2000. Data for Grotta Polesini are from Radmilli 1974; for Grotta Palidoro, Cassoli 1976–1977. OIS chronology follows Martinson et al. 1987.

NOTE: Total game counts are NISP for vertebrates but MNI for mollusks, in order to correct for significant differences in mean fragment sizes. Birds are mostly partridges, along with doves, waterfowl, and large passerines. Edible shellfish are mainly mussels, limpets, and turbans (Stiner 1994, 1999; Kuhn and Stiner 1998a).

KEY: (EP) Epipaleolithic; (UP) Upper Paleolithic; (MP) Middle Paleolithic. Site codes are (RM) Riparo Mochi; (GPo) Grotta Polesini; (GPa) Grotta Palidoro; (GB) Grotta Breuil; (GM) Grotta dei Moscerini.

loprovincialis, Monodonta turbinata, Ostrea edulis, Callista chione, Glycymeris spp.) and tortoises (Testudo sp. and Emys orbicularis), appear early in the sequence, and shellfish persist in human diets thereafter. Birds become important only with the beginning of the Upper Paleolithic (fig. 8.2); they are mostly gray partridges (Perdix perdix) and quail (Coturnix coturnix), along with lower frequencies of doves (Columba livia) and aquatic (Anseriform) birds, among others. Small mammals, mainly hares *(Lepus capensis)* and rabbits *(Oryctolagus cuniculus)*, become important somewhat later, mainly in the Epipaleolithic phases. The diachronic changes in the contents of the small game fractions at these sites are directional, whereas variation in the contribution of small animals to total game intake based on NISP counts is not.

The preponderance of shellfish in the Late Epigravettian (Layer A) of Riparo Mochi is enigmatic relative to the Upper Paleolithic phases before it and best resembles the shellfish component of the Middle Paleolithic assemblage from Grotta dei Moscerini. However, inland Late Epigravettian samples from Grotta Palidoro and Grotta Polesini in Latium indicate that birds and lagomorphs figured prominently in small game fractions elsewhere at this time (see also Cassoli 1976–1977; Radmilli 1974). Unlike the residential camps rich in diverse cultural materials that characterize the series overall, the final Paleolithic occupation at Riparo Mochi almost certainly represents a special use camp (Stiner 1999) at a time when sea level transgressed nearly to the foot of the shelter.

The Middle Paleolithic sample from Grotta dei Moscerini, with its moderately large percentage of small game animals (table 8.1), is peculiar in its own right, because many sites of this period in Italy lack substantial small game components (Stiner 1994). Where small game remains are present and clearly attributable to Middle Paleolithic humans, the focus nonetheless is on sessile types (see also Blanc 1958–1961; Palma di Cesnola 1969; Stiner 1993, 1994:176–192; on North Africa, Klein and Scott 1986). At Moscerini, shellfish and tortoise collection (46 percent of total game NISP) was combined with an odd pattern of terrestrial foraging—occasional collecting (scavenging) of ungulate head parts (Stiner 1994). Small mammal remains have also been reported in the Middle Paleolithic layers of Grotta di Sant'Agostino (Tozzi 1970), but taphonomic evidence associates most of these materials with denning wolves, not humans (Stiner 1994:166–171). The wolf den of Sant'Agostino nonetheless is important for establishing the abundant presence of lagomorphs in coastal Italy during Middle Paleolithic times, despite their rare occurrence in Middle Paleolithic archaeofaunas.

The faunal series from Italy and Israel represent distinct ecogeographical zones, so the trends cannot be dismissed as local phenomena. The changes in small game use occurred in the context of relatively stable biotic communities in which small animals were consistently diverse and abundant. Variation in animal

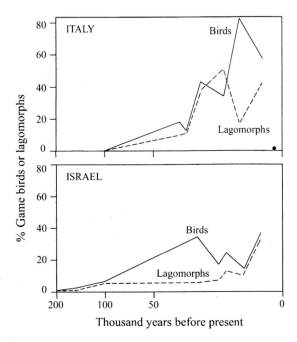

Figure 8.2
Relative abundances of birds and lagomorphs in the small game fractions of the Paleolithic faunal series from the Wadi Meged and coastal Italy. The isolated point represents Mochi Layer A, as in figure 8.1.

community content, based on the numbers of species recruited and lost, was minor during the Late Pleistocene in Italy (Stiner 1994:68–77) and Israel (Tchernov 1981, 1992b). Greater variation in living species availability may have occurred during the Middle Pleistocene (Tchernov 1992b, 1994), but the most pronounced shifts in humans' use of small game took place in the Late Pleistocene.

More significant than any expansion of the taxonomic spectrum in human diets was a rising emphasis on a few highly productive taxa that were less easily caught by hand. Each small game series begins in the early Middle Paleolithic with nearly exclusive use of sessile or slow-moving prey. This is followed in the early Upper Paleolithic by major proportional increases in quick-flying common game birds and, by the Epipaleolithic, in fast-running lagomorphs as well. Highly ranked prey animals, as defined by Middle Paleolithic exploitation, were the slow-moving tortoises and shellfish. Use of these animals continued through the Upper Paleolithic and Epipaleolithic, but agile types supplemented the diet in ever greater proportions, despite their lower ranking on grounds that their capture costs were higher.

The high productivity of galliform birds and lagomorphs is well known; here they are termed "high turnover" populations. Humans continued to collect slow-moving, slow-growing prey in the later periods, but exploitation may have provoked declines in the local abundances of some of these animals. Two separate lines of evidence concerning early favorites on the small game menu lend credence to the idea that their availability was conditioned by the interplay between predator densities and prey population resilience or turnover rates.

CORROBORATION FROM SIZE DIMINUTION IN SLOW-GROWING PREY

Differences in population turnover rates are important for modeling the long-term outcomes of population interactions, a concept first articulated by MacArthur and Wilson (1967), Odum (1971), and Pianka (1978), among others. Intensive harvesting over intervals shorter than a population's regeneration time is known to reduce mean individual age in a great variety of modern vertebrate and invertebrate species (Caughley 1977; Dye et al. 1994; Keck et al. 1973; Koslow 1997; Lambert 1982; Levinton 1995:94–95; McCullough et al. 1990; Russell 1942). Species with slow rates of development and prolonged reproductive careers are particularly sensitive to losses of mature adults (on shellfish, see Botkin 1980; on tortoises, Blasco et al. 1986–1987; Hailey et al. 1988; Lambert 1982; on fish, Koslow 1997; Russell 1942). These are "low turnover" species. Of interest is the relation between the pace of prey population recovery and foragers' needs and options (see also Bayham 1979, 1982; Broughton 1994; Stiner 1990, 1994; Winterhalder et al. 1988). If a prey species is slow growing, then recruitment from neighboring areas cannot easily compensate for local losses in less than a few years.

When humans collect sessile small-bodied animals, they should and apparently do prefer adults, because adults represent the largest packages of their type (e.g., Yesner 1981). There is an appreciable difference, for example, between mature tortoises or mussels and young ones in terms of food units gained for the effort. Heavy exploitation easily alters the age and size structure of some of these species, and a predator's preference for larger individuals accelerates the effect. Because age corresponds to body size in species that continue to grow during their adult years, a reduction in mean age also brings about a reduction in average individual size for the population as a whole. This kind

of pressure will not necessarily destroy the prey population, so long as the population's intrinsic potential for growth (r) is not exceeded; more often, the prey type just becomes rarer in the environment, forcing the predator to switch to other resources. Most prey populations have some tolerance for predator pressure, because younger/smaller adults can reproduce as the intraspecific competition for mates relaxes, though these adults tend to produce fewer young (Hailey and Loumbourdis 1988; Levinton 1995:90, 94–95). Size diminution in spur-thighed tortoises *(Testudo graeca)* from the Middle Paleolithic through the Epipaleolithic was examined in chapter 7: measurements of tortoise humeri from the Wadi Meged sites and Kebara Cave revealed a clear trend toward size reduction. The tortoises collected by Middle Paleolithic foragers were large on average and considerably larger than those collected by later humans after about 44 KYA.

Commercially important shellfish populations (mussels, oysters, and various clams) are also known to experience sudden declines in average age and size if exploited too intensively, and some species are particularly sensitive (e.g., Dye et al. 1994; Hockey 1994; Levinton 1995). These facts about modern shellfish ecology have been used in research on human predation intensities in prehistory (e.g., Botkin 1980; Clark and Straus 1983; Jerardino 1997; Klein 1979; on fish, see Broughton 1997). Only limited consideration of the subject is possible in the Italian cases, because whole shells are comparatively few, due to shattering from trampling, heat, or both (Kuhn and Stiner 1998a; Stiner 1994, 1999). At Riparo Mochi, limpet size reduction occurred abruptly between the Gravettian (Upper Paleolithic) and the Early Epigravettian (Epipaleolithic) periods (fig. 8.3). The samples form only two size groups, and an analysis of variance shows that the size differences among assemblages deviate significantly from random (F-ratio = 75.92, $p < .001$, df = [4, 403]). Because suppression of mean limpet size is constant through both the Last Glacial Maximum (Mochi C) and the warm conditions of the terminal Pleistocene (Mochi A), the diminution trend seems not to be explained by climate change. Nor is it a result of variation in the relative frequencies of the constituent species (predominantly *Patella caerulea* throughout). Changes in habitat quality associated with the rise and fall of sea level could in principle account for size diminution in the limpet populations independently of Paleolithic forager effects (e.g., Bailey 1983; Jerardino 1997). However, this is unlikely in the case of Riparo Mochi, because the steep coastal topography of the Balzi Rossi and its widespread rocky surfaces lent unusual stability to habitat configurations there (Stiner 1999).

Whatever the importance of shellfish to Middle Paleolithic diets at sites such as Grotta dei Moscerini, humans' net impact on the age and size structures of shellfish colonies was minimal throughout the Middle and early Upper Paleolithic. Most mussel, clam, and oyster shells in the Middle Paleolithic samples are unmeasurable—at least in terms of complete dimensions—but it is clear from the fragments that the individual animals were quite large on average.

CRITICAL VARIABLES IN HUMAN PREDATOR–SMALL PREY INTERACTIONS

The foregoing observations do not present a simple story of subsistence change in the later Paleolithic, but together they suggest increasing predator pressure on prey. This implies that human population densities were rising, probably in several demographic pulses. Which aspects of predator-prey relationships are most critical to understanding the process of change? The archaeological findings and background information on the common prey types represented in the Mediterranean series indicate that two variables were of crucial importance: the ways in which small animals eluded predators and small prey maturation rates.

PREDATOR AVOIDANCE MECHANISMS

The means by which prey animals avoid predators can greatly affect their relative rank in a predator's foraging regimen. In the Mediterranean study areas, we have two broad categories of small prey—those that are easily caught by hand and those that are not. These differences in "catchability" translate into distinct work-of-capture costs in the absence of special tools or physiological structures. For humans, this is principally a matter of technology. A tortoise's defense against predators combines cryptic habits, slow movement, and a portable fortress. Safety in numbers replaces hiding in some mollusks. The attractiveness of these animals to humans is of course their ease of collection in combination with (in many cases) relatively low processing costs. Humans should have collected these resources whenever they were encountered. Also significant is that modern tortoises and shellfish can exist at very high densities in the absence of human

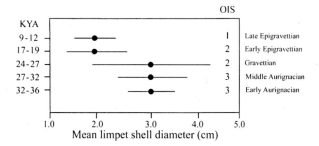

Figure 8.3

Size reduction trend in limpets (predominantly *Patella caerulea*) from time-ordered Upper Paleolithic layers of Riparo Mochi, based on mean shell diameter and standard deviations. No substantial skewing occurs in the size distribution for any assemblage (<10 percent difference between mean and median). Oxygen isotope stages (OIS) follow Martinson et al. 1987.

disturbance, because of their low metabolic rates, high subadult and adult survival rates, and potentially long life spans (e.g., Hailey 1988; Shine and Iverson 1995). Mussel colonies and limpet patches are relatively easy to find along rocky shorelines. Terrestrial tortoises are less clustered in the environment, because they lead solitary lives, but adults are quite visible seasonally.

The small species that increase in later Paleolithic diets are the quick-fleeing types. Hares, rabbits, partridges, and other birds are given to bolting from cover at high speed and generally are more difficult to catch than tortoises or shellfish without the help of nets, snares, or other traps. Because of these animals' higher pursuit and handling costs, Paleolithic humans' incentives to switch to quick-moving types would have to have been quite strong. Prolonged scarcity of easily collected small prey ultimately would also increase the selective advantage of any technology that reduced the cost of capturing quick, agile animals.

PREY MATURATION RATES

Predator-prey relationships are mediated by the life history characteristics of prey. Because slow-growing small taxa dominate the earlier portions of each Paleolithic sequence and fast-growing types become important later, the periodicity or intensity of predation relative to prey maturation rates must have been important. Ground birds, lagomorphs, tortoises, and most shellfish may produce many young per year (table 8.3). However, tortoises require roughly a decade to mature, and many large-bodied shellfish, up to half that time (minimally two to five years in the wild; e.g.,

Table 8.3

Life history and predator defense characteristics of common small prey types in the Mediterranean Paleolithic faunas

Prey	Offspring Production Rate	Maturation Rate	Predator Defense Mechanisms
Shellfish	High	Slow–moderate	Safety in numbers, armor, some cryptic
Tortoises	Moderate	Slow	Freezing, hiding, armor
Lagomorphs	Moderate	Fast	Hiding, bolting and rapid running
Game birds	Moderate	Fast	Hiding, bolting and rapid flight

Epifanio and Mootz 1976; Keck et al. 1973; Little and Kitching 1998). Lagomorphs and game birds such as partridges reach reproductive age within a single year, which accounts in large part for the high turnover rates of these populations.

A harvesting bias favoring adults, and adult females in particular, may amplify the sensitivity of slow-maturing populations to predator pressure and possibly accelerate size diminution. In tortoises and edible mollusks, older females may produce more eggs on average, owing to their larger body size (e.g., Hailey and Loumbourdis 1988). Moreover, the sexes in tortoises are size dimorphic, with females being significantly larger than males of equivalent age (Lambert 1982). Large body size also confers a reproductive advantage on females in many shellfish species (Levinton 1995:90–95). Selective removal of adults may erode the reproductive core of any sort of prey population, but its effect is especially strong among slow-maturing species, because mature females lost to predation are replaced only slowly. Human predators have developed some distinctive search windows, including a strong attraction to hard-shelled prey in a variety of environmental substrates. The relationship between slow-maturing prey species and predatory humans appears to be particularly delicate, sustainable only if harvesting rates remain very low.

Table 8.3 summarizes the main life history and predator avoidance characteristics of edible mollusks, tortoises, lagomorphs, and partridges in the Mediterranean faunal series. These generalizations liken tortoises to mollusks, and lagomorphs to partridges, at least from the human perspective. It is clear that prey birth rates alone cannot explain the differences in prey population turnover rates or the trends in small game

use of the later Paleolithic. The explanatory power of maturation rate and predator avoidance mechanisms is much more promising: work of capture influences prey rank, and birth rate and maturation rate should together determine a population's potential resilience to heavy predation.

SIMULATIONS OF PREDATION ON TORTOISES, HARES, AND PARTRIDGES

Differences in small prey productivity amount to differences in the ways animals occupy environments and stock them with progeny. Humans should be affected by the life history characteristics of any prey animal that, for other reasons, is ranked highly in the foraging spectrum. Contrasting life history strategies suggest that low-turnover prey species should respond quite differently to human predation than high-turnover species. What is poorly understood at present is the magnitude of difference in productivity among the subject prey animals. The predator-prey simulations described here were designed around the life history traits of three small prey items that are common in Mediterranean Paleolithic sites—tortoises, hares, and partridges. The simulation models were constructed with two questions in mind: First, what is the maximum annual "yield" that predators can take from a subject prey population without destroying it over the long term, or, more importantly, what is the threshold for a stable (sustainable) predator-prey relationship? Second, how much more resilient are hare and partridge populations than tortoises to similar increases in predator density?

SIMULATION MODEL DESIGN: HIGH AND LOW GROWTH CONDITIONS

The parameters for the simulations were taken from a variety of modern wildlife studies, preferably but not exclusively for the species identified in the Mediterranean archaeofaunas. Figure 8.4 summarizes the life history parameters used for tortoises *(Testudo)*, hares *(Lepus)*, and partridges *(Alectoris* and *Perdix)*. Not all wildlife studies are equally suitable sources of simulation parameters. Cases involving substantial habitat loss or catastrophic population decline are compelling ammunition for conservation issues, for example, but they are not necessarily appropriate standards for prehistoric prey population dynamics. Data from long-term studies of viable populations, with good control over birth rates, mortality rates, and their causes, were favored for modeling purposes (table 8.4).

To investigate the interplay of life history traits in predator-prey systems, we modeled two extremes of population growth for each kind of small prey animal—a high growth model (HGM) and a low growth model (LGM). Truly average conditions are rare in the life of any individual, and most or all years in that individual's lifetime will likely fall between the curves defined by our high and low growth models. Because prehistoric prey and predator densities cannot be

---

Individual variables by sex
> Male
>> Age
>> Mass (tortoises only)
> Female
>> Age
>> Mass (tortoises only)
>> Next age of reproduction
>> Litter size

Fertility parameters
> Female minimum age of reproduction
> Birth interval (spacing)
> Minimum number of offspring
> Maximum number of offspring

Natural mortality parameters
> Maximum potential life span
> Age of onset of adult mortality rate
> Annual adult mortality rate
> Juvenile mortality rate

Hunting parameters
> Minimum age or size to hunt
> Annual kill percentage

Figure 8.4
The structure and variables of the predator-prey simulation model. (Reproduced from Stiner et al. 2000.)

Table 8.4
Fertility and mortality parameter values for tortoises, hares, and partridges in the high growth (HGM) and low growth (LGM) models

Parameter	Tortoises		Hares		Partridges	
	HGM	LGM	HGM	LGM	HGM	LGM
Fertility						
Female age at first reproduction (years)	8	12	0.75	1.0	0.75	1.0
Birth interval (days)	365	730	365	365	365	365
Minimum number of offspring per annum	7	7	9	7	11	9
Maximum number of offspring per annum	14	14	11	9	13	11
Mortality						
Maximum potential life span (years)	60	60	12	12	8	8
Age of adult-level mortality onset (years)	1	1	0.5	0.5	0.2	0.2
Annual adult-level mortality rate	0.053	0.093	0.4	0.5	0.5	0.6
Annual (base-level) juvenile mortality	0.70	0.85	0.6	0.7	0.42	0.6

SOURCES: For tortoises, Blasco et al. 1986; Castanet and Cheylan 1979; Hailey and Loumbourdis 1988; Lambert 1982, 1984; Meek 1989; Shine and Iverson 1995; Wilbur and Morin 1988; Willemsen and Hailey 1989. For partridges, Ali and Ripley 1969; Alkon 1983; Ash 1970; Bannerman and Bannerman 1971 cited in Alkon 1983; Blank and Ash 1962; Bohl 1957; Christensen 1970; Dement'ev and Gladkov 1967; Galbreath and Moreland 1953; Harper et al. 1958; Jenkins 1961; Mackie and Buechner 1963; Middleton 1935; Potts 1986; Robbins 1984. For hares, Bronson and Tiemeier 1958; Broekhuizen 1979; Flux 1981; Frylestam 1979; Hansen 1992; James and Seabloom 1969; Keith 1981; Kovacs 1983; Lechleitner 1959; Petruesewicz 1970; Pielowski 1971, 1976 cited in Broekhuizen 1979; Raczynski 1964; Swihart 1983.

known absolutely, our strategy was to compare the relative resiliencies of tortoise, hare, and partridge populations under favorable and lean conditions for prey reproduction and growth. The simulations were written by Todd Surovell as Visual Basic macros in Microsoft Excel 7.0. Populations were modeled as sets of actual individuals, each characterized by age, sex, and, in tortoises, body mass. Additionally, females were assigned values for next age of reproduction and annual litter size. Individual age increased by a fixed value per unit of time elapsed.

Population dynamics were governed largely by fertility and mortality variables (fig. 8.4). Fertility was controlled by three parameters in the model: female minimum reproductive age and the minimum and maximum number of offspring per annum. When a female was born, her next age of reproduction was set to the minimum age at which she could begin reproducing and to normal birth spacing thereafter. A predetermined number of offspring, between the minimum and maximum values in table 8.4, was added to the population each year (except for LGM tortoises, as we discuss shortly). An even sex ratio at birth was maintained on the basis of empirical evidence presented in chapter 7 (fig. 7.18)

Mortality was controlled by four parameters in the model: maximum potential life span, annual juvenile mortality, annual adult mortality, and age of onset of adult-level mortality. Mortality effects were divided between only two age groups, juveniles (including newborns) and adults, an approach justified by available wildlife data. Adult mortality randomly removed a fixed percentage of adults from the population each year, in addition to removing any individuals lucky enough to have exceeded the maximum potential life span. Density-dependent mortality from nonhuman causes affected only juveniles, because young animals are most likely to suffer under conditions of high prey population density. Thus, juvenile mortality was allowed to vary as a linear function of population density:

$$m_{jt} = m_{j0} + [(pop_t/pop_k) (1 - m_{j0})]$$

where m_{jt} is juvenile mortality at time t; m_{j0} is base-level juvenile mortality; and pop_t/pop_k is population density at time t. Therefore, $m_{jt} = m_{j0}$ when $pop_t = 0$, and $m_{jt} = 1$ when the population is at environmental carrying capacity.

Hunting by humans was controlled by two constants in any given run—minimum age (or size) to hunt (a selectivity factor) and annual kill percentage. As long as individual prey above a given age or body size threshold were available, it was assumed that humans would be attracted to them. If individuals above the threshold were no longer available, then humans would target the oldest available individuals below that age threshold. Parameter definitions and sources used in modeling the tortoise, partridge, and hare populations are presented separately in the following sections.

TORTOISES

Testudo, the common genus of tortoise in the archaeofaunal series, provides an ideal standard for comparing small game use in the Mediterranean Paleolithic. Unfortunately, little modeling work has been done on tortoises in general, making it necessary to begin from scratch (but see Doak et al. 1994). In doing so, we note several important insights from wildlife studies of modern *Testudo graeca* and *T. hermanni* in the Mediterranean Basin. First, the illegal pet trade, which favors large specimens for international markets, has rapidly driven down mean individual size in affected tortoise populations in North Africa (Lambert 1982; Stubbs 1989) and Spain (Blasco et al. 1986–1987). Second, immature tortoises generally are much more difficult to find than are adults in Mediterranean habitats (Lambert 1982). Third, adult female tortoises tend to be larger than males of the same age (Blasco et al. 1986–1987; Lambert 1982), making the reproductive core of the population that much more vulnerable to size-dependent predation by humans. Our model takes into account the steeper growth curve of females relative to males, because size-biased collecting should affect females and males somewhat differently. Tortoises over about 0.3 kg were considered adults, on the basis of curve fitting, corresponding to 10 years of age for females and 12 years of age for males. Figure 8.5 shows idealized growth curves for male and female tortoises created by gross estimation and curve-fitting of the data from Blasco et al. 1986–1987 and Lambert 1982. Lambert (1982) documented considerable variation in the degree of sexual size dimorphism in *Testudo*, but to be conservative our model minimized the potential adult size differences between the sexes.

Testudo graeca and *T. hermanni* populations can be modeled as one taxon for our purposes, because they respond in nearly identical ways to variations in food supply and to human-caused disturbances in areas where their distributions overlap, and they have similar reproductive rates as measured by annual egg mass production, clutch sizes, and laying frequencies (Blasco et al. 1986–1987; Hailey and Loumbourdis 1988; Hailey et al. 1988; Stubbs 1989). *T. graeca* is the more widespread species (Ernst and Barbour 1989; Stubbs 1989), and a maximum adult weight of about 1–2 kg is typical today in the Wadi Meged study area.

Information on the population dynamics of Mediterranean and other tortoises is scarce, and the available data on fertility and mortality are coarse grained, making it necessary to broaden the taxonomic scope to estimate the ranges for certain parameters. Fortunately, tortoises differ little in terms of the variables employed here, especially if compared with most mammals and birds. Tortoise life histories are characterized by high hatchling mortality but very low subadult and adult mortality, in addition to long life spans and delayed reproductive maturation (Hailey 1988; Shine and Iverson 1995; Wilbur and Morin 1988).

Adult mortality in *Testudo* varies among populations and across years, but composite study results show that survival tends to be continuously high after the first year of life (Hailey 1988; Lambert 1982; Meek 1989). Although tortoises are far from mature at this stage, the age of onset of adult-level mortality was set at one year in both models. We set hatchling mortality (i.e., for the first year of life) at 70 percent in the HGM (females produce 2.1 to 4.2 yearlings per annum) and at 85 percent in the LGM (0.7 to 1.4 yearlings produced per annum), partly on the basis of estimates by Doak et al. (1994) for hatchling survival in desert tortoises *(Gopherus)*. High adult survivorship is essential to the health of tortoise populations (chapter 7), whereas hatchling survival rates can vary much more without detracting from the long-term fate of those populations (Doak et al. 1994; Heppel, Crowder, and Crouse 1996; Heppel et al. 1996); these observed characteristics are reflected in our models. Because egg production depends partly on female body size (Hailey and Loumbourdis 1988), the number of offspring (eggs) produced per annum was allowed to vary linearly with body mass within the specified range. Because wild individuals of the genus *Testudo* seldom live beyond 60 years (Lambert 1982), this value served as the maximum potential life span. It allowed 53.5 and 48.0 years of reproductive activity in the HGM and LGM, respectively.

A strong negative correlation exists between age at sexual maturity and the adult mortality rate in chelonians (turtles and tortoises) and many other reptiles (Shine and Iverson 1995). We used the regression line associated with this correlation to control the covariance of these parameters. To account for published variation in age at first reproduction (cf. Blasco et al. 1986–1987; Castanet and Cheylan 1979; Hailey 1990), values for *T. graeca* and *T. hermanni* were set at 12 years in the HGM and at 8 years in the LGM. These corre-

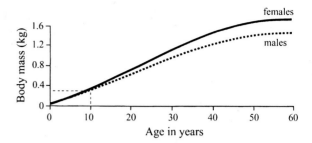

Figure 8.5
Idealized curves of tortoise body growth (mass) by sex. Adulthood is set at a body weight of 0.3 kg (fine dashed line), which is reached by females at around 9 to 10 years of age and by males at 11 to 12 years of age.

spond to adult mortality values of 5.3 percent and 9.3 percent, respectively, well within the range documented for modern wild populations (Hailey 1988, 1990; Lambert 1982; Meek 1989). Annual egg production for *T. graeca* varies between 7 and 14, according to Hailey and Loumbourdis (1988). Birth spacing was set to 365 days in the HGM but at 730 days in the LGM, on grounds that as few as half the adult females in a tortoise population might reproduce in a given year (Wilber and Morin 1988).

Figure 8.6 presents the simulated outcomes of incremental increases in predation on tortoises over 200 years under high growth (HGM) and low growth (LGM) conditions. Adult tortoises were assumed to be preferred wherever available to the predators. It is clear from this exercise that tortoise populations cannot tolerate annual losses of more than 4 to 7 percent (LGM and HGM respectively) of reproductively mature individuals without crashing. In comparison with hares and birds, tortoise populations are exceptionally sensitive to predation and are easily destroyed. Sustainable harvesting is possible only below these thresholds. The same may have been true for certain shellfish (e.g., limpets, *Thais*) that Paleolithic humans in the Mediterranean area depended upon for food, although they are not modeled here.

PARTRIDGES

The faunal series from Israel and Italy include three partridge species, chukar *(Alectoris chukar)*, gray partridge *(Perdix perdix)*, and quail *(Coturnix coturnix)*. The simulations emphasized the parameters available for chukars and gray partridges, which are widely distributed in the Mediterranean area and recently have been introduced into numerous other habitats worldwide (Alkon 1983; Potts 1986). The gray partridge is better

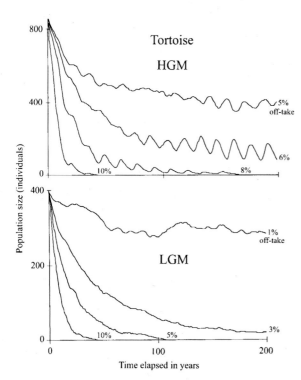

Figure 8.6
Simulated predation on tortoise *(Testudo)* populations under high growth (HGM) and low growth (LGM) conditions. Percentages refer to annual off-take (mortality) from the total population, with adults weighing more than 0.3 kg taken preferentially. The undulations (chattering) in some of the HGM curves are due to alternating focuses on male and female prey, a product of model design. (Reproduced from Stiner et al. 2000.)

adapted to cold winters, and the chukar to arid conditions, but otherwise these species have similar needs and population characteristics.

Normally gregarious, partridges and chukars disperse into breeding pairs in the spring, and most hens lay one clutch of eggs per year (Alkon 1983; Bohl 1957; Christensen 1970; Mackie and Buechner 1963). The timing of egg laying and hatching is not fully synchronized among reproducing females, however, particularly in milder climates. The average number of hatchlings per female chukar per year is 11, on the basis of nine separate studies (Ali and Ripley 1969; Alkon 1983; Bannerman and Bannerman 1971, cited in Alkon 1983; Bohl 1957; Christensen 1954, 1970; Dement'ev and Gladkov 1967; Galbreath and Moreland 1953; Harper et al. 1958). Gray partridges' reproductive output is similar. One standard deviation was therefore added to and subtracted from the figure for mean annual production to produce the HGM and LGM birth rates, respectively.

Juvenile mortality in partridges is greatest during the first and second week after hatching and declines rapidly as the chick approaches adult size (Blank and Ash 1962; Potts 1986). Winter mortality for chukars and gray partridges is comparable where their ranges overlap in southern Europe. Adult mortality from predation is especially high for hens during the spring laying period (Potts 1986) and can be as much as 10 times the normal level for adults at large. Our adult average mortality rate value of 0.55, plus or minus one standard deviation, relies heavily on Potts's (1986) long-term study of gray partridges and his summaries of more than 50 other sources on partridges in England, mainland Europe, and the United States (e.g., Ash 1970; Blank and Ash 1962; Jenkins 1961; Middleton 1935). This average incorporates the risks unique to nesting hens. The onset of the adult mortality rate is two months of age for both the HGM and LGM, following Jenkins 1961 and Potts 1986. Juvenile and adult mortality rates significantly influence the growth and maintenance of partridge populations (see also Caughley 1977 on birds in general). Robbins (1984) reported that female partridges were capable of breeding through four to seven reproductive seasons. Eight years thus serves as the maximum potential life span for both the HGM and LGM, with a maximum of one year devoted to individual development.

Figure 8.7 presents the simulated outcomes of incremental increases in predation on partridge populations over 200 years under high growth (HGM) and low growth (LGM) conditions. Individuals aged two months and older were taken preferentially. It is clear that partridges are very resilient to sustained heavy predation, and their populations are difficult to destroy, even where off-take is consistently high. The simulation results indicate that partridge populations can tolerate up to about 65 percent annual losses of adults in the HGM, and about 22 percent in the LGM. Though not explored here, certain other game birds, such as common dove species, probably also fit this pattern.

HARES

The brown, or cape, hare *(Lepus capensis)* is the most common lagomorph in the Paleolithic samples, but little reliable information could be found for populations of this animal in modern western Asia. The closely related European hare *(Lepus europaeus)* is widespread and well studied in Europe, however, including the northern Mediterranean Rim. Some biologists argue that it is the same species as the brown hare; virtually all others agree that they are closely related. Because the adult mortality reported for modern European hares varies greatly among studies, largely because of the hares' popularity with sport hunters, we distinguished hunted from nonhunted populations in our use of mortality parameters. The literature on the North American jackrabbit (*L. californicus* of grassland–desert scrub habitats; *L. alleni* and *L. townsendi* of grassland habitats) was also consulted.

The reproductive seasons of hares last eight to nine months each year in warm temperate environments. Age at first reproduction normally varies between nine months (the value for our HGM) and one year (the LGM value), although females are physically capable of reproducing at six months of age. Our use of reported birth rates is conservative and relies mainly on observed rather than estimated values, because many biologists' techniques for estimating birth rates assume extraordinary rather than normal reproductive conditions (cf. Bronson and Tiemeier 1958; Gross et al. 1974; Hansen 1992; Lechleitner 1959). A doe's annual production of young (leverets) in the HGM and LGM (table 8.4) is based on standard deviations around an average from nine studies, six on the brown hare and three on North American jackrabbits (Bronson and Tiemeier 1958; Flux 1981; Hansen 1992; James and Seabloom 1969; Keith 1981; Petrusewicz 1970; Pielowski 1976, cited in Broekhuizen 1979; Raczynski 1964). As with partridges, one standard deviation was added to and subtracted from the mean annual production value for hares in order to estimate the HGM and LGM birth rates, respectively.

The maximum potential life span was set at 12 years for both the HGM and LGM, a compromise based on studies by Pielowski (1976), Abildgard et al. (1972), and Broekhuizen (1979). Three studies conducted in areas minimally affected by humans suggest that adult mortality rates apply to all individuals past the minimum reproductive age of six months (HGM and LGM) (Abildgard et al. 1972; Marboutin and Peroux 1995; Pielowski 1976, cited in Broekhuizen 1979). Mortality in recreationally hunted hare populations is consistently higher (cf. Broekhuizen 1979; Frylestam 1979; Kovacs 1983; Lechleitner 1959; Marboutin and Peroux 1995; Pepin 1987; Petruesewicz 1970) and cannot be taken to represent the human-independent dynamics of hare populations.

Juvenile mortality rates in hares are high, ranging between 0.60 and 0.89 per annum (Frylestam 1979; Gross et al. 1974; Hansen 1992; Petruesewicz 1970). However, these figures represent stable hare populations existing at high densities; juvenile mortality nat-

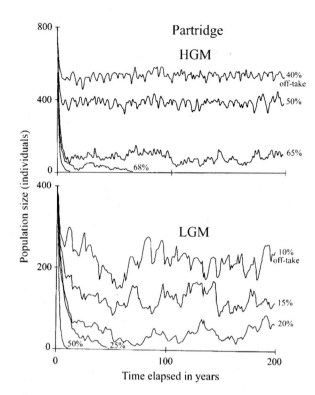

Figure 8.7
Simulated predation on partridge (*Alectoris* and *Perdix*) populations under high growth (HGM) and low growth (LGM) conditions. Percentages refer to annual hunting off-take (mortality) from the total population, with adults over two months of age taken preferentially. (Reproduced from Stiner et al. 2000.)

urally increases as a population approaches equilibrium. In our model the effects of population density on juvenile mortality rate were such that the hare population reestablished equilibrium at a reduced density and with an associated reduction in juvenile mortality. The variation produced by our simulations matches well the variation observed by Gross et al. (1974) for a real hare population that shifted over several years between high and low density conditions.

Figure 8.8 illustrates the simulated outcomes of incremental increases in predation on hares over 200 years in high growth (HGM) and low growth (LGM) conditions. Clearly, hare populations are resilient in the face of heavy predation and are difficult to destroy. These populations rebound easily after heavy harvesting, provided that no more than about 53 percent (HGM) to 18 percent (LGM) of mature or nearly mature individuals are removed from the population in a given year. Essentially similar results were obtained for rabbits (*Oryctolagus* and *Sylvilagus*) in an unpublished study by Natalie Munro.

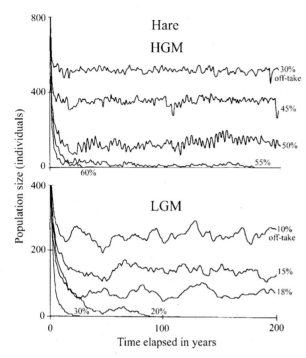

Figure 8.8
Simulated predation on hare *(Lepus)* populations under high growth (HGM) and low growth (LGM) conditions. Percentages refer to annual hunting off-take (mortality) from the total population, with adults over six months of age taken preferentially. (Reproduced from Stiner et al. 2000.)

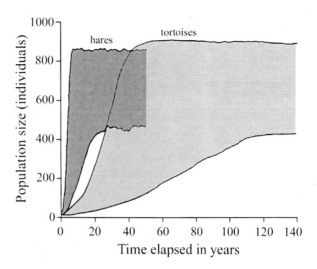

Figure 8.9
Comparison of areas between the high and low growth curves for simulated tortoise and hare populations. The upper line represents the high growth model (HGM) and the lower line the low growth model (LGM) for each kind of prey. Initial population size was 10 individuals. Carrying capacity was set at 1,000 for tortoises and 1,250 for hares in order to render population sizes comparable in the graph. (Reproduced from Stiner et al. 2000.)

DIFFERENCES IN PREY POPULATION RESILIENCE

Folk wisdom tells us that lagomorphs are exceptionally productive. What it does not tell us is how game birds and tortoises compare with them. Figure 8.9 compares the areas between the high and low growth curves for tortoises and hares in our simulations of population growth. The area enclosed by the HGM and LGM curves for tortoises does not overlap at all with that for hares during the years of population growth, despite our rather puritanical limits on hare productivity. In our simulations, hare populations reached equilibrium between about 7 (HGM) and 25 (LGM) years, whereas tortoise populations reached equilibrium between about 50 (HGM) and 125 (LGM) years. Partridge populations may be even more resilient than hares, reaching equilibrium by about 5 years in the HGM and by 10 years in the LGM (not shown). Apart from this qualification, the areas enclosed by the HGM and LGM curves for partridge and hare populations at equilibrium overlap almost completely and may be considered to be about the same.

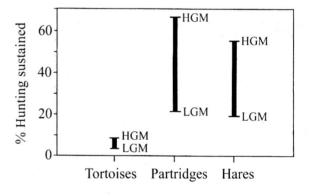

Figure 8.10
Comparison of hunting tolerance thresholds for tortoise (4–7 percent), partridge (22–66 percent), and hare (18–53 percent) populations under high (HGM) and low (LGM) growth conditions. Upper horizontal bars represent thresholds above which predators' dependence on the designated prey type is no longer sustainable. Vertical bars represent natural variation in population resilience as defined by the LGM and HGM. (Reproduced from Stiner et al. 2000.)

Figure 8.10 summarizes the resilience ranges of the three types of small prey animals common in the Paleolithic series. The simulations confirm major differences in the scale at which humans could possibly hope to depend on tortoises, hares, and partridgelike birds for meat. Other things being equal, hare populations can support proportionally 7 times greater off-take by predators than tortoises can support, and partridges can support 10 times greater off-take than tortoises. This means that humans' reliance on tortoises is sustainable only if human population densities are very low. Humans' reliance on partridges and hares is sustainable in both low- and high-density conditions.

Differences in small animal productivity make greater economic sense of Upper Paleolithic and Epipaleolithic humans' increasing use of birds and hares when large tortoises were in short supply. One can also argue that partridges and hares represented more stable or reliable sources of small meat packages as human population densities increased. However, the high-turnover prey species in the two Mediterranean study areas are also quick and thus more difficult to catch by hand. It is for this reason that they may have been ranked lower in Middle Paleolithic foraging systems. Humans could have overcome the disadvantages of these prey items only with greater technological investment.

LINKS BETWEEN PREDATOR PRESSURE AND PREDATOR POPULATION DENSITY

Inherent differences in prey population resilience are significant if human populations are experiencing stress from territorial circumscription and rising densities. As it happens, prey "catchability" correlates closely with differences in prey population resilience, at least among the small animal species that were important to Paleolithic foragers in the Mediterranean Basin. Tortoises and some shellfish require years to mature, their populations are relatively unresilient, and they seldom can elude humans once discovered. The high rank of these animal resources in humans' eyes relates to the relatively low costs of capturing and processing them.

Partridges, rabbits, and hares are very different from tortoises and certain shellfish in that they mature in well under one year and their populations rebound rapidly despite heightened harvesting pressure. Partridges and lagomorphs would have been less attractive to early foragers because they are more difficult to catch by hand. By the Upper Paleolithic, it seems that people had no choice but to pursue greater proportions of quick prey, to meet their need for dietary protein. Some of the radiations in Upper Paleolithic and Epipaleolithic foraging technologies no doubt evolved on

the heels of demographic increases as ways to reduce the cost of acquiring agile prey (Stiner 2002a; Stiner et al. 2000).

Small animal species vary tremendously in predator defense mechanisms and population resilience, in contrast to the ungulates that were commonly hunted by prehistoric humans. In addition, small animal species vary less in body weight relative to humans than do large game animals (see table 6.1). It is for these reasons that data on small game exploitation can reflect subtle changes in Paleolithic demography. Adult tortoises and shellfish ideally must contribute young to their population for many years if they are to enjoy any measure of reproductive success. Heavy harvesting by humans reduces prey population viability and, soon, the frequency with which human foragers can find suitably large individuals of the affected species (Botkin 1980; Christenson 1980; Earle 1980; Mithen 1993; Pianka 1978). It therefore is remarkable that up to about half of all identifiable animal remains (NISP) in some early Middle Paleolithic assemblages of Hayonim Cave are from a reptile that is exceptionally sensitive to predation. What is more, the sizes of the individual tortoises taken during the Mousterian were larger on average than those of the later Paleolithic periods. High archaeological frequencies of a low-resilience prey species, along with large individual body sizes, imply that the early human populations that depended upon the species were small and highly dispersed. Middle Paleolithic populations may simply not have experienced the sorts of stresses that would have made agile, fast-growing small animals attractive. Low human population densities during the early Mousterian also imply small social groups and networks, certainly limiting the numeric scope of individual interactions. Under these conditions, the possibilities for evolution of complex sharing and exchange behavior as a way to counter the effects of unpredictable resource supplies would also have been limited.

The analyses presented in this chapter suggest more than one pulse of growth in human population density during the later Paleolithic in the Mediterranean Basin. Tortoise diminution in Israel is evidenced by at least 44,000 years ago and probably earlier, whereas mollusk diminution in Italy first occurred about 23 KYA. Increasing reliance on birds and lagomorphs during the Upper Paleolithic and Epipaleolithic was almost certainly a response to the declining availability of higher-ranked prey types relative to the number of human consumers. Although one would expect the natural abundances of shellfish, tortoises, game birds, and lagomorphs in the environment to have shifted somewhat with global climate, the trends in small prey emphasis bear a strong human signature. This is not to say that climatic shifts of short periodicity, such as the Younger Dryas, had no effects on human life. Climatic oscillations almost certainly changed effective latitude, precipitation, resource patchiness, and available life space for humans in some way (Keeley 1995; see also Bar-Yosef 1995, 1996; Binford 1968, 1999; Flannery 1969). These factors were not, however, the main determinants of change in small game use by Paleolithic foragers between 200 and 9 KYA.

Predator-prey simulation modeling illustrates how rising human population density and associated preda-tor pressure may alter prey abundances and thereby select for changes in the small species emphasized by foragers. The findings uphold the importance of population growth for subsistence and social change in the later Paleolithic. More surprising is the evidence that resource intensification began so early in the story of subsistence revolution in Eurasia. The results suggest a notable expansion in dietary breadth with the onset of the early Upper Paleolithic and an even greater expansion during the later Upper Paleolithic and Epipaleolithic. How do the data on small game exploitation stand up in a formal analysis of prey diversity and dietary breadth following the predictions of classic foraging theory? That is the subject of chapter 9.

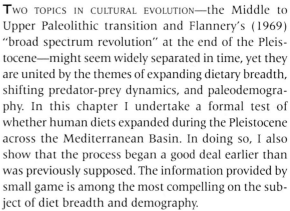

CHAPTER NINE

Prey Diversity and Changes in Human Dietary Breadth

TWO TOPICS IN CULTURAL EVOLUTION—the Middle to Upper Paleolithic transition and Flannery's (1969) "broad spectrum revolution" at the end of the Pleistocene—might seem widely separated in time, yet they are united by the themes of expanding dietary breadth, shifting predator-prey dynamics, and paleodemography. In this chapter I undertake a formal test of whether human diets expanded during the Pleistocene across the Mediterranean Basin. In doing so, I also show that the process began a good deal earlier than was previously supposed. The information provided by small game is among the most compelling on the subject of diet breadth and demography.

The approach taken here is diversity analysis, the classic means for testing hypotheses about human dietary breadth (Stiner 2001). A case was made in chapter 8 for humans' ranking of prey on the basis of predator defense behaviors and differences in absolute body size. I further explore these points here via manipulation of the criteria for categorizing prey in diachronic comparisons of three distinct faunal series from the northern and eastern rims of the Mediterranean Basin. The analysis benefits from the extended accumulation times that are typical of cave faunas and that might result from multiple visits to a site by humans. Time averaging, however, was never great enough to destroy the overall chronological integrity of the faunal series. Constants in this approach are geographic setting, high data quality, the general range of prey types represented, and attribution of the faunas to human collectors. Thus, patterns in the data are most likely to reflect climate-induced shifts in faunal content or changes in human predatory behavior.

HISTORICAL REVIEW

All Paleolithic hominids lived by hunting and collecting wild foods, an aspect of existence that began to disappear only with the emergence of farming and herding societies of the Neolithic, 10,000 years ago or less. The roots of this remarkable economic transformation lie in equally revolutionary changes that took place in certain Stone Age cultures several millennia beforehand. L. Binford (1968) and Flannery (1969) first recognized links between the expanding diets of Late Pleistocene foragers in Eurasia and culture change. Binford described substantial diversification of human diets in middle- and high-latitude Europe at the end of the Paleolithic, or the Mesolithic, roughly 12–8 KYA. Rapid diversification in hunting practices, food processing, and food storage equipment generally

accompanied the dietary shifts, which he took to be symptoms of intensified use of habitats and fuller exploitation of the potential foodstuffs they contained. Some of this behavior was directed toward grinding, drying, and storing nuts, but it also involved small animals (see also Clark and Straus 1983; Coles 1992; Jochim 1998; Keeley 1988; Price and Gebauer 1995).

Flannery pushed these observations further in 1969 with his broad spectrum revolution (BSR) hypothesis, proposing that the emergence of the Neolithic in western Asia was prefaced by local increases in dietary breadth among foraging societies of the late Epipaleolithic. He argued that subsistence diversification, mainly through the addition of new species to the diet, raised the carrying capacity of an environment that was increasingly constrained by climate instability at the end of the Pleistocene. The two authors suggested that local imbalances in human population density relative to available food were somehow integral to the remarkable changes that took place in human societies just prior to the forager-farmer transition. Binford's and Flannery's papers have stimulated much archaeological research and many debates since their publication, not least because they offered some explicit predictions about subsistence changes. One of the most widely used of these predictions is Flannery's suggestion that the actual number of species in human diets increased with the BSR.

These arguments about human subsistence evolution were directly influenced by early works in the science of population ecology, including works on what later came to be known as foraging theory and diet breadth models. Inspired particularly by Odum and Odum (1959), Emlen (1966), and MacArthur and Pianka (1966), Binford and Flannery argued that economic change could have resulted from demographic crowding in certain regions of the world and presumably from the conditions of selection on human societies. Most archaeologists continue to think of demography as one of several ingredients necessary to the forager-farmer transition (e.g., Bar-Yosef and Meadow 1995; Binford 1999; Davis et al. 1994; Keeley 1988; Redding 1988; Watson 1995). We know that density-dependent effects can play decisive roles in shaping the evolutionary histories of predator-prey systems in general (Boutin 1992; Gavin 1991; Pianka 1978; Sinclair 1991), and humans should not be altogether immune to these effects in principle (Harpending and Bertram 1975; Winterhalder and Goland 1993). Changes in human population density certainly influence rates of interspecific and intraspecific contact and the availability of critical foodstuffs. With increases in population density, people's solutions for

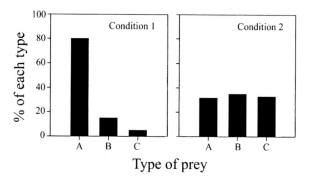

Figure 9.1
Predicted differences in evenness among three hypothetical types of prey taken by predators under distinct foraging conditions. Condition 1 is high availability of the highest-ranked prey types; condition 2 is declining availability of the highest-ranked types such that predator choice diversifies to include more of the lower-ranked types.

getting enough to eat are bound to shift as well. Rapid technological change and the greater densities of archaeological sites during the later Paleolithic (e.g., Mirazón Lahr and Foley 2003) lend some credence to this position.

THEORETICAL EXPECTATIONS FOR CHANGES IN DIETARY BREADTH

Evidence of increasing dietary breadth is expected to take the form of more species in the diet and/or greater proportional evenness between high-ranked and low-ranked prey items in response to the declining availability of preferred types. A predator can afford to ignore lower-quality prey at little cost if the chance of finding a superior type in the near future is high, fostering a narrower diet that emphasizes a favored type disproportionately to its availability in the environment (condition 1 in fig. 9.1). As the supply of preferred prey dwindles, broadening the diet to include common but lower-yield prey types maximizes a predator's returns per unit of expenditure by reducing search time (Pianka 1978). This second set of conditions therefore encourages a more diverse diet in the sense that the predator's emphasis is spread more evenly between two or more prey types that occur in the environment (condition 2).

A broadening of Paleolithic diets in Eurasia certainly is apparent in the greater exploitation of energy-rich nuts and large seeds from the Epipaleolithic onward. Because the nutritional benefits of these resources require considerable work and equipment to extract (Keeley 1988), the broadening trend is most readily apparent in the proliferation of milling tools after the Last Glacial Maximum (Wright 1994) and, to a lesser extent, in increasing evidence of storage facilities and preserved plant parts (Hillman et al. 1989; Miller 1992) and the rise of commensal rodents (Tchernov 1984a). Under lean conditions, people should also have become less selective about which animals to hunt rather than go hungry. Yet measures of diversity in game use based on Linnean taxonomic categories (counts of species or genera) register only one clear economic transition—that from foraging to farming in the early Neolithic, when there was a gradual *decline* in dietary breadth (Davis 1982; Edwards 1989; Horwitz 1996; Neeley and Clark 1993). Any variation that could be found in the taxonomic diversity of archaeofaunas across the Middle and Upper Paleolithic and Epipaleolithic periods was more easily explained by climate-driven environmental changes or geographical variation in animal and plant community composition (Bar-Oz et al. 1999; Simek and Snyder 1988; Stiner 1992). Until recently, there seemed to be no zooarchaeological support for the BSR hypotheses of expanding diet breadth in the Paleolithic at all.

MEASURING VARIATION IN DIETARY BREADTH

The basic idea behind the BSR hypothesis remains a good one. Discrepancies between the results for plant and animal exploitation seem to stem from the way zooarchaeologists tend to categorize prey animals (Stiner et al. 2000). Because the cultures of interest are extinct, prey ranking systems cannot be inferred by watching people make decisions. The relative values (payoffs) of prey species must instead be evaluated from knowledge of modern variants of the animals whose bones occur in archaeological deposits. Species and genera present the most obvious and accessible analytical categories, and the most literal expectation of Flannery's BSR hypothesis is indeed more species in the diet, a more balanced emphasis on those species, or both. Thus, zooarchaeologists normally examine diet variation in terms of indexes of taxonomic richness (N-species or N-genera relative to total NISP) and taxonomic evenness (proportionality in abundance) (Edwards 1989; Grayson and Delpech 1998; Neeley and Clark 1993; Simek and Snyder 1988). Such analyses employ either Kintigh's simulation-based technique (Kintigh 1984) or a longer-standing regression approach (Grayson 1984) that apparently grew out of the work of Fisher, Corbet, and Williams (1943) for problems of sampling in modern community ecology.

The main weakness of diversity approaches that rely on Linnean taxonomic units is their insensitivity to the physical and behavioral differences among prey animals. The only qualification normally added to such analyses is prey body size, since all game animals are composed of similar tissues and large animals yield more food than small ones, even if they are more difficult to catch. The logic of this practice is fine as far as it goes, but it potentially overlooks great differences in prey handling costs among animals that are broadly equivalent in food content and package size. Indeed, some distantly related taxa are nearly equivalent from the viewpoint of handling costs because of their locomotor habits or ways of avoiding predators. Tortoises and rock-dwelling marine shellfish, for example, are sluggish or immobile. Hares and partridges, though similar in body weight to tortoises or an armful of shellfish, are quick and agile. Humans, because of their generalist dietary tendencies, should exploit a wider range of prey as the resident species diversity in ecosystems increases, which further increases the potential for prey type interchangeability.

TRENDS IN DIETARY BREADTH IN MEDITERRANEAN ITALY, TURKEY, AND ISRAEL

Differing prey classification systems greatly affect archaeologists' perceptions of change in prehistoric diet breadth. More to the point, those systems bound most strictly by taxonomy seem to obscure critical information. This can be demonstrated in three faunal assemblage series from the Mediterranean Basin using a simple measure of diversity, the Reciprocal of Simpson's Index, or $1/\Sigma(\rho_i)^2$, where ρ represents the proportion of each prey type for array $_i$ in an assemblage (Simpson 1949; Levins 1968). In three treatments of the same data set in the following sections, the prey categories are manipulated and progressively simplified, from categories rooted strictly in biological systematics to those defined by independent energetic criteria such as prey body size and, among small-bodied prey, predator avoidance strategies and speed.

The Study Samples

The Mediterranean faunal series include a total of 32 assemblages (appendixes 15–17) from shelter sites in the Wadi Meged, Israel (200–11 KYA) (Stiner et al. 2000; Munro 2001), the western coast of Italy (110–9 KYA) (Stiner 1994, 1999), and the south-central coast of Turkey (42–17 KYA) (Güleç et al. 2002; Kuhn et al.

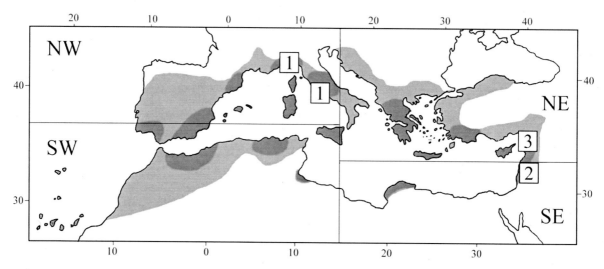

Figure 9.2
Geographical origins of the three Mediterranean faunal series: (1) western coast of Italy, with 16 assemblages; (2) Wadi Meged, inland Galilee of Israel, with 9 assemblages; and (3) Hatay coast of south-central Turkey, with 7 assemblages. Each faunal series originated in a distinct Mediterranean quadrant defined by somewhat distinct arrays of endemic species and environments. (Base map after Blondel and Aronson 1999:8.) Light shading represents modern Mediterranean vegetation; dark shading represents Pleistocene refugia.

1999, 2001; Stiner et al. 2002). NISP was the basic counting unit except for shellfish, for which MNI was used in order to render fragment sizes comparable to the invertebrate remains. Three quadrants of the Mediterranean Basin (fig. 9.2), each with its own set of endemic species (chapter 6), are represented. The faunal series coincide geographically with Mediterranean vegetation refugia during peak glaciations (see Blondel and Aronson 1999:28). Thus, the chances for stability in community composition in the study areas are as high as they can be for any Mediterranean ecosystem. The spectrum of animal taxa eaten by Paleolithic peoples in the Mediterranean Basin did not vary much: tortoises, marine shellfish, large legless lizards, ostrich eggs, hares, rabbits, and game birds such as partridges, in addition to ungulates such as deer, gazelles, goats, and wild cattle.

Time-averaging effects were a concern for analyses of faunal series in any one site, since evidence of trends was being sought. This problem was addressed in several ways, the main tools being rooted in a knowledge of the stratigraphy and sediment formation processes of the sites in question. An understanding of the faunal series from the sites in Italy and Turkey benefited from the fact that sterile or semisterile layers separated major cultural horizons there. In these cases, sorting discrete components from those that might not have been discrete was a relatively straightforward endeavor (for the Italian series, see Stiner 1994, 1999; Kuhn and Stiner 1992, 1998a). The data from Üçağızlı Cave in

Turkey are part of a new project and are as yet only partly published (Güleç et al. 2002; Kuhn et al. 2003, 2004; Stiner et al. 2002). Most of the cultural components in the cave, however, are clearly separated stratigraphically by sterile red clays (Kuhn et al. 1999, 2001). Hayonim Cave in Israel poses some challenges on the question of time averaging, because, although its major stratigraphic divisions (Layers B, C, D, E, and F) were relatively clear, its very thick Layer E was subdivided on the basis of subtler variations, a combination of sediment characteristics and variations in the vertical distributions of stone artifacts (chapters 1 and 2). One cannot claim that Mousterian (MP) stratigraphic units 1–7 represent wholly discrete events, but they certainly represent a coherent time series suitable for studies of long-term trends (chapter 4). Meged Rockshelter poses similar challenges (Kuhn et al. 2004), but at least two distinct lithic industries have been identified there from the morphology of cores and blanks and from tool frequencies.

Results as a Function of Prey Type Groupings

Application of the Reciprocal of Simpson's Index to assemblages that each potentially could contain about 20 Linnean genera yields consistently low levels of evenness in dietary breadth for the three faunal series (fig. 9.3). There is only a weak correlation between this index of evenness and time ($n = 32$, Pearson's $r = .386$, $r^2 = .15$, $p = .05$), and there is no correspondence

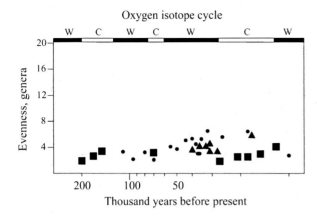

Figure 9.3
Evenness in the representation of Linnean genera in the faunal series from Italy (circle), Israel (square), and Turkey (triangle), using the Reciprocal of Simpson's Index (20 = most even). Time is expressed on a logged scale, as are oxygen isotope climate cycles (following Martinson et al. 1987; Shackleton and Opdyke 1973). *C* denotes a cold stage, and *W*, a warm stage. (Reproduced from Stiner 2001.)

between the index and the 6–7 climatic oscillations indicated by oxygen isotope data from deep sea cores (Martinson et al. 1987; Shackleton and Opdyke 1973). Although sample sizes vary among the assemblages (table 9.1), that does not explain the pattern in figure 9.3. This treatment of the data provides no support for the BSR hypothesis within the Paleolithic, just as Edwards (1989), Neeley and Clark (1993), and Horwitz (1996) observed from their studies of other Old World faunas. Most investigations of diet breadth changes have stopped here.

Regrouping the data into three categories based on a simplified combination of prey size and defense traits (large and small, and slow and quick, among the small types) reveals much temporal variation (fig. 9.4) but no overarching trend ($n = 32$, $r = .035$, $r^2 = .001$, $p = 1$). The Turkish series from the Hatay coast does trend toward greater dietary evenness from beginning to end, but perhaps only by virtue of its brevity. Cyclical rather than unidirectional change could be suggested in the sense that warm climate stages might increase evenness in human diets, independently of adaptation changes, by favoring the expansion of small species in animal communities. Figure 9.4, however, indicates both a lack of sensitivity to climate cycles and a tendency for minor increases in dietary breadth after 40–45 KYA. A shortage of data for 80 to 50 KYA from Israel, at the lowest latitude in the sample, is not necessarily to blame, because the exploitation of slow, collectable small game appears to have continued

through the late Middle Paleolithic in the Levant (e.g., Kebara Cave, Speth and Tchernov 2002, n.d.). Combining body size and defense traits in categorizing prey yields a provocative pattern, certainly more interesting than the one displayed in figure 9.3, but it also appears to conflate some crucial information that needs unpacking.

What about small animal exploitation? This was where Binford and Flannery expected to see the greatest changes in game use. Variation within the small game fraction of each series reveals a clear trend toward more even dependence on high-ranked and low-ranked small prey (fig. 9.5), confirmation of expanding dietary breadth during the later part of the Mediterranean Paleolithic. This regrouping of the data distinguishes only between slow-moving, easily collected types (tortoises and shellfish), fast-running mammals (mostly lagomorphs), and quick-flying game birds. The number of assemblages in the Mediterranean series that contain small game components large enough to be compared is reduced to 18 (table 9.1), with four from the early Middle Paleolithic of Italy collapsed into one to increase sample size. Removing large game from the comparison allows clear expression of expanding dietary breadth in small game exploitation ($n = 18$, $r = 0.606$, $r^2 = .37$, $p = .01$). Though not part of the statistical calculations, estimated values of evenness are shown for the segments of the late Mousterian sequence of Kebara Cave, based on general information provided by J. D. Speth.

Much of the dietary expansion took place during a cold climate stage (OIS 2). This is the opposite of what is usually expected to result from climate-driven changes in animal community composition, because the number of small animal species tends to be higher in warmer environments (Blondel and Aronson 1999; Pianka 1978). The evidence indicates a categorical, or stepwise, change in the way humans interacted with small animal populations after about 40–50 KYA. It reveals surprisingly little change in the way humans interacted with populations of large mammals.

IMPLICATIONS OF "SENSITIVE SPECIES" FOR PREY RANKING SYSTEMS

If different ways of categorizing prey in studies of Paleolithic diet breadth yield contradictory results, which approach is more appropriate? The answer depends on how we think foragers should have ranked prey according to expected energy returns. Linnean taxonomy is a

Table 9.1
Indexed proportions of animal prey in Mediterranean Paleolithic diets, 10–200 KYA

| Assemblage | Culture | KYA | Count | N-Genera | Index of Evenness | | | % | |
					Linnean	Size and Defense	Small Only, Defense	Ungulates	Slow Small Game
Italy									
R. Mochi A	EP	10	901	13	2.74	1.24	1.02	11	99
G. Polesini	EP	10	—	—	—	—	1.95	97	0
G. Palidoro	EP	15	—	—	—	—	1.39	99	0
R. Mochi C	EP	18	1,860	24	6.45	2.36	2.49	43	14
R. Mochi D	UP	26	1,549	23	5.58	1.94	2.74	67	19
R. Mochi F	UP	32	691	23	6.52	2.34	1.65	46	76
R. Mochi G	UP	35	860	22	5.27	2.16	1.85	32	71
G. Breuil 3/4	MP	36	351	10	3.01	1.00	—	100	—
G. Breuil br	MP	37	290	7	3.02	1.01	—	99	—
R. Mochi I	MP	38	335	9	4.49	1.02	—	99	—
G. St'Agostino 0	MP	40	771	10	5.34	1.07	—	96	—
G. St'Agostino 1	MP	44	771	10	5.06	1.05	—	97	—
G. St'Agostino 2	MP	50	355	10	3.72	1.03	—	98	—
G. St'Agostino 3	MP	55	164	9	4.09	1.05	—	98	—
G. Moscerini 1-2	MP	70	193	15	2.08	1.61	│	21	│
G. Moscerini 3	MP	80	471	16	3.23	1.65	│	73	│
G. Moscerini 4	MP	95	185	13	2.17	1.37	1.02	84	99
G. Moscerini 6	MP	110	237	14	3.33	1.66	│	73	│
Israel									
Hayonim B	EP	12	6,010	19	4.22	2.91	2.75	27	44
Hayonim C	EP	15	2,022	16	3.05	1.68	1.58	73	78
Meged Shelter	EP	18	1,063	12	2.68	2.19	2.04	51	65
Meged Shelter	UP	21	334	10	2.69	2.01	1.60	60	77
Hayonim D	UP	27	9,123	15	2.06	1.41	2.09	83	60
Hayonim E.1	MP	70	63	6	3.36	1.44	—	82	—
Hayonim E.2	MP	150	809	15	3.43	2.08	1.25	50	89
Hayonim E.3	MP	170	4,214	14	2.81	1.98	1.06	38	97
Hayonim E.4	MP	200	3,385	11	2.03	1.75	1.04	31	98
Turkey									
Üçağızlı Epi	EP	17	321	9	5.79	2.91	2.55	25	52
Üçağızlı B	UP	28	795	11	3.33	2.20	1.20	41	91
Üçağızlı B1-4	UP	31	1,247	11	3.33	2.22	1.20	42	91
Üçağızlı C-D	UP	32	40	7	4.33	1.34	—	85	—
Üçağızlı E-E2	UP	33	176	9	4.14	1.08	—	97	—
Üçağızlı F-F2	UP	36	146	9	4.13	1.11	—	96	—
Üçağızlı G-I	UP	41	265	9	3.60	1.04	—	98	—

SOURCES: Natufian data (Hayonim Layer B) are from Munro 2001; Aurignacian large mammal data (Hayonim Layer D) are from Rabinovich 1998.

NOTE: Counts are for specimens identified to the genus or finer taxonomic category only. Percentages of ungulates were calculated relative to total assemblage count; percentages of slow small game were calculated within the small game fraction only. Reciprocal of Simpson's Index ranges are 1–20 for Linnean categories; 1–3 for size and defense criteria of second and third indexes.

KEY: (MP) Middle Paleolithic; (UP) Upper Paleolithic; (EP) Epipaleolithic. Vertical lines indicate assemblages that were lumped to increase sample size.

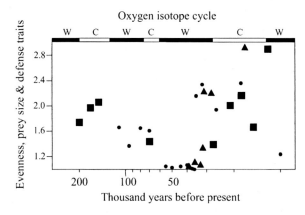

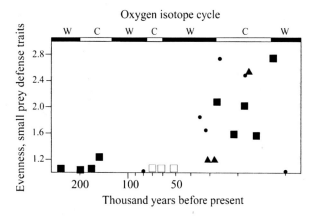

Figure 9.4
Evenness in the representation of three prey categories based on body size and defense mechanisms—large game, small slow game, and small quick game—using the Reciprocal of Simpson's Index (3 = most even). Symbols are as in figure 9.3. (Reproduced from Stiner and Munro 2002.)

Figure 9.5
Evenness among three prey categories within the small game fraction only, based on prey defense mechanisms—slow game, quick-running terrestrial mammals, and quick-flying birds—using the Reciprocal of Simpson's Index (3 = most even). Symbols are as in figure 9.3. Open squares represent estimates for the late Mousterian of Kebara Cave. (Reproduced from Stiner 2001.)

powerful tool in biology and zooarchaeology, not least because it reflects considerable agreement about what animals should be called and how they are related to one another phylogenetically. Foragers' perceptions of

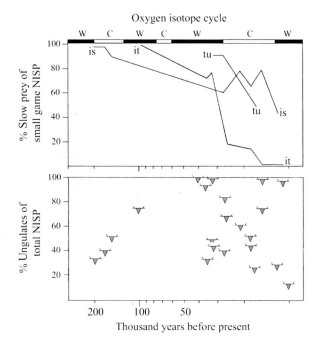

Figure 9.6
Trends in the percentages of slow small prey (lines) in the small game fraction of each assemblage from Israel (is), Italy (it), and Turkey (tu), together with ungulate remains (inverted triangles) in the total count of each assemblage. (Reproduced from Stiner 2001.)

prey, however, seldom if ever follow the rules of biological systematics.

Prey body size should be a valuable nontaxonomic criterion for ranking the potential returns of prey (see chapter 7), but it, too, has its limits because of the complications of capture costs and, in some cases, processing costs (Madsen and Schmitt 1998; Stiner et al. 2000). The large-to-small body size contrast in the three Mediterranean series, expressed as the percentage of ungulates in the total assemblage, is largely trendless (fig. 9.6; $n = 18$, $r = .276$, $p = 1$). In contrast, the proportion of slow animals within the small game fraction of each assemblage clearly declines with time ($n = 18$, $r = .572$, $.02 < p < .01$), the converse of which is increasing reliance on small quick animals. Prey body size must also have had economic significance, but it seems that the absolute differences in prey size were recalibrated to some extent by the differing capture costs among small prey animals.

The distinction between "backup" resources and total meat volume acquired is important from strategic and nutritional points of view. Returning to the topic of prey biomass of an earlier chapter, figure 9.7 reveals a decline in size across the entire prey spectrum in all three of the Mediterranean series. The timing of change and some of the dominant prey taxa differ among the regions considered. The trends are not gradual but rather represent dramatic shifts toward the end of each sequence as small game meat volume increases noticeably in the Epipaleolithic. But even before small game gained such prominence in forager diets, downward shifts took place within the ungulate body size continuum: species such as gazelles and roe deer gained greater importance before small game biomass contributions increased. The proportional contribution of ungulate biomass to Paleolithic diets was nevertheless consistently high (usually 97 percent or more) until late in each archaeofaunal sequence, when it fell abruptly to 83 percent or lower (fig. 9.8).

Detailed information about the broad types of small animals exploited indicates that significant changes took place well before the Epipaleolithic period in each region, most obviously the greater use of quick-moving, quick-reproducing small prey with the onset of the Upper Paleolithic. The biomass-corrected counts across the entire prey body size spectrum suggest greater use of faster-reproducing small ungulates, in addition to lagomorphs and game birds, over time. Unfortunately, trends in the ungulate data are a good deal subtler than the signals of economic and strategic change in the small game data.

Experimenting with small prey categories, albeit with attention to independently documented characteristics of the subject species, allows the evidence for early increases in diet breadth to spring into focus. Independent standards for prey classification in the study were isolated from wildlife data and linked to demographic increase by predator-prey simulation modeling and examination of diminution effects in certain slow-growing species (chapter 8). That small prey animals differ tremendously in their development rates also permits an unusually clear view of the way increases in Paleolithic dietary breadth shifted with local demographic growth. This is not simply a matter of how much small game animals contributed to total game intake; it is more a matter of how certain sensitive species serve as symptoms of threshold effects in predator-prey systems—like fume-sensitive canaries carried into coal shafts by nineteenth-century miners.

EVIDENCE FOR THE BROAD SPECTRUM REVOLUTION

The Mediterranean Basin is a textbook case of high diversity in animal and plant communities. This quality changed little over the course of the Pleistocene in comparison with the rest of Europe and northern Asia (e.g., Blondel and Aronson 1999). The Mediterranean Basin is thus an ideal laboratory for testing ideas about dietary expansion in human evolution, with more

small species to choose from and a high potential for shifting predator-prey dynamics to be expressed in small game data.

Contrary to the results of prior studies, the data on small game use in southern Europe and western Asia during the late Pleistocene support Flannery's broad spectrum revolution hypothesis of expanding dietary breadth in response to demographic packing. The findings for the three Mediterranean faunal series, however, present an interesting twist to Flannery's original predictions with respect to timing. It does seem that the BSR began in the eastern end of this vast region, but the earliest evidence of dietary expansion and demographic pulses is associated with the emergence and spread of the earliest Upper Paleolithic cultures from Asia into Europe, the same basic path as the spread of Neolithic adaptations after 10 KYA (Ammerman and Cavalli-Sforza 1984; Hewitt 2000; Reich and Goldstein 1998). The dietary shifts identified by Binford and Flannery between 12 and 8 KYA were merely the last in a longer series of economic changes. The fact that these changes began earlier in the eastern Mediterranean Basin than at its northern and western ends reinforces the likelihood that prehistoric human populations were largest in the semiarid subtropical to tropical latitudes of Asia and Africa (Harpending and Bertram 1975; Keeley 1988). Demographic pulses emanated from southwestern Asia into Europe several times.

Demographic forces were part of the evolutionary arena within which changes in human dietary breadth and society occurred during the Late Pleistocene. This observation need not represent backpedaling to the days of simplistic or linear explanations of population pressure as the engine of culture change. It simply shows us that demographic processes were somehow part of the substrate upon which selective factors may have operated on human societies, regardless of whether the question is about the emergence of Upper Paleolithic lifeways, social complexity in hunter-gatherers, or the eventual origins of plant cultivation and animal husbandry. In most periods, human populations adjusted to environmental stresses in a classically demographic way, without substantive changes in behavioral adaptations. In rare instances, there is evidence of behavioral evolution. Although the zooarchaeological data on diet breadth do not by themselves explain unprecedented human responses to declining resource availability or why human populations were increasing, they greatly refine the questions we can ask of archaeological records.

Figure 9.7
Percentages of total prey biomass obtained by Paleolithic hunters across periods for size-ordered prey species in three Mediterranean faunal series from Israel, Italy, and Turkey. Key: (u) total ungulate percentage; (sg) total small game percentage. Mochi A in the Italian series represents an extreme situation but is still fairly typical of coastal occupations for the period.

Any factor that reduces the seasonality or unpredictability of food supplies for human foragers can lead to a substantial population increase. Perhaps one early shift in this regard occurred through the use of a wider range of meat sources, some of which were a good deal more resilient than others in the face of heavy exploitation. Of course by the later Upper Paleolithic and Epipaleolithic, food storage (plants in lower latitudes, but also dried or smoked meat and rendered fat in other areas) was also part of the subsistence equation. The latter development is quite familiar to archaeologists who work on Late Pleistocene cultures. But returning to the first point, an interesting quality of certain warm-blooded small prey populations that rebound quickly is their greater reliability as a food source, if capture costs can be reduced artificially through technology.

A forager population that can grow faster on lower-value but more resilient food species will have a demographic advantage over competing populations. In the late Pleistocene this involved a subtle lowering of humans' position in the food chain in some arid, low-latitude regions, largely because greater plant use was also part of the BSR. Large-seeded plants permit more direct access to primary production, despite higher collection and processing costs, and thus may support humans at higher population densities (Harpending and Bertram 1975; Keeley 1988). Small quick animals may present this possibility to a lesser extent, but toward the same end and often complementary to the intensified use of plant seeds in the absence of domesticated ungulates. A parallel issue is the acceleration in social and technological complexity among Paleolithic cultures that seems to have accompanied demographic packing (Keeley 1995; chapter 12). The early indications of expanding diets in the eastern Mediterranean seem to precede rather than follow the evolution of the kinds of tools (specialized projectile tips, nets, and other traps) needed to capture quick small animals efficiently (Kuhn and Stiner 2001; Stiner et al. 2000).

SUMMARY

Formal dietary breadth models hold that resource-ranking systems will vary with the type of consumer. Because game animals are composed of similar tissues, simple body size gradients are widely used for ranking taxa in zooarchaeological research on subsistence evolution, and the ranks are used in conjunction with numbers of Linnean taxa. Oddly, neither approach exposes significant changes in diet breadth in the Pale-

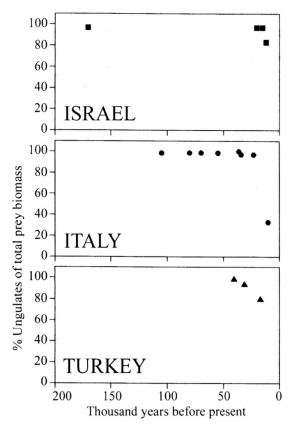

Figure 9.8

Percentages of total ungulate biomass obtained by Paleolithic hunters across periods in each of the three Mediterranean faunal series from Israel, Italy, and Turkey.

olithic sequence of Mediterranean Eurasia, despite radical transformations in other aspects of the archaeological record. The interplay of archaeological anomalies and wildlife data for disparate Mediterranean faunal series reveals that absolute differences in prey size often were qualified from the foragers' point of view by the costs of capturing certain kinds of small prey. It is striking that most of the shifts in predator-prey dynamics noted here concern small animals, not big ones. These insights into prehistoric diet breadth could not have been gained without careful control of other potential sources of variation in archaeofaunas, such that other explanations could be excluded systematically. The spatial distribution of sites served as a general but essential constant in this study.

Slow-growing, slow-moving tortoises and marine mollusks dominate the small game fractions of the Mediterranean Middle Paleolithic record, constituting up to half of all identified specimens in several of the assemblages. The timing and duration of associated diminution effects in terrestrial tortoises and marine limpets are largely independent of global climate trends—the other potential cause of diminution—and thus point to a human cause. With evidence of harvesting pressure on "low-turnover" prey populations during the later Paleolithic comes evidence for a corresponding increase in the exploitation of agile, warm-blooded small animals, mainly lagomorphs and birds such as partridges, populations that can rebound rapidly from heavy hunting by humans. Paleolithic foragers' emphasis on slow (highly ranked) and quick (lower ranked) small prey grew more "even" with time, the predicted outcome of hunting pressure and demographic increase in the absence of a correlation with climate warming. The pattern of expansion in dietary breadth is directional and covers many thousands of years and large areas of the Mediterranean Basin, indicating changes in the nature of human adaptations rather than simple adjustments to environmental variation. This process of change also began a good deal earlier than previously supposed, with the earliest Upper Paleolithic cultures known in the eastern Mediterranean Basin.

CHAPTER TEN

Food Utility, Attrition, and Transport of Ungulate Body Parts

Pᴿᴇʏ ʙᴏᴅʏ ᴘᴀʀᴛ ʀᴇᴘʀᴇꜱᴇɴᴛᴀᴛɪᴏɴ ᴘʀᴏᴠɪᴅᴇꜱ ᴀ ʙᴀꜱɪꜱ for comparing the transport strategies and contexts of food consumption by humans at different times and places. Interspecific comparisons have previously shown that the spatial distribution of carcasses on a landscape and the nutritional quality of body parts either limit or foster a predator's strategic emphasis on hunting, confrontational scavenging, or itinerant scavenging in the context of gathering (Stiner 1991b, 1994). Much discussion has been devoted to the way Mousterian hominids obtained large game—mainly, whether they hunted it in the sense in which later foragers certainly did or whether they relied principally on scavenging. This dichotomy is too artificial to be of much heuristic value, because most predators engage in both behaviors. It is fair, however, to suggest that more than one kind of predator-prey relationship could have evolved among divergent hominid populations, as they have evolved among other predator species, and that evidence of food transport strategies might reflect differences in adaptation.

Bone transport is most interesting for what it might say about how prehistoric people staged carcass dismemberment and processing across landscapes and seasons. This point has been demonstrated most vividly by Binford's (1978) ethnoarchaeological study of the Nunamiut in the high arctic, by Yellen's (1977, 1991a) research among the Bushmen in the arid subtropics, and, more recently, by the work of two research teams among the Hadza (Bunn et al. 1988; O'Connell et al. 1988a). There are many potential ways to butcher and distribute meat from large prey, depending on the contingencies of social group size, food dispersion, aridity, availability of fat in the environment, and so on.

In this chapter I address the extent to which nonhuman biases and human-induced biases shaped the patterns of ungulate body part representation observed in the Paleolithic faunas of the Wadi Meged. The analyses emphasize a body part profiling technique I developed using "anatomical regions" (Stiner 1991a). As demonstrated in what follows, this profiling technique removes most of the potential distortion that could arise from density-mediated bone attrition (Stiner 2002b). I also examine portion-by-portion patterns according to the recommendations of Lyman (1994) and Grayson (1989). Body part representation is considered against published food utility indexes for meat, marrow, and grease content (sensu Binford 1978) and relative to variation in structural density. Last, the implications for foraging strategies are discussed. The Wadi Meged assemblages used in this study, as well as the comparative assemblages from other Mediterranean regions, came from shelter sites. This fact helps to limit the potential contexts of assemblage formation, in that body parts must have been carried into the sites from other locations. Nonetheless, potential conflicts of interpretation remain, because much can happen during and after the deposition of bones in a shelter.

IN SITU ATTRITION AND BODY PART PROFILING TECHNIQUES

Bone destruction from mechanical processes such as crushing in sediments, ravaging by carnivores, and marrow processing by humans is thought to be conditioned by the "structural density" of bone tissues (*sensu* Lyman 1994:235–238), principally the mineral component. Human-collected faunas tend to be highly fragmented, so skeletal element counts must be estimated from the frequencies of unique morphological features that can be recognized from partial specimens, such as the head of a femur, the nutrient foramen of a humerus, an occipital condyle of a cranium, a prezygopophysis of a lumbar vertebra, or the medial face of the distal epiphysis of a tibia. Because the macrostructure of mammal skeletons, particularly those of large mammals, is heterogeneous, the many identifiable components of a skeleton may not resist decomposition forces equally well. Zooarchaeologists therefore work in steps, beginning with questions about the agents of bone collection, modification, and destruction. Later, and assuming that biases introduced by nonhuman agents can be excluded or controlled for, analysts may take on questions about prehistoric human behavior.

Attempts to infer how prehistoric foragers handled carcasses are made difficult by the potentially equifinal consequences of other processes, particularly density-mediated bone attrition. Showing that density-mediated attrition could account for variation in skeletal representation is important to the analytical process, but it does not demonstrate that postdepositional attrition was the sole or even a partial cause of anatomical biases (Beaver 2004; Lyman 1994). A positive correlation between skeletal survivorship and bone density points only to the possibility of two separate causes for anatomical biases in an assemblage, causes that might have acted alone or together. Fine-scale comparisons, such as between large and small taxa in the assemblage (Munro 2001) or among spongy tissues distributed within the skeletons of prey, are needed to resolve the matter, if it can be resolved at all.

Techniques for analyzing body part profiles on the basis of fragmented faunal material must either address the differential survivorship of the full range of structural density classes in the skeleton or stick to comparisons of parts that fall within a narrower density range that is widely represented in the vertebrate skeleton. Both kinds of approaches have found their way into the zooarchaeological literature, usually for different applications by different research groups. The term "body part profiling" refers to almost any systematic comparison of anatomical representation in archaeofaunal assemblages to independent standards based on a natural skeleton model. The results of body part analyses can be portrayed in many ways, ranging from standardized bar charts to anatomical indexes and correlation statistics. It is, however, the *relation* between the archaeofaunal body part representation and an accurate anatomical model that determines the reliability of the profiling approach.

COMPARISONS ACROSS MAJOR BONE MACROSTRUCTURE CLASSES

The more common body part profiling approach requires reasonably complete knowledge of structural variation throughout the vertebrate skeletal anatomy. Skeletal tissues normally are grouped on the basis of bulk density (= fragility) into major macrostructure types—compact bone, cancellous bone, tooth enamel, dentine—and researchers focus their classification efforts on recognizable portions of elements composed of these macrostructures that may persist in fragmented assemblages. This practice sounds simple in principle but proves to be complex. Currently there is much variation among investigators in their working characterizations of structural density, the structural scale at which they model resistance to destruction, and the anatomical standards they use as controls.

The extent to which skeletal density explains the loss of bone recognizability as defined by zooarchaeological practice is unknown. Many investigators nonetheless have demonstrated a relation between some measure of bone density and observed biases in vertebrate body part representation in faunal assemblages (e.g., Behrensmeyer 1975; Binford and Bertram 1977; Brain 1967, 1969, 1981; Lyman 1984). It is widely assumed, therefore, that variation in bone macrostructure has some explanatory power in questions of skeletal survivorship (for a thorough review, see Lyman 1994:235–293). Considerable information exists on the mineral, or "structural," density of the skeletons of many vertebrates, but two very different sets of standards have been proposed (compare Lyman 1984 and Lam et al. 1998). In what follows, I use Lyman's (1984) photon densitometry standards, noting that measurements of bone density by this technique are generally analogous to the *unadjusted* measurements obtained by Lam et al. (1999, BMD_1 as opposed to BMD_2) using the computed tomography technique (Stiner 2004).

The contrast between compact and cancellous (spongy) macrostructures of bone has received the most attention, and skeletal survivorship in these two macrostructure classes normally is compared to assess the potential severity of in situ attrition (e.g., Grayson 1989; Lyman 1984; Rogers 2000). The often significant correlations between bone mineral density and bone survivorship are compelling, but wedge-shaped distributions of points when body part survivorship is plotted against structural density (Beaver 2004) testify to the fact that one cannot fully distinguish between human economic causes and density-mediated attrition.

COMPARISONS WITHIN A SINGLE BONE MACROSTRUCTURE CLASS

Given that the relations among observable types of skeletal macrostructures, countable morphological features, and the processes by which skeletal structures break down are still something of a mystery, profiling techniques that rely on fewer assumptions about how all of this works have some obvious benefits. I developed the second approach to body part profiling in order to address questions about niche evolution in Pleistocene humans and the comparative ecology of food transport and processing behaviors of several species of ungulate predators (Stiner 1991a, 1994). The technique has come under fire recently, mainly with the claim that differential attrition resulting from variations in mineral density may so thoroughly obscure the original (culturally determined) patterns in ungulate body part representation that any research findings based on the technique are meaningless (Bartram and Marean 1999; Grayson 1996; Marean and Kim 1998). The next section offers a systematic review of the strengths and limitations of the "anatomical regions" technique. By applying most or all of the same control data that have been said to invalidate it, I hope to show where critics have gone wrong.

This approach is distinct from the "cross-macrostructure" technique described earlier in that it relies principally on compact bone, which is widely distributed throughout the vertebrate skeleton. The idea

is to control for the possible effects of density-mediated bone attrition on estimates of the minimum number of skeletal elements (MNE) by narrowing the tissue density range to features dominated by compact bone. Teeth, the densest of all skeletal elements because of the exceptionally high degree of mineralization of the enamel (ca. 95 percent), are confined to the skull. Spongy bone structures (a.k.a. cancellous or trabecular bone) dominate much of the axial skeleton, making axial members the least dense, or "least resistant," of elements in general. Compact bone is prevalent in the skeletons of all terrestrial mammals and birds. Limb elements such as the humerus, radius, femur, tibia, and metapodials contain large tracts of compact bone (figs. 10.1 and 10.2), as do some of the nondental components of the skull such as the petrous complex. Certain smaller features of the vertebrae (zygopophyses) and ribs (proximal heads) are also fairly dense, though less so than many limb and cranial features.

The thickness of compact bone varies among skeletal elements, but this variation is of a smaller order than that observed between compact bone and the other skeletal tissue classes named earlier. One simply needs to know the locations of compact bone tissues relative to the distribution of the unique morphological features ("portions") normally used to estimate MNE for each kind of element. Available literature suggests that there is enough information with which to do this and that a reasonably good correspondence exists between visible and measured variation in the distribution of compact bone in ungulate skeletons (e.g., Elkin and Zanchetta 1991; Kreutzer 1992; Lyman 1984, 1994; Lyman et al. 1992).

THE "ANATOMICAL REGIONS" PROFILING TECHNIQUE

An anatomical-region–based approach to ungulate body part representation circumnavigates much of the variation in mineral density within skeletons by focusing on compact bone. Pooling MNE counts by anatomical region evens out variation in structural density further still. It should be noted, however, that the technique is directed toward interassemblage comparisons from which only the most robust differences in patterns of body part representation are sought.

The minimum number of elements (MNE) is estimated for each skeletal member of a given taxon on the basis of the most common morphologically unique "portion" or feature in the assemblage. Some portions

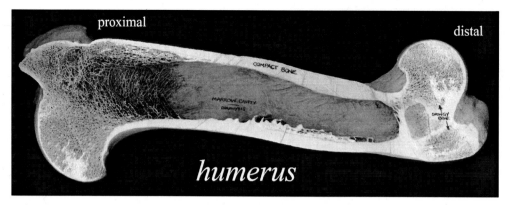

Figure 10.1
Cross-sectioned humerus of an artiodactyl ungulate, showing the distribution of compact and spongy macrostructures along the shaft (diaphysis) and ends (epiphyses). The proximal epiphysis of this element is dominated by spongy bone, but the distal end and the shaft are dominated by compact bone. Oblique pilastering and the internal outlet of the foramen are visible inside the medullary chamber.

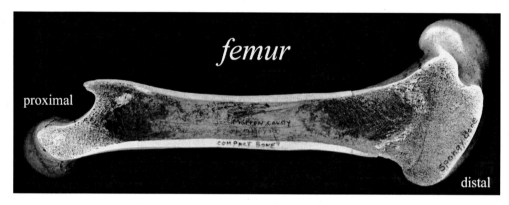

Figure 10.2
Cross-sectioned femur of an artiodactyl ungulate, showing the distribution of compact and spongy macrostructures along the shaft (diaphysis) and ends (epiphyses). The proximal and distal epiphyses of this element are dominated by spongy bone. Smooth walls and the tubular outlet of the foramen (with wire insert) are visible inside the medullary chamber.

tend to yield higher counts than others, presumably because of their greater inherent resistance to mechanical destruction. Limb end (epiphyseal) and shaft features (e.g., foraminae, see fig. 5.5) are considered in MNE estimates (appendix 2), and all fragments in an assemblage, including limb shaft splinters, are examined systematically. For the skull, only bony portions are used in comparisons with postcranial MNEs, because tooth enamel is so much denser than any kind of bone (Currey 1984; see chapter 3). Small, compact features are favored for counting, and many of these portions coincide with Lyman's photon densitometry scan sites (1994:234–250; see table 5.1).

The MNE counts are condensed into nine anatomical regions (fig. 10.3): (1) the horn/antler set, (2) the head, (3) the neck, (4) the rest of the axial column including the ribs and pelvis, (5, 6) the upper and lower front limbs, (7, 8) the upper and lower rear limbs, and (9) the feet (Stiner 1991a, 1994:240–245). Species-specific identifications are pooled with more general identifications of specimens of the appropriate ungulate body size group in order to increase sample size and overcome the fact that some elements and portions of elements are more diagnostic of taxon than are others. For the purposes of bar chart comparisons, body part representation can be standardized against a whole skeleton model (Stiner's [1991a] standardized

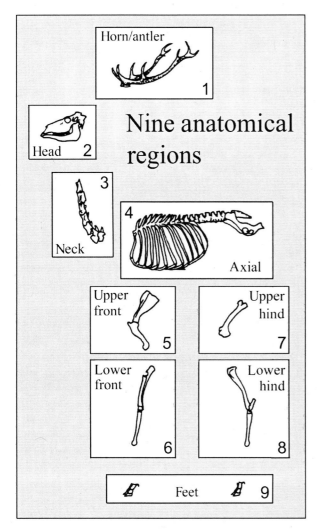

Figure 10.3
Nine anatomical regions for the ungulate skeleton.
(Reproduced from Stiner 1991a.)

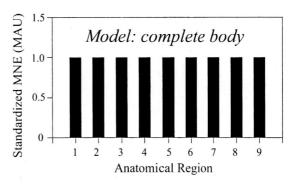

Figure 10.4
Standardized bar chart for the nine anatomical regions of a theoretically complete skeleton: (1) antler/horn, (2) head, (3) neck, (4) axial skeleton, (5) upper front limbs, (6) lower front limbs, (7) upper hind limbs, (8) lower hind limbs, (9) feet. Dental elements are not used to calculate the frequencies of head parts. Standardized MNE (observed MNE divided by expected MNE for one complete skeleton) is equivalent to Binford's (1978) MAU. (Reproduced from Stiner 1991a.)

TEST OF DENSITY RANGE CORRESPONDENCE AMONG ANATOMICAL REGIONS

How much variation in observed MNE values among anatomical regions in the density control data can be explained by variation in the density of the bone portions used for counting? This question was addressed through a statistical comparison that considered all the possible portions listed in appendix 2 for which photon densitometry control data were available, irrespective of measured density. I then narrowed the list of variables to include only those portions most commonly represented in my prior analyses of diverse Mediterranean faunas (Italy and Israel) from 1985 to the present. Control data for deer served as the skeletal density standards (Lyman 1994:table 7.6), because intraskeletal variation seems to follow a similar pattern in artiodactyl and perissodactyl species (Lam et al. 1999).

Density value midpoints and ranges in the control data (fig. 10.5) and pairwise statistical comparisons of density values (table 10.1) indicate that the chances for reduced recognizability of bone portions are about the same for the head region and various limb regions in the ungulate skeleton. An F-ratio statistic confirms that there are no major differences among the pooled cranial, limb, and foot regions ($n = 32$, $r^2 = 0.27$, $p = 0.124$). Upper front limbs and foot bones have a somewhat lower probability of preservation than heads and other limb regions (table 10.1), but these differences are minor (table 10.2).

MNE, equivalent to Binford's [1978] minimum animal unit, MAU) by dividing the observed MNE for a skeletal element or group of elements by the expected MNE for the same element or group in one complete skeleton. If skeletal representation is complete, then standardized values for all body regions will be equal (fig. 10.4), making major anatomical biases among regions easy to detect. The observed MNEs for anatomical regions can also be indexed relative to one another, and the total element count (tMNE) can be compared to the number of individual animals (MNI) in the ratio tMNE/MNI (Stiner 1991a).

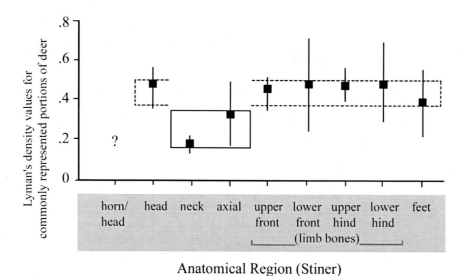

Figure 10.5
Ranges and midpoints of variation in bone structural density for nine anatomical regions of the artiodactyl skeleton, using Stiner's (1991a) profiling method and Lyman's (1994) photon densitometry data for deer. (Reproduced from Stiner 2002b.)

Table 10.1
Pairwise differences in density between deer cranial, limb, and foot regions

Anatomical Region Pair	t	df	p	Difference in Means
Head–upper front limb	2.251	9.0	0.051	0.138
Head–lower front limb	−0.347	8.7	0.737	−0.023
Head–upper hind limb	1.075	2.9	0.362	0.100
Head–lower hind limb	−0.037	9.3	0.972	−0.003
Head–feet	2.292	7.3	0.054	0.158

Table 10.2
Mean photon densitometry values for bony portions of deer cranial, limb, and foot regions

Anatomical Region	No. Portions Considered	Mean Density	SD
Head	5	0.52	0.09
Upper front limb	6	0.38	0.11
Lower front limb	6	0.54	0.13
Upper hind limb	3	0.42	0.14
Lower hind limb	7	0.52	0.18
Feet	5	0.36	0.12

SOURCE: Lyman 1994.

The chances for reduced recognizability among cranial and limb regions are closer still for the portions I most commonly used to estimate MNE in archaeo- logical assemblages from Mediterranean shelter sites ($n = 23$, $r^2 = 0.330$; F-ratio = 1.671, $p = 0.195$) (table 10.3). A more stringent nonparametric version of

Table 10.3
Pairwise differences in mean structural density for bone portions by anatomical region in the Mediterranean cave faunas

Anatomical Region and Code	Anatomical Region					
	2	5	6	7	8	9
All portions for which density parameters were available						
Head (2)	—					
Upper front limb (5)	−0.138	—				
Lower front limb (6)	0.023	0.162	—			
Upper hind limb (7)	−0.100	0.038	−0.123	—		
Lower hind limb (8)	0.003	0.141	−0.020	0.103	—	
Feet (9)	−0.158	−0.020	−0.181	−0.058	−0.161	—
Portions most commonly used to estimate MNE						
Head (2)	—					
Upper front limb (5)	−0.114	—				
Lower front limb (6)	−0.029	0.085	—			
Upper hind limb (7)	−0.067	0.048	−0.037	—		
Lower hind limb (8)	−0.033	0.082	−0.003	0.034	—	
Feet (9)	−0.195	−0.080	−0.165	−0.128	−0.162	—

NOTE: Anatomical regions are numbered as in figure 10.3. None of the relations between region pairs is statistically significant.

ANOVA, a Kruskal-Wallis statistic, yielded essentially the same answer as did the tests already described (8.393, df = 5, $p = 0.136$).

The conceptual basis for the profiling technique, therefore, is well supported by Lyman's and others' estimates of variation in bone structural density. The risks of overinterpretation in this technique actually center on the vertebral column ("neck" and "axial skeleton"), anatomical regions that are deemphasized in this discussion on grounds that their representation is likely to be ambiguous. Questions about why vertebrae are underrepresented in some assemblages must be addressed in other ways (see chapter 5). The relative abundances of cranial and major limb bones are suitable for investigating the food acquisition strategies of humans and other ungulate predators. The anatomical index (H + H)/L (Stiner 1991a, 1994)—the sum of horn-antler MNE and head MNE divided by total MNE for major limb elements excluding phalanges ("feet")—and a related index, HEAD/L, are also largely unaffected by potential variation in bone tissue density.

The assertion that head-dominated patterns or head-and-foot-dominated patterns in ungulate remains are likely to be the products of density-mediated attrition is not supported by documented variation in structural density in the skeletons of vertebrates (*contra* Marean and Kim 1998). If anything, foot bones as a group are slightly less dense than the countable compact bone portions of the head and limb groups (table 10.2), and the latter regions are composed of many portions of roughly equivalent densities. None of the differences in bone density of head and limb regions is of the order needed to greatly bias MNE estimates in the region-by-region profiling technique.

To summarize so far, portions of elements composed of compact bone can be thick or thin, but many of the components of the skull and limbs have relatively similar chances of resisting mechanical sources of in situ destruction. Anatomical indexes that compare head to various limb region frequencies in the manner just described are largely immune to the biasing effects of density-mediated attrition. Certain other anatomical indexes may be affected more, such as the horn-antler index HORN/L, because they combine skeletal elements representing a greater range of tissue densities. Yet horn-antler biases abound in modern hyena dens of Africa and Pleistocene dens of Italy, despite the spongy macrostructure of bovid horn cores and cervid antlers (Stiner 1991a, 1994).

RAMIFICATIONS

There is a considerable fallacy in the assumption that limb bone shafts are more persistent in Paleolithic archaeofaunas than either end of the same bones

(Marean and Kim 1998). Most indications are that mean shaft persistence is about the same or nearly the same as that of one end but often greater than that of the other end of the element. The maximum loss differential between soft ends and harder portions of limb bones appears to be 2:1 or at most 3:1 (compare, for example, cases in Lyman 1994; Stiner 2004). Photon densitometry data predict this fairly well. Perhaps with the advent of milling and boiling technologies in later archaeological periods, spongy bone fared much worse than compact bone (e.g. Bar-Oz and Dayan 2002; Brink 1997; Munro 2001), but this situation does not apply to the Lower, Middle, and much of the Upper Paleolithic in this study.

Marean and Kim (1998) argued that two other sources of bias might work together to obscure archaeologists' perceptions of prey body part representation. These sources are selective deletion by hyenas or other carnivores and human observer error arising from the practice of ignoring shaft fragments during data collection. The latter is irrelevant to this study, because all specimens recovered from the sites were examined. The selective deletion concept is based on actualistic experiments, some of them ethnoarchaeological and others experiments in which investigators break and discard bones as they think early hominids would have done in open settings and then observe what local predators (mainly hyenas) carry off. Hyenas' choices are said to follow the compact-cancellous tissue distinction for prebroken bones; when bones are broken and still greasy, hyenas are attracted to spongy parts (Bartram and Marean 1999; Capaldo 1997; Lupo 1995; Marean et al. 1992). Not all limb bone ends are spongy or greasy, however; some are quite dense, such as the distal tibia and humerus and the proximal radius and metapodials (see figs. 10.1 and 10.2). From a paleontological point of view, at least for the Mediterranean region, where spotted hyenas were once common, there are no indications that hyenas spirited off large or disproportionate quantities of limb bone ends of all sorts from human sites. Instead, there is good quantitative agreement between MNE estimates for the shaft and at least one end of most limb bone elements (chapter 5; Stiner 2002b).

The larger point is that the experiments cited by these authors explain, at most, missing portions of particular elements. They explain less about why certain elements are missing and much less about disproportionate representation of most regions of the ungulate skeleton other than the vertebral column. Anatomical regions are arguably closer approximations than portions of elements to what people and other large predators generally decide to carry away from procurement sites, outside of some rather special foraging circumstances.

BODY PART PROFILES FOR THE WADI MEGED UNGULATES

The ungulate body part profiles for the Wadi Meged series are shown in figures 10.6–10.8 (data in appendixes 18–29). The profiles are generally similar to one another irrespective of culture period: one sees a close-to-natural balance in the representation of head and limb bones in the Natufian, Kebaran, pre-Kebaran, and Mousterian periods. Vertebral elements, particularly below the neck region, are underrepresented, as one might expect from their greater inherent vulnerability to mechanical destruction of any sort, although differential transport is not refuted. Indeed, the observed incidence of neck and axial elements often falls well below the predicted level of susceptibility for backbone elements shown in figure 10.5. This pattern occurs in several assemblages for which density-survivorship comparisons indicate no significant correlation. What is more, table 5.20 (chapter 5) shows that vertebrae are no more or less fragmented across periods or taxa on the basis of the MNE/NISP ratio. Some of the explanation for the low frequencies of neck and axial bones in the Wadi Meged ungulate assemblages therefore must lie in prehistoric humans' decisions not to carry them to the shelters.

Foot bones are also underrepresented in many of the Wadi Meged assemblages, especially in those dating to the Mousterian period. Foot bones should be only slightly more susceptible to mechanical destruction than the major elements of the four limb regions, judging from skeletal density reference data (fig. 10.5). Yet because the Epipaleolithic and Upper Paleolithic assemblages tend to be somewhat less well preserved than the Mousterian assemblages (chapter 5), the lower frequencies of foot bones in the Mousterian assemblages represent the opposite of what poorer preservation would predict.

Variation among the four limb regions of ungulates in the assemblage-by-assemblage comparisons seems nearly random (figs. 10.6–10.8) and may not be very meaningful given the small sizes of some of the assemblages. However, the composite results for nonvertebral anatomical regions presented in figure 10.9 indicate mild overrepresentation of limb elements and foot bones in the Epipaleolithic and Upper Paleolithic periods. As before, foot bones appear to be

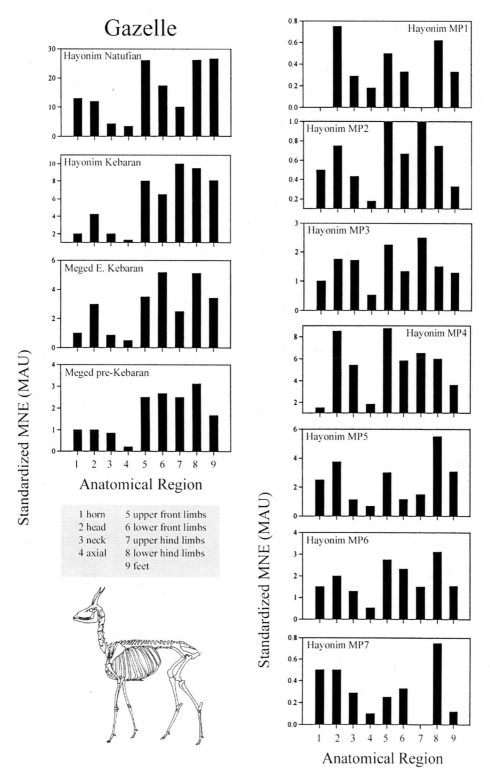

Figure 10.6

Gazelle body part profiles for Natufian, Kebaran, pre-Kebaran, and seven Mousterian assemblages in the Wadi Meged faunal series.

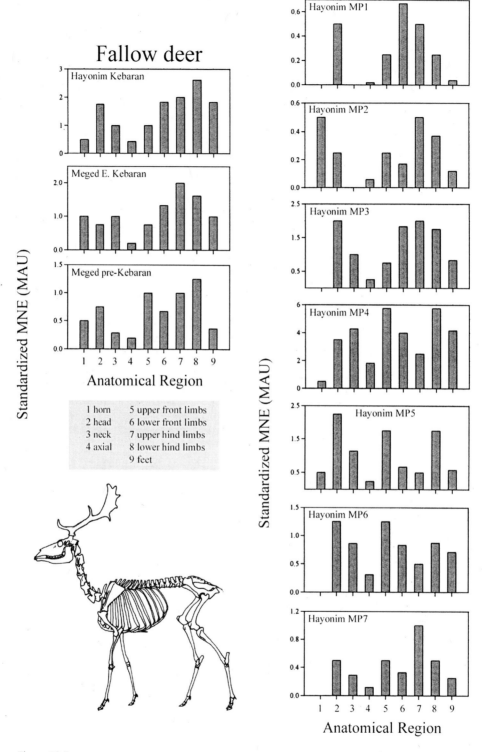

Figure 10.7
Fallow deer body part profiles for Natufian, Kebaran, pre-Kebaran, and seven Mousterian assemblages in the Wadi Meged faunal series.

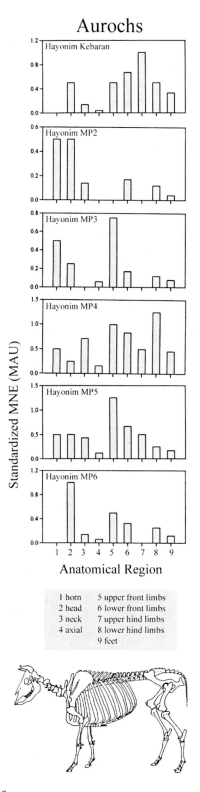

Aurochs

Standardized MNE (MAU)

Anatomical Region

1 horn	5 upper front limbs
2 head	6 lower front limbs
3 neck	7 upper hind limbs
4 axial	8 lower hind limbs
	9 feet

Figure 10.8
Aurochs body part profiles for Natufian, Kebaran, pre-Kebaran, and seven Mousterian assemblages in the Wadi Meged faunal series.

underrepresented in the Mousterian relative to their expected durability, whereas the incidence of head and major limb bones is nearly balanced. Finally, figure 10.9 shows that there are fewer differences among ungulate body size groups within each major culture period than there are for any ungulate size group between the periods.

Comparisons of bone-based and tooth-based MNE estimates for the head region, described in chapter 5 (see fig. 5.4), indicate close agreement between the two, although bone-based counts are somewhat lower, with slopes of 0.882, 0.559, and 0.832 for the Natufian, Epipaleolithic–Upper Paleolithic, and Mousterian periods, respectively. Teeth are the densest elements of the skeleton, the most easily identified, and the most resistant to decomposition, yet head counts in the Wadi Meged assemblages, whether based on teeth or bones, are significantly outnumbered by counts based on certain limb elements, whose mineral density is less than that for any fully formed tooth element. The underrepresentation of head parts generally is not an artifact of poor preservation—and in any case, heads should be as persistent as limbs (fig. 10.5). The mild limb bias therefore bears a human signature: certain limb elements were brought to and deposited in the shelter somewhat more often than head parts and a good deal more often than foot bones and axial vertebrae.

Another question is whether the mandibles and crania of prey were separated at acquisition sites and transported to different extents. This can be tested through simple comparisons of element abundance (MNE) by ungulate body size group. Mandibles are somewhat more prevalent than crania in the Wadi Meged assemblages in general (table 10.4), indicating some transport bias. However, there is considerable variation among the assemblages, some of which is not explained by differences in sample sizes. Specifically, the number of mandibles well exceeds the number of crania for small and medium ungulates in the pre-Kebaran assemblage from Meged Rockshelter and in the Mousterian assemblage from MP unit 3 of Hayonim Cave. Imbalances in the numbers of crania and mandibles for the large ungulates display the opposite pattern, with crania outnumbering mandibles in Mousterian (MP) units 4 and 6–7. In the other assemblages, the ratio of crania to mandibles appears to be balanced, irrespective of ungulate body size, which is consistent with transport of entire skulls to the shelter sites. The variation in the relative representation of crania and mandibles is not explained by culture period. Rather, it seems to reflect fine-scale circumstantial (and unknown) variation in foraging conditions.

Head parts are relatively more common in the Mousterian assemblages than in assemblages dating to the later Paleolithic periods—that is, they approach the closest natural balance with limbs. Head parts are less well represented in the Kebaran and Natufian assemblages and may be as few as half the number expected in relation to limb elements. Interestingly, gazelle horn counts are on a par with head counts only for the Natufian, but head parts are much less common than postcrania, suggesting a tendency to preferentially transport only adult male skulls to Hayonim Cave during that period (Munro 2001:159–160). There was no substantial sex bias in head transport by the same criterion in the earlier periods, including the Kebaran at Hayonim Cave (compare figs. 10.6–10.8; see also chapter 11).

The patterns in body part profiles for the Wadi Meged ungulates can be compared more broadly with patterns observed for Upper and Middle Paleolithic faunas in Mediterranean Italy and elsewhere, as well as with the patterns commonly generated by other predators of ungulates under a wider variety of conditions (following Stiner 1991a). Figure 10.10 shows that the differences between the Epipaleolithic–Upper Paleolithic and early Mousterian faunas of the Wadi Meged series are dwarfed by this interspecific comparison. All of the Wadi Meged cases fall within a narrow range. There were no great differences in the ways Paleolithic people treated carcasses in the Wadi Meged during the Epipaleolithic–Upper Paleolithic and Middle Paleolithic periods, although some (rare) Middle Paleolithic cases elsewhere on the Mediterranean Rim (head-dominated assemblages from coastal Italy) are quite different and are associated with shellfish exploitation (Stiner 1994). It seems that early Mousterian humans of the Wadi Meged area had full, uninhibited access to the most valuable parts of ungulate prey. These findings imply that these large prey animals were obtained principally or exclusively by hunting.

INDEXED FOOD UTILITY AND DENSITY CORRELATIONS

It is zooarchaeologists' custom to compare observed patterns of mammal body part representation with independently formulated indexes of food utility and bone density. *Food utility* refers to measurements of relative food values obtained from tissues external to and inside bones, including the major muscle masses associated with certain portions of each skeletal element (meat utility index), bone marrow from the medullary cavities of the limb bones and mandible (marrow utility index), and bone grease obtained from the fine

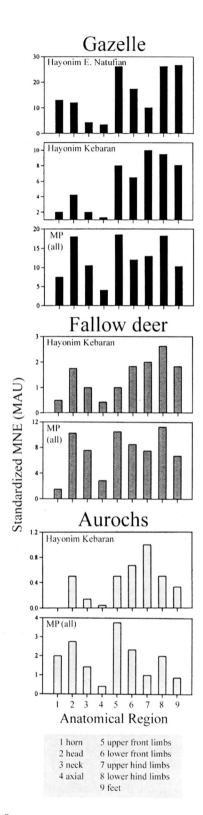

Figure 10.9

Summary comparison of ungulate body part profiles for the combined Mousterian layers and later periods at Hayonim Cave.

Table 10.4
Ungulate crania and mandible representation (MNE) in the Wadi Meged faunal series

Element	SU	MU	LU
By assemblage			
Kebaran, Hayonim Layer C			
Half cranium	9	4	1
Hemimandible	8	4	1
Early Kebaran, Meged Rockshelter			
Half cranium	5	3	—
Hemimandible	7	3	—
Pre-Kebaran (UP), Meged Rockshelter			
Half cranium	1	1	—
Hemimandible	4	3	—
Mousterian MP units 1–2			
Half cranium	3	1	1
Hemimandible	3	2	2
Mousterian MP unit 3			
Half cranium	2	2	1
Hemimandible	5	8	—
Mousterian MP unit 4			
Half cranium	16	15	4
Hemimandible	18	17	1
Mousterian MP unit 5			
Half cranium	8	5	1
Hemimandible	7	6	1
Mousterian MP units 6–7			
Half cranium	4	4	3
Hemimandible	6	4	1
By period			
All Kebaran assemblages			
Half cranium	15	8	1
Hemimandible	19	10	1
All Mousterian assemblages			
Half cranium	33	27	10
Hemimandible	39	37	5

NOTE: "Half cranium" refers to the fact that identifiable portions come in pairs and are separable with fragmentation, making MNE counts for right and left sides separate.

KEY: (SU) Small ungulate; (MU) medium ungulate; (LU) large ungulate.

pores of spongy bone tissues (grease utility index). This concept was first systematically applied by Binford in his 1978 publication on the ethnoarchaeology of the Nunamiut Eskimo. From these observations he also proposed a modified general utility index (MGUI) that combined information for meat, marrow, and bone grease in relation to specific skeletal elements and portions of elements.

For the Wadi Meged study, photon density parameters were taken from Lyman (1994:234–247) for *Antilocapra* (applied to *Gazella*) and *Odocoileus* (applied to *Dama*), and from Kreutzer (1992) for *Bison* (applied to *Bos*). Food utility indexes for meat, marrow, and bone grease, as well as the modified general utility index (MGUI), were taken from Binford (1978; reviewed in Lyman 1994:223–234), with data on wild *Ovis* applied

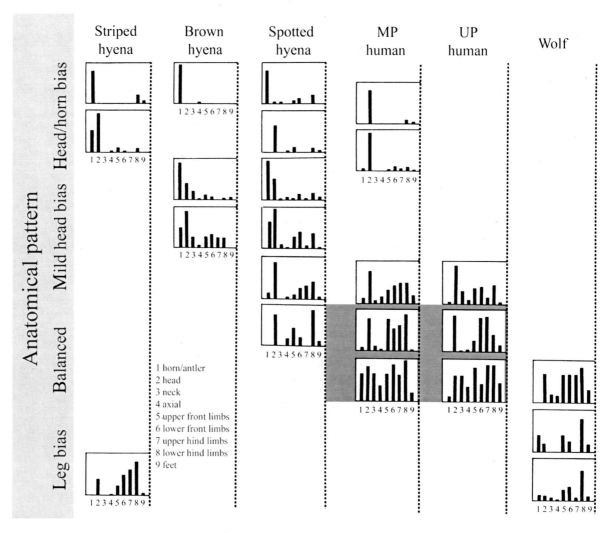

Figure 10.10
The full extent of variation in anatomical patterns in medium ungulate remains collected in shelters by six types of predators, with the observed range of variation in the Paleolithic faunas from the Wadi Meged (dark shaded area). (Base chart from Stiner 1991a.)

to *Gazella* and data on *Rangifer* applied to *Dama* and *Bos*, except for *Bison* MGUI, which was taken from Emerson (1990) and applied to *Bos*. The utility indexes were more problematical than the parameters for skeletal bulk density for this study, because the reference taxa for food utility indexes are less well allied morphologically or metabolically with Near Eastern ungulates. The applications for deer were the least problematical, because all deer are of the Cervidae family and possess similar anatomies. It should also be noted that the tests presented here for the relation between skeletal survivorship and density are less thorough than those presented in chapter 5, because the comparisons had to be limited to parts for which reference data on food utility were also available.

INTERCORRELATIONS AMONG REFERENCE DATA SETS

Table 10.5 explores the degree to which the various food utility and bone density indexes are intercorrelated (following Lyman 1994:258–265). Generally speaking, meat utility and MGUI are strongly linked to one another and appear to measure roughly the same thing. Likewise, marrow and grease are correlated, albeit less strongly, and this pair of parameters is largely independent of the meat and MGUI indexes. Independence between the meat-focused and bone nutrient utility indexes is a large part of the justification for their application to problems of carcass transport and processing strategies in humans and other predators (e.g., Binford 1978, 1981; Brain 1981; Bunn et al. 1988; O'Connell et al. 1988a; Yellen 1977).

Table 10.5

Spearman's correlation statistics for relations between food utility and skeletal density indexes for all major skeletal elements, excluding limb shafts

Index	Meat	Marrow	Grease	MGUI	Skeletal Density
Small ungulates (sheep/pronghorn)					
Meat	—				
Marrow	0.307	—			
Grease	0.012	**0.505**	—		
MGUI	**0.958**	0.200	0.173	—	
Skeletal density	**0.672**	**0.639**	0.041	**0.657**	—
Medium ungulates (caribou/deer)					
Meat	—				
Marrow	0.130	—			
Grease	0.095	**0.586**	—		
MGUI	**0.867**	0.089	0.183	—	
Skeletal density	0.309	**0.557**	0.271	0.184	—
Large ungulates (caribou/bison)					
Meat	—				
Marrow	0.298	—			
Grease	0.103	0.379	—		
MGUI	**0.930**	0.287	0.037	—	
Skeletal density	0.230	0.162	0.029	0.300	—

NOTE: Figures set in boldface and underscored indicate a very significant relation.

Unfortunately, skeletal density is also strongly linked to the meat, marrow, and MGUI indexes in small ungulates and to the marrow index in medium ungulates, making it difficult to separate the effects of human economic behaviors from postdepositional destruction of bones. When correlations are run between density and the same utility indexes for large paired limb elements only, with shaft features omitted, the main relations between skeletal density and meat and MGUI become stronger for the small and medium ungulates (table 10.6). The existence of significant intercorrelations between food utility and density indexes is bound to be somewhat problematical in any attempt to interpret archaeological patterns of body part representation (Grayson 1989; Lyman 1984, 1991).

RESULTS FOR THE WADI MEGED UNGULATES

Spearman's correlation statistics for the relations between skeletal survivorship and the published food utility and bone density indexes are summarized in table 10.7. The results for utility indexes are vague and scattered, and only a few of the relations are signifi-

cant. A stronger or more selective meat focus *may* be suggested for the Kebaran assemblages, and perhaps a slightly greater focus on marrow plus meat for some of the early Mousterian assemblages. Density-mediated attrition is most prominent for the Kebaran and pre-Kebaran assemblages from Hayonim Cave and Meged Rockshelter and for the base of the early Mousterian sequence in Hayonim Cave. The degree to which density explains body part representation nonetheless is quite limited in all but a few cases; only between 9 and 16 percent of all variation can be explained by density (chapter 5). A narrower, limb-focused analysis ("Paired major limb elements only" in table 10.7) reveals fewer relations, although the results are generally consistent with the foregoing observations. Again the Kebaran assemblages in both Hayonim Cave and Meged Rockshelter may have been affected the most by in situ attrition of some kind, though not badly overall. The possibility that humans engaged in more intensive bone processing cannot be rejected for the Natufian (Munro 2001). Milling instruments existed in the material culture of the region following the Last Glacial Maximum and were particularly common from the Natufian onward (Wright 1994), and

Table 10.6
Spearman's correlation statistics for relations between food utility and skeletal density indexes for paired major limb elements only

Index	Meat	Marrow	Grease	MGUI	Skeletal Density
Small ungulates (sheep/pronghorn)					
Meat	—				
Marrow	0.343	—			
Grease	0.219	0.068	—		
MGUI	**0.931**	0.374	0.363	—	
Skeletal density	**0.720**	0.342	0.507	**0.775**	—
Medium ungulates (caribou/deer)					
Meat	—				
Marrow	0.317	—			
Grease	0.097	0.036	—		
MGUI	**0.868**	0.258	0.192	—	
Skeletal density	**0.782**	0.373	0.441	**0.683**	—
Large ungulates (caribou/bison)					
Meat	—				
Marrow	0.366	—			
Grease	0.108	0.071	—		
MGUI	**0.903**	0.452	0.364	—	
Skeletal density	0.459	0.242	0.196	0.356	—

NOTE: Figures set in boldface and underscored indicate a very significant relation.

evidence for density-depending attrition in the gazelle remains of Hayonim Cave is not also apparent for hares in the same faunal assemblages (Munro 2004). For periods prior to the Natufian, however, vertebrae—likely targets for heat-in-liquid and/or pulverization techniques of bone processing but usually ignored in "cold" marrow processing regimes (see chapter 5)—are no more fragmented in the Kebaran and pre-Kebaran assemblages than they are in the early Mousterian assemblages overall. So grease rendering may not be the explanation for higher levels of attrition in the Kebaran and pre-Kebaran periods.

Overall, the comparisons of food utility indexes and bone density do little to improve on the quality of information provided by the profile-based comparisons presented earlier. With behavioral phenomena, particularly those involving a versatile predator, protracted site formation histories, and prey states, perhaps we should not be surprised that the results for potential biases toward density, meat, and fat are ambiguous. Part of the problem was predicted by the intercorrelations of the food utility and density parameters (tables 10.5 and 10.6). Equally important and less widely recognized are the effects of long accumulation times, which promote fuzzy relations between utility indexes

such as MGUI and the marrow index and observed patterns of skeletal part representation. Utility indexes are constructed from modern situations with exceptionally short time frames and of necessity are compiled from single or a relatively small number of procurement events. This certainly is not the situation represented by the early Mousterian faunas of the Wadi Meged. Information is gained nonetheless from body part profiles on an evolutionary time scale.

CONCLUSIONS ABOUT PROCUREMENT AND TRANSPORT STRATEGIES

The ungulate body part profiles of the early Mousterian of Hayonim Cave indicate procurement primarily by hunting (as opposed to scavenging). The larger question about hominid adaptations of the Mousterian culture period is not whether these people could hunt large game but rather how their hunting techniques, schedules, and priorities were configured within a larger foraging system. Body part representation for gazelles and deer is relatively complete in the early Mousterian assemblages, except for the neck and vertebral column and some foot bones, and carcasses of the much larger aurochs are only somewhat less complete. The mild underrepresentation of vertebral elements in principle could be due to the transport decisions of Mousterian foragers in combination with mild rates of decomposition in situ. Cross-examination of the data indicates that the perceived bias against vertebrae and ribs is indeed due mainly to the transport decisions of prehistoric humans, because, for example, significant indications of density-mediated attrition occur in far fewer assemblages than those lacking in vertebral elements and ribs. It seems that axial parts were often left at acquisition sites or elsewhere on the landscape before foragers returned to small base camps in Hayonim Cave and Meged Rockshelter. Both the near balance between head and limb parts in the Mousterian assemblages and the lower representation of heads relative to the natural number of limbs in the Epipaleolithic and Upper Paleolithic assemblages are typical of uninhibited access to large prey. This normally means hunting, judging from interspecific comparisons between various members of the order Carnivora and recent human societies. Transport decisions appear to have become only somewhat more selective with time; head parts did not always make it back to camp.

The differences that exist between the Epipaleolithic–Upper Paleolithic and Middle Paleolithic ungulate assemblages—somewhat fewer head and foot bones in the later periods—are interesting and perhaps important from a strategic point of view. These differences are subtle, however, when considered in the larger universe of predator-generated body part profiles. Early Mousterian people brought bulky parts of high-quality prey back to shelters in the Wadi Meged, places where food could be taken for delayed processing. Some of the highest-quality foods that could be obtained in this environment were not eaten immediately upon procurement but rather transported to a camp and likely shared. The frequency of site visits must have been low, but the large sizes of the game animals argue for considerable sharing during site occupations. In this regard, Mousterian society was very "human." The rate of bone accumulation accelerated and the number of sites increased globally in the late Mousterian (Mirazón Lahr and Foley 2003)—for

example, at Kebara Cave on Mount Carmel (Speth and Tchernov 1998, 2001)—but the basic formula of bringing high-quality food back to camp may have remained unchanged between the early and late Mousterian in the Levant.

The other major conclusion to be drawn in this chapter is a methodological one. Differential preservation of specific element portions need not translate to differential representation or preservation at the scale of anatomical regions, particularly if the question is about the bony components of heads versus limbs versus feet. Some misconceptions about this topic appear to stem from the assumption that shafts of limb elements resist decomposition far better than *any* limb end, hard or soft. The density parameters considered in this study do not support this in principle or empirically. The other source of confusion, and perhaps the more profound one, seems to arise from analysts' tendency to focus upon portion representation *within elements* in their assessments of attrition effects and then to attempt to generalize their observations to larger patterns of anatomical representation across the skeletal anatomy. In fact, pooling elements by anatomical regions eliminates much of the potential ambiguity originating from differential bone density.

Exploration of the profiling method used here—its strengths and weaknesses—in terms of independent referents of skeletal density demonstrates that the risks of misinterpretation vary greatly with different ways of grouping the data. Experiments and many zooarchaeological studies concerned with the relative rates of loss of head and limb parts have tended to focus on why certain portions of elements might be missing from an assemblage. Such analyses may say little about the probability of imbalances at the level of anatomical regions in prey. The latter are much coarser divisions of the anatomy of large prey animals, yet they are arguably closer approximations, however imperfect, of the butchering and transport units that humans commonly create. The anatomical-region approach to body part profiling is simpler than most, requiring less information about density variation and its relation to portion identifiability, and thus it offers fewer ways to be wrong in first-order interpretations of the data.

Table 10.7
Spearman's correlation statistics for relations between percentage of bone portion survivorship and body part utility indexes for common ungulate taxa

Assemblage (n)	Meat Utility Index		Marrow Utility Index		Grease Utility Index		MGUI		Survivorship	
	rho	p	rho	p	rho	p	rho	p	rho	p
All major skeletal elements (excluding limb shafts)										
Gazelle										
Hay Kebaran (24)	0.438	.05	0.337	>.1	0.001	>.1	0.468	.05 > p > .02	0.505	.02 > p > .01
Meg Kebaran (24)	0.446	.05 > p > .02	0.482	.02	0.257	>.1	0.393	.1 > p > .05	0.455	.02
Meg pre-Keb (23)	0.356	>.1	0.129	>.1	0.059	>.1	0.431	.05	0.473	.05 > p > .02
Hay MP1 (17)	0.202	>.1	0.579	.02 > p > .01	0.460	.1 > p > .05	0.171	>.1	0.395	>.1
Hay MP2 (19)	0.172	>.1	0.409	.1 > p > .05	0.356	>.1	0.178	>.1	0.214	>.1
Hay MP3 (22)	0.017	>.1	0.213	>.1	0.024	>.1	0.010	>.1	0.091	>.1
Hay MP4 (24)	0.007	>.1	0.037	>.1	0.139	>.1	0.075	>.1	0.020	>.1
Hay MP5 (24)	0.380	.06 > p > .05	0.194	>.1	0.038	>.1	0.387	.06 > p > .05	0.471	.05 > p > .02
Hay MP6 (24)	0.278	>.1	0.455	.05 > p > .02	0.012	>.1	0.239	>.1	0.453	.05 > p > .02
Hay MP7 (15)	0.201	>.1	0.483	.05	0.519	.05 > p > .02	0.111	>.1	0.365	>.1
Fallow deer										
Hay Kebaran (24)	0.529	.01	0.386	.05	0.310	>.1	0.389	.05	0.709	.001
Meg Kebaran (20)	0.374	.1	0.114	>.1	0.344	>.1	0.365	.1	0.221	>.1
Meg pre-Keb (18)	0.076	>.1	0.309	>.1	0.396	>.1	0.001	>.1	0.503	.05 > p > .02
Hay MP1 (7)	0.206	>.1	0.612	>.1	0.408	>.1	0.206	>.1	0.206	>.1
Hay MP2 (11)	0.108	>.1	0.670	.02	0.464	>.1	0.094	>.1	0.558	.1 > p > .05
Hay MP3 (25)	0.346	.1	0.449	.05 > p > .02	0.265	>.1	0.290	>.1	0.389	.1 > p > .05
Hay MP4 (25)	0.418	.05	0.120	>.1	0.104	>.1	0.392	>.1	0.288	>.1
Hay MP5 (23)	0.050	>.1	0.275	>.1	0.357	.1	0.123	>.1	0.277	>.1
Hay MP6 (22)	0.135	>.1	0.081	>.1	0.148	>.1	0.265	>.1	0.059	>.1
Hay MP7 (12)	0.172	>.1	0.275	>.1	0.225	>.1	0.171	>.1	0.437	>.1
Aurochs										
Hay Kebaran (13)	0.072	>.1	0.211	>.1	0.274	>.1	0.150	>.1	0.202	>.1
Hay MP3 (6)	0.143	>.1	0.551	>.1	0.428	>.1	0.232	>.1	0.086	>.1
Hay MP4 (23)	0.231	>.1	0.268	>.1	0.248	>.1	0.288	>.1	0.391	.1 > p > .05
Hay MP5 (13)	0.123	>.1	0.384	>.1	0.146	>.1	0.039	>.1	0.359	>.1
Hay MP6 (12)	0.001	>.1	0.181	>.1	0.039	>.1	0.133	>.1	0.664	.02

Continued on the next page

Table 10.7, continued

Assemblage (n)	Meat Utility Index		Marrow Utility Index		Grease Utility Index		MGUI		Survivorship	
	rho	p	rho	p	rho	p	rho	p	rho	p
Paired major limb elements only										
Gazelle										
Hay MP5 (15)	—	—	—	—	—	—	—	—	0.590	.05 > p > .02
Hay MP6 (15)	—	—	—	—	—	—	—	—	**0.665**	.01 > p > .001
Fallow deer										
Hay Kebaran (14)	**0.698**	.01 > p > .001	—	—	—	—	0.589	.05 > p > .02	**0.730**	.01 > p > .001
Hay MP3 (15)	0.528	.05	**0.644**	.01	—	—	—	—	—	—

SOURCES: Percentage of survivorship follows Binford 1978 and Lyman 1994. Photon density parameters are for scan sites listed in Lyman 1994 for *Antilocapra* (applied to *Gazella*) and *Odocoileus* (applied to *Dama*) and from Kreutzer 1992 for *Bison* (applied to *Bos*). Utility indexes are from Binford 1978 (reviewed in Lyman 1994) for wild *Ovis* (applied to *Gazella*) and *Rangifer* (applied to *Dama* and *Bos*), except for *Bison* MGUI, which is from Emerson 1990 (applied to *Bos*).

NOTE: Correlation tests for the relation between survivorship and density are less stringent than those presented in chapter 5; this set of tests is limited to parts for which reference data on food utility are available. Application of published utility indexes is more problematical than that of density parameters, because the model taxa are less well allied genetically or metabolically with Near Eastern ungulates in the Wadi Meged series. An underscore indicates a mildly significant relation; boldface and underscore together indicate a very significant relation.

CHAPTER ELEVEN

Evidence of Large Game Hunting from Ungulate Mortality Patterns

■

UNGULATE REMAINS DOMINATE PLEISTOCENE archaeofaunas nearly worldwide. From ungulate remains one can trace shifts in hominid predator-prey relationships over at least one million years—from what may have been desultory hunting and scavenging to specialized hunting and eventually domestication of some species. These ecological transitions in human lifeways were shaped partly by the risks of interference competition from other meat eaters. Meat is among the few sources of "complete" protein in the environment, and hominids frequently sought meat from large game animals. These large resource packages of high quality were relatively difficult to obtain, and so the potential for interference competition from other large predators could be great. A theoretical outcome of interspecific competition is ecological differentiation, or character displacement, which relieves the stresses of conflict between consumers that coexist in an ecosystem (MacArthur and Levins 1967; Pianka 1978: 189–199, 260).

When did hominids become the sort of ungulate predator that we see among recent humans? Specifically, did prime-adult-focused hunting, a behavior now commonly discussed in the zooarchaeological literature, get an early start in the Middle Pleistocene or even earlier? And how different were Mousterian people from later foragers, particularly if all of these early humans were capable of bringing down large game? In this chapter I address the antiquity of large game hunting practices and the possibility of trends in predator-prey interactions involving humans and ungulates from the late Middle Pleistocene to the Pleistocene-Holocene transition. The evidence comes principally from the age structures of ungulate death assemblages—mortality patterns inferred from data on tooth eruption and wear and bone fusion. Information on ungulate sex ratios is also considered, although this information is less robust in the Wadi Meged samples. The methods of analysis are presented alongside the results for dental remains, bone fusion, and sex ratios, respectively, because each is distinct. Of special interest is evidence of hunting pressure—or lack of it—in the common artiodactyl (cloven-hoofed ungulate) species. Although this chapter is about more than the hunting-scavenging debate, the evidence also lays to rest any suggestion that meat acquisition by early Mousterian humans was confined to scavenged sources. Hayonim Cave provides one of the earliest and best-preserved examples of prime-age-focused hunting, the evolutionary implications of which reach beyond the Near East.

MORTALITY PATTERNS FROM DENTAL ERUPTION AND WEAR

The methods and models used in studying the ungulate dental remains in the Wadi Meged assemblages were ones I developed more than a decade ago (Stiner 1990, 1994). Since then this approach has been applied widely in research on Paleolithic faunas (e.g., Enloe 1997; Gaudzinski 1995; Lyman 1994; Pike-Tay et al. 1999; Speth and Tchernov 1998; Steele and Weaver 2002). Interestingly, and reassuringly, the results have been fairly consistent among studies, at least in the larger patterns observed. For readers unfamiliar with the three-cohort system and models, I first review the technique, its assumptions, and its potential sources of bias (see Stiner 1990 and 1994 for detailed treatment).

Ungulate mortality patterns can be constructed from observations of tooth eruption and wear, measurements of crown height, or both, depending on a species' dental morphology and analysts' preferences (e.g., Davis 1980b, 1983; Deniz and Payne 1982; Gifford-González 1991; Grant 1982; Klein 1978, 1979; Klein and Cruz-Uribe 1984; Lowe 1967; Payne 1973; Roettcher and Hofmann 1970; Severinghaus 1949; Stiner 1990). The wear stages for deer, cattle/bison, and sheep/goats are well documented (reviewed by Hillson 1986; Payne 1973) and are the basis of the criteria used here. Because there is less precedent for mountain gazelles, age criteria for them are based on a combination of Payne's (1973) and Davis's (1980b) systems and Stutz's (2002) study of known-age individuals in the comparative collections of Tel Aviv University and the Hebrew University of Jerusalem. Only eruption and wear observations are used, although I generally find good agreement between crown height measurement data for hipsodont species and occlusal wear stage data. Regardless, a nonlinear relation between tooth wear and age in real years must be recognized for all wear-based and crown height methods (Gifford-Gonzalez 1991; Hillson 1986; Severinghaus 1949).

The archaeofaunal patterns are presented initially as fine-grained (nine-cohort) data and then are collapsed into three age groups as determined from a combination of modern wildlife data, principles of mammalian population dynamics, and the life history characteristics of the prey species. The three-cohort system facilitates comparisons of archaeofaunal patterns with natural variation associated with each of a variety of mortality factors.

POTENTIAL TAPHONOMIC BIASES

Fully formed, unresorbed teeth are very dense structures and for this reason resist decomposition forces exceptionally well. Teeth that are only partly formed at the time of death are a good deal less resistant. This fact raises the question of whether observed biases in ungulate mortality patterns could be explained by variable states of dental development in combination with poor preservation conditions. Considerable debate has taken place over this question, but only some of the claims about biases in mortality data are justified by the structural biology of teeth. In this section I address the relative resistance of deciduous and permanent teeth to decomposition factors of the sort that might distort mortality profiles. As argued in what follows, the primary concern should be with the developmental state of a tooth at the time of death; whether it is a deciduous or a permanent tooth is of less concern.

Mammalian tooth enamel is very porous during formation but only briefly so relative to the full potential use-life of the tooth. This is true for permanent teeth and many deciduous teeth, whose enamel compositions and microstructures are similar if not identical. A tooth's susceptibility to decomposition is controlled foremost by its porosity and its proportions of the three main constituents of mammalian tooth enamel—the mineral hydroxyapatite, a protein matrix, and water—which differ greatly between partly and fully formed elements. The mineral fraction in a developing tooth may be as low as 37 percent (about half that of mature bone), the organic fraction 19 percent, and water as high as 44 percent (Wainwright et al. 1976:224). The proportions of the three components change radically with maturation, when enamel is composed of about 94–97 percent mineral, 2 percent organic matrix, and 4 percent water. In addition to a very high inorganic/organic ratio, the apatite crystals of fully developed tooth enamel are roughly 10 times larger (about 40 nm across) than the apatite crystals of bone (about 4 nm on average) (Currey 1984; Weiner and Traub 1992). The lower ratio of surface area to volume of the fully formed crystalline structure of enamel makes it denser, less porous, and much less soluble than bone.

The absence of a hardened dentin backing in partly formed teeth also makes them fragile. The difference between a developing tooth and a mature one is basically that between a semihollow shell and a solid entity with a stony covering. Dentin is a distinct structural variant of bone that lacks cells but, like bone, is composed of about 70 percent hydroxyapatite. Dentin is somewhat harder than bone but considerably softer

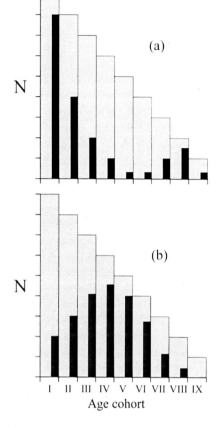

Figure 11.1
Idealized stable living-structure of a mammalian population (large gray bars) contrasted with (a) U-shaped mortality pattern (black bars); (b) prime-dominant mortality pattern (black bars).

than enamel. Enamel is an essentially nonelastic material, but in combination with a semielastic dentin backing, which serves as a shock absorbance system, the tooth crown is a very dense and relatively closed structure (e.g., Hillson 1986; Wainwright et al. 1976; Weiner and Wagner 1998). In whole form, teeth therefore resist destruction better than virtually any other part of the mammalian skeleton.

The period of dentin and root development represents a fraction of the total time in which the tooth may exist in the mouth of a mammal. This point applies to both deciduous and permanent elements. Of concern, then, is the amount of time during which a particular element is vulnerable to destruction, should the animal die young. Many authors have suggested that the visibility of juveniles in death assemblages declines at a faster rate than the visibility of adult animals under suboptimal preservation conditions, particularly in carnivore-ravaged faunas,

because juvenile teeth are as a rule more fragile (Binford and Bertram 1977; Levine 1983; Marean 1995; Munson 1991, 2000). By this reasoning, some argue, the apparent absence or scarcity of juveniles in archaeofaunal mortality patterns is a product of differential preservation, not of selective hunting. Specifically, small numbers of juveniles in the mortality profiles are taken to represent missing data rather than as an indication of the original age structure of a death population (Marean 1995). Although this assessment can be defended in some instances (e.g., Hoffecker et al. 1991; Stiner 1998b), it is inappropriate in most other instances where prime-adult biases or other distinctive mortality patterns are observed.

In the case of artiodactyl faunas, two sources of information contradict the more extreme positions that have been taken on attritional biases in archaeological mortality patterns. One is the great variation in mammal age structures observed in assemblages attributed to different bone collecting agents in stratigraphic series from a single shelter site, where mortality patterns may vary from U-shaped or L-shaped to staircaselike or other patterns when viewed in multicohort bar charts (fig. 11.1). Deeply stratified Pleistocene shelter sites in Italy, for example, yield prime-dominant ungulate age structures in human-collected material but U-shaped patterns for the same ungulate species in another layer attributed to denning spotted hyenas (Stiner 1994), animals famous for destroying bone (Brain 1981; Kruuk 1972; Marean and Spencer 1991; Sutcliffe 1970). Because there were no significant differences in preservation chemistry among the stratigraphic levels of the Italian sites (Stiner 1994), the contrasting age structures in the large mammal remains must be explained in terms of prey selection by the agents responsible for bone collection and modification.

Another source of confusion about differential in situ destruction of teeth seems to be the equation of the immaturity of a specific tooth element with youth in the individual animal. This is only partly correct. Most mammals develop two functional sets of teeth in succession, a deciduous set followed by a permanent set. Dental elements develop sequentially within each set as well. The developmental schedule effectively offsets loading within the dentary row, with adjacent hardened teeth absorbing more of the masticatory strain as other teeth form. The lowest level of mineralization and the greatest susceptibility to destruction exist before eruption, or gingival emergence. At this stage, dentin formation lags well behind enamel formation, so the fragile enamel has a hollow core. During this short interval a young animal might not be "visible" in a death assemblage, at least on the basis of that particular tooth, but it should be visible—and countable—on the basis of an adjacent element. A straightforward solution to the developmental wrinkle therefore is to examine adjacent (in fact, all) dental elements in any assemblage and to cross-check the total number of individual animals represented by each. Further, because the aging system, as discussed later, is designed to avoid the double counting of individuals by blocking out certain eruption-wear intervals, it also avoids most of the potential biases from mechanical decomposition (e.g., Stiner 1994:288–292, 1998b:310–314).

Finally, tooth size is not a proxy for mineral density. In ungulates, a fully developed deciduous tooth often is no less mineralized than a permanent tooth of the species. Extensive resorption may occur toward the end of a deciduous tooth's use-life in some mammals, such as bears (Torres 1988) and perhaps equids (Levine 1983), but this cannot be generalized to most artiodactyl ungulates. All ungulate teeth—deciduous and permanent—are more fragile and thus more susceptible to destruction prior to complete mineralization and, only in certain mammalian groups, following substantial resorption at the end of the deciduous tooth's use-life. A permanent tooth crown worn to its base with advanced age may also break relatively easily and thus be more difficult to recognize in faunal assemblages (e.g., Stiner 1998b). But this liability in mortality analysis is tempered by the fact that very old adults are rare in populations to begin with, whereas young animals can be abundant in living populations and potentially a more abundant source of food to predators. It should be understood that deciduous teeth generally are smaller than permanent teeth in a given species, but deciduous teeth of one taxon can be a good deal larger than the permanent teeth of another species, and all may survive the rigors of mechanical destruction if they are fully mineralized. Mass-mediated dissolution in which smaller specimens suffer more than bulky ones may occur in some sedimentary regimes, but this mode of differential destruction requires that acidic liquid recharge be quite limited or, alternatively, that mechanical size sorting cause deciduous teeth to be deposited in chemical environments distinct from those in which permanent teeth are deposited. Neither of these situations is common in sedimentary environments, and certainly not in the limestone shelter sites under discussion.

NINE-COHORT MORTALITY PROFILES

The mortality patterns of the Wadi Meged ungulates appear as nine-cohort bar charts in figure 11.2, based on age at death as determined from eruption and wear for the deciduous fourth premolar (dP4) and the permanent fourth premolar (P4) or permanent third molar (M3). The lower dentition is preferred for this technique, but its upper counterpart has a fairly similar eruption and wear schedule and can be substituted as necessary. Some of the age structures in figure 11.2 are dominated by prime adult animals, and others are not. The patterns are not explained by the counting methods used nor by decomposition, because independent taphonomic evidence indicates favorable preservation environments for the Wadi Meged assemblages (chapters 4 and 5). Although the mortality patterns are interesting, it is difficult to evaluate them in this format against a set of alternative mortality models, because the variation associated with each model is not apparent. This was one impetus for the development of the three-cohort system.

THE THREE-COHORT SYSTEM

The three-age system—consisting of juvenile, prime adult, and old adult cohorts—is keyed to the principal life stages through which long-lived individuals pass (Stiner 1990). Because few age cohorts are involved, this system handles small assemblages better than most. The method accommodates isolated teeth as well as those in intact dental rows. The eruption-and-wear stage criteria stem from the systems developed by Payne (1973), Grant (1982), and Lowe (1967). The technique targets the deciduous fourth (or third) premolar followed by the fourth premolar or, alternatively, the third molar (fig. 11.3), with cross-checks to adjacent elements. The dP4–P4 or dP3–M3 sequence in artiodactyls has three attractions: these teeth are easily distinguished from other cheek teeth, they track a lifetime of food intake and occlusal abrasion, and the fourth premolar can be completely worn away in advanced age (and the third molar nearly so). This means that the dental set collectively represents the full potential lifetime of an ungulate. The fully formed dP4 is not much more fragile than the P4 in artiodactyl ungulates, so preservation should not favor the mature permanent teeth of adults much more than it does the fully formed deciduous tooth.

Juveniles are defined as individuals younger than the age at which the fourth deciduous premolar is normally shed and replaced by the permanent fourth pre-

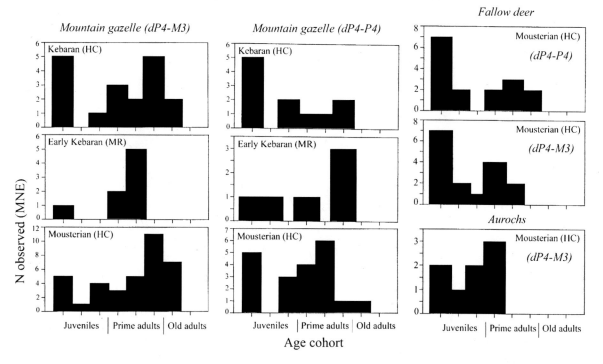

Figure 11.2
Age structures (mortality patterns) of mountain gazelles, fallow deer, and aurochs based on dental remains from the Kebaran, early Kebaran, and early Mousterian assemblages in the Wadi Meged series, using nine age cohorts. In the notation dP4–M3, dP4 represents the juvenile cohorts, and M3, as an alternative to P4, the adult cohorts.

molar. The timing of this developmental juncture varies by species (table 11.1) but is predictable relative to the time of birth. The transition between juvenile and prime adult status is especially clear, because the permanent premolar supplants the deciduous counterpart by pushing it up and out of the jaw (fig. 11.4). A permanent premolar, however, must show some evidence of occlusal wear in order to be counted, because occlusal wear is the only reliable indicator that the permanent tooth was "exposed" in the jaw at the time of death. Although the third molar may be substituted for counts of the permanent fourth premolar in the subject artiodactyl species, it should be noted that the M3 generally erupts just before the loss of dP4.

The prime adult stage begins at the deciduous-permanent boundary and continues to about 65 percent of the maximum potential life span. The prime age–old age boundary in fossil assemblages is based on how much of the tooth is left relative to the complete form and the state of the occlusal surface. The maximum potential life span (MPL) is defined by destruction of the tooth crown in long-lived individuals of the species, preferably across multiple death populations in a given region. Old adults are indicated when more

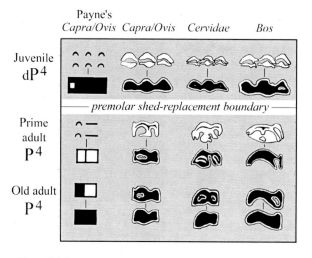

Figure 11.3
Tooth eruption and wear stages according to the three-age-cohort system for deciduous and permanent fourth premolars of sheep/goats, deer, and cattle. M3 can be substituted for P4. (Adapted from Stiner 1990.)

than half the height of the crown has been worn away. The MPL theoretically ends when the crown is destroyed, which happens to the P4 in advanced age.

At this point, feeding becomes difficult or compromised, and the animal soon perishes (see Stiner 1990 on wildlife reference cases.)

No attempt is made in this system to estimate the absolute ages of individual animals except at the deciduous-permanent tooth replacement boundary, for which reliable biological standards are available. The three age cohorts—juveniles, prime adults, and old adults—do not represent equal fractions of the maximum potential life span but instead are tied to the realities of population dynamics and life history characteristics in mammals. Thus, the three age cohorts correspond to the differing susceptibilities of individuals to attritional death factors such as disease, accidents, and malnutrition, as well as to significant changes in body composition. Fat content in adult individuals—a factor of great importance to predators—may vary radically with season and according to somewhat different schedules of maturation and reproduction, depending on the sex of the animal. Prior to the very late stages of development, juveniles are generally poor in fats, because so much of their energy is devoted to growth.

MORTALITY MODELS AND EXPECTED RANGES OF VARIATION

The number of individuals in each of the three age cohorts can be converted to percentages of the total sample and plotted on tripolar graphs (fig. 11.5). Areal divisions of the graph represent different classes of

Figure 11.4
Fallow deer deciduous third and fourth premolars astride emerging permanent third and fourth premolars.

mortality patterns and the range of variation associated with each. The two most common mortality patterns in nature occupy the two lower-center panels: these are U-shaped, or attritional, mortality on the left and "living-structure," or nonselective mortality, on the right. The latter pattern refers both to the typical proportions of young, prime, and old animals in the living population on which predators depend and the set of animals a predator may take nonselectively from that population over time. The three corners of the graph correspond to biases toward each of the three designated age groups. The range of variation for each mortality pattern is defined on the basis of theoretical expectations from population ecology and empirical

Table 11.1
Known-age ranges in months for the replacement of dP4 by P4/M3 in artiodactyl ungulates

Taxon	dP4 in Place	P4 Erupting	P4/M3 in Place	Source
Gazelle (*Gazella gazella*)	Birth–1	~10	~15–18	Davis 1980b
	—	~10	~16	Stutz 2002[a]
Red deer (*Cervus elaphus*)	1-2	23–27	27	Hillson 1986:210
	—	25–26	27	Lowe 1967:138
Fallow deer (*Dama dama*)	At birth	19	22	Chapman and Chapman 1975
	At birth	17	20	Hillson 1986:211
Aurochs (*Bos primigenius*)	Birth–1	30	36	Hillson 1986:206

NOTE: dP4 is the deciduous fourth premolar; P4 is the permanent fourth premolar; M3 is the permanent third molar. P4 and M3 emerge at roughly the same time, although M3 is generally slightly ahead of P4.

[a] Wear stages for gazelles were developed by Stutz (2002) from comparative collections of mountain gazelles of known ages at Tel Aviv University and the Hebrew University of Jerusalem, Israel, following Payne's (1973) system for sheep/goats.

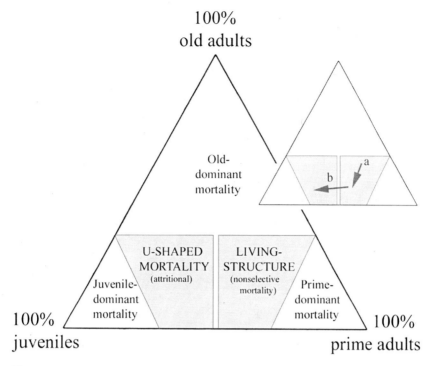

Figure 11.5
Living-structure and mortality models in tripolar format. Inset shows expected trends caused by moderate (a) and high (b) levels of hunting pressure.

comparisons of living-structures and death patterns from nonviolent causes in recent contexts (Stiner 1990, 1994).

The population structures of living mammals, encompassed by the "living-structure" panel of the tripolar plot, normally vary between stable, growing, and declining states (Caughley 1977). Predator pressure on prey populations can steepen the age structure pyramid when it is viewed in bar chart format. Under these conditions, the proportions of juveniles and young adults in the population rise somewhat relative to the proportions of mid-prime and old adults, and are more likely to be killed because they have become more common in the population. A population may be pushed into sustained growth mode by hunting pressure, and the reproductive prospects for young adults may improve in spite of their having less life experience (see Caughley 1977, among others). One signal of this effect in tripolar format is a decline or "compression" on the old adult axis (*a* in fig. 11.5; Stiner 1994:307–315). This concept has been modeled further by Wolverton (2001:44) using projections from Taber et al.'s (1982) simulation-based study of American elk (*Cervus canadensis*) population dynamics with increased hunting pressure. Mortality patterns theoretically may tend toward the lower part of the "attritional" (U-

shaped) area of the pattern map as the juvenile bias from hunting pressure becomes extreme *(b)*. Hunting pressure, however, is not the only or even the most widespread cause of U-shaped mortality, and such patterns therefore must be interpreted carefully while holding predator species in mind.

VARIATION IN THE MORTALITY PATTERNS GENERATED BY PREDATORS

Figure 11.6 illustrates the tendencies of various non-human predators, recent humans, and Paleolithic humans (Italy only) that have hunted ungulates. The points represent broad averages of multiple cases framed against natural variation in living prey population structures and the mortality patterns arising from nonviolent causes. The ungulate mortality patterns generated by cursorial (long-chase) predators such as spotted hyenas, wolves, cheetahs, and African wild dogs overlap completely with the death patterns resulting from disease, malnutrition, and senescence. Predators that ambush their prey, such as lions and tigers, tend to hunt nonselectively with respect to prey age unless the prey species is exceptionally large or small. The death patterns created by these ambush specialists tend to resemble the age structures of the living prey populations as well as patterns resulting from mass deaths from catastrophic floods or fire. The extent of scavenging for ungulate carcasses seems to push the mean value higher on the old age axis for cursorial and ambush predators alike, as can be seen for Indian tigers in this comparison (also spotted hyenas, Stiner 1990). Obviously, a mortality pattern can be tied to a particular predator only by appeal to independent taphonomic evidence, in light of the complete overlap observed between predator-induced and nonviolent sources of mortality.

Late Middle Paleolithic and Upper Paleolithic cases from the Italian peninsula and recent human cases from multiple world regions tend to fall on the right half of the distribution. The full range of cases spans the entire living-structure and prime-dominant areas of the pattern map but averages to a mild bias toward prime adult prey. That the Upper Paleolithic (UP) cases are positioned lower with respect to the old age axis than are the hunted Mousterian faunas (MP1) in figure 11.6 may suggest somewhat greater hunting pressure on ungulate populations in the later period. Anomalies in the Mousterian data involve apparent scavenging in a few cases from coastal Italy (MP2 in fig. 11.6); these are head-dominated assemblages representing small amounts of meat acquired per carcass, and several are associated with the collection of slow-moving small game (chapter 9; Stiner 1994). Even without these cases, the Mousterian mean from Italy is situated somewhat higher on the old adult axis than the Upper Paleolithic–Epipaleolithic mean from the same region.

The averaged patterns for the four "recent" human cultures are similar to one another and to the Paleolithic results. These cases date throughout the Holocene period, come from different parts of the world, and involve different suites of ungulate species and distinct weapons systems, from bows and arrows, atl-atls, and spears to firearms. Because the mortality patterns are essentially the same, yet quite distinct from those created by nonhuman predators, we must be seeing a property of the human species, evidence of niche separation as a result of hominids' long-standing membership in ungulate predator guilds (Stiner 1990, 1994). The results are surprisingly unaffected by the weapons technology used. Furthermore, only subtle differences are apparent for the Middle Paleolithic versus the Upper Paleolithic.

RESULTS FOR THE WADI MEGED AND OTHER MEDITERRANEAN SITES

The comparisons made so far have concerned cases outside the Levant. Figure 11.7 presents the mortality models and observed patterns of variation for ungulate prey in the Middle and Upper Paleolithic and Epipaleolithic assemblages from Hayonim Cave, Meged Rockshelter, and Kebara Cave (Mount Carmel) in the Galilee (tables 11.2–11.4). There are no obvious differences between the total distributions of the Epipaleolithic–Upper Paleolithic and Mousterian cases; both span the living-structure and prime-dominant pattern areas. There is no substantive difference, moreover, between the early Mousterian from Hayonim Cave and Speth and Tchernov's (1998) results for the late Mousterian from Kebara Cave. Figure 11.7 includes for comparison a wider range of cases involving artiodactyl prey from early Upper Paleolithic through terminal Epipaleolithic assemblages from Mediterranean Italy (tables 11.3–11.4), Levantine Turkey, and Lebanon. The data from Riparo Mochi in Italy and from Üçağızlı Cave on the Hatay coast of Turkey are from ongoing projects (Kuhn et al. 1999, 2001; Kuhn and Stiner 1992, 1998a; Stiner 1999; Stiner et al. 2002). Other Italian data are from Lazio (Stiner 1990, 1994). The Epipaleolithic cases from Lebanon are taken from Kersten's (1987) detailed study of the ungulates from Ksar 'Akil. Though equid species are present in some of the faunas, they generally are too rare for their mortality patterns to be considered.

There are very few outliers to the Upper Paleolithic–Epipaleolithic distribution, despite the diverse prey species, regions, and habitats compared. The only anomalies, isolated in figure 11.8, are the terminal Epipaleolithic cases, which display much stronger biases toward juveniles than expected. These are the Natufian gazelle assemblages from Hayonim Cave and Hayonim Terrace (from Munro 2001), one Epipaleolithic chamois assemblage from Grotta Polesini (10 KYA, Stiner 1990), and one Mesopotamian fallow deer assemblage from Ksar 'Akil (Kersten 1987).

Detailed information is available for the Natufian cases, which are distinctive for their remarkable emphasis on fawns (Munro 2001; Davis 1983; Davis et al. 1994) in comparison with earlier gazelle assemblages from the Galilee (fig. 11.9). This effect in the Natufian is consistent with the modeled consequences of excessive hunting pressure (prediction b in fig. 11.8). Seasonal procurement in ethnographic contexts is also known to produce extreme juvenile biases; an

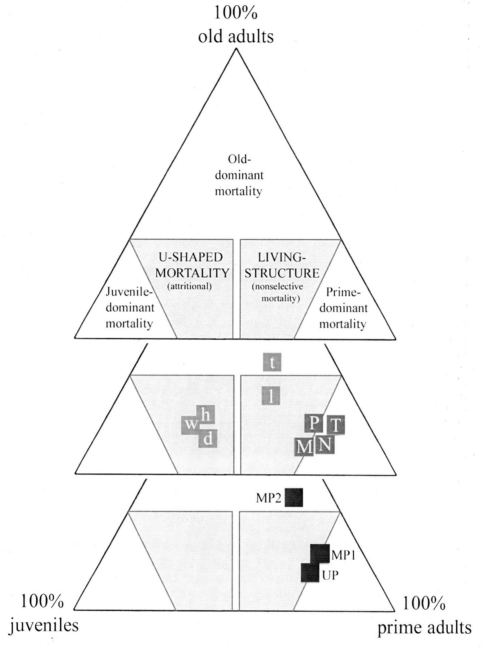

Figure 11.6
Models and averaged values for observed ungulate mortality patterns created by various human and nonhuman predators. Shaded panels represent natural variations in the age structures of living ungulate populations and thus also nonselective mortality patterns (right panel) and mortality patterns caused by attritional factors such as disease, accidents, and malnutrition (left panel). Each corner of the diagram represents a strong bias toward the designated prey age group. Ambush predators are tigers (t) and lions (l); cursorial, or long-chase, predators are spotted hyenas (h), wolves (w), and African wild dogs (d). Holocene human predators are Paleoindian-Archaic (P), Mississippian farmers (M), Nunamiut Eskimo (N), and trophy hunters in modern game parks (T). MP[1] is the mean for most Middle Paleolithic hunted faunas from Italy (100–33 KYA); it closely resembles that for the Upper Paleolithic (UP). MP[2] represents a handful of cases from coastal Italy dominated almost exclusively by head parts from old adult prey. (Data from Stiner 1990, 1994.)

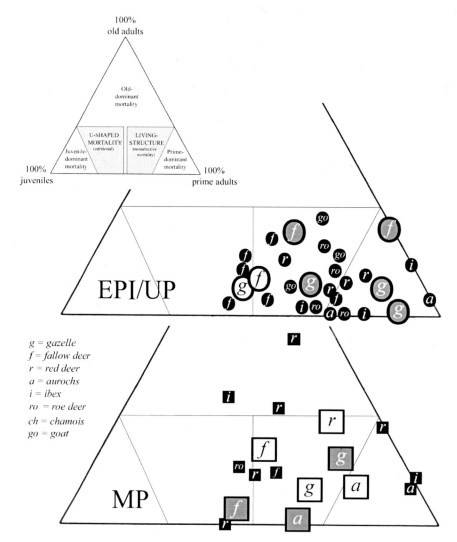

Figure 11.7
Models and observed mortality patterns in artiodactyl prey from Mediterranean Upper Paleolithic–Epipaleolithic and Mousterian assemblages from Israel, Lebanon (Ksar 'Akil), Turkey (Üçağızlı Cave), and Italy. Large symbols represent Galilee sites: Hayonim Cave and Meged Rockshelter in the Wadi Meged (gray infill); Kebara Cave on Mount Carmel (white infill). Italian sites are Riparo Mochi in Liguria and Grotta Polesini, Grotta Palidoro, Grotta Breuil, Grotta di Sant'Agostino, Grotta Guattari, and Grotta dei Moscerini in Lazio.

example is the periodic, concerted procurement of caribou calves in the spring by Nunamiut hunters to obtain very soft hides (Binford 1978; Stiner 1990). But these episodes represent only a small portion of the total number of animals the Nunamiut procure in a year, and the consequences of the practice are invisible in the averaged value for this culture in figure 11.6.

In the case of high residential mobility, juvenile prey remains may be concentrated at one site but not another. Such a scenario could be proposed for the exploitation of highland chamois at Grotta Polesini in Italy, since an extreme juvenile bias is not apparent in

several other artiodactyl species from the same site. The situation is less clear for Mesopotamian fallow deer in the final Epipaleolithic layer of Ksar 'Akil, although this assemblage contrasts markedly with fallow deer assemblages in all of the earlier Epipaleolithic layers recorded also by Kersten (table 11.3). In the Natufian of Hayonim Cave, however, occupation times are known to have been quite long in any given year (Munro 2001; see also Bar-Yosef and Belfer-Cohen 1989; Belfer-Cohen 1988a, 1988b, 1991; Belfer-Cohen and Bar-Yosef 2000), so another explanation for the extreme juvenile bias in the gazelles must be found.

Table 11.2

Three-cohort system results for artiodactyls in the Wadi Meged assemblages, based on the dP4–P4 or dP4–M3 dental eruption and wear sequence

Assemblage (Site)	Total MNE	% Juveniles	% Prime Adults	% Old Adults
Gazelle				
Kebaran (HC)	18	28	61	11
Early and pre-Kebaran (MR)	8	12	88	0
Aurignacian (HC)	128	~11	~79	~10
Early Mousterian (HC)	36	16	64	19
Fallow deer				
Kebaran (HC)	4[a]	25	50	25
Early and pre-Kebaran (MR)	4[a]	0	75	25
Early Mousterian (HC)	16	56	44	0
Aurochs				
Early Mousterian (HC)	8	37	62	0

NOTE: Total MNE refers only to the total number of dental specimens that could be age-scored. Data for the Aurignacian are estimated from Davis 1989:59.

KEY: (HC) Hayonim Cave; (MR) Meged Rockshelter.

[a] Sample was too small for reliable calculation.

EVIDENCE OF HUNTING PRESSURE

Figure 11.10 encapsulates information from chapters 7 and 9 on trends in ungulate prey biomass in three Mediterranean regions—the Galilee, Hatay (Levantine Turkey), and coastal Italy. Marked juvenile biases in the ungulate faunas of these Mediterranean series appear around the same time as a precipitous drop in the total ungulate biomass acquired by Paleolithic foragers, not long before a transition to Neolithic or Mesolithic lifeways, depending on the region. The biomass data lend credence to the idea that the juvenile biases in some terminal Epipaleolithic cases, particularly but not exclusively the large Natufian sites in the Levant, are about more than seasonal episodes of specialized hunting. Some of these cases must signal a fundamental shift in human foraging ecology, probably linked to chronic territorial circumscription.

All of the other Upper Paleolithic and Epipaleolithic cases shown in figure 11.7 occur in the living-structure and prime-dominated areas of the graph. The total distribution for these periods appears slightly compressed on the old age axis (*a* in fig. 11.5). The effect is highlighted for Epipaleolithic–Upper Paleolithic gazelles in figure 11.9, where points trend to the lower left area of the graph with the end of the Paleolithic. In two-dimensional format, this distribution would correspond to a steeper age structure in which a somewhat larger proportion of individuals is composed of younger adults or juveniles. It must be admitted, how-

ever, that the Levantine Mousterian distribution differs little from the later cases except in being slightly more scattered along the old-adult axis. There are no discernible trends in the *mean* values by Paleolithic period and subperiod in the Levantine sequence. Greater variation in the Middle Paleolithic distribution might be explained by the greater span of time represented by this sample, but it could also suggest less overall pressure on artiodactyl prey populations.

That nearly all of the European and Mesopotamian fallow deer cases hug the left margin of the distribution in the Galilee and elsewhere (fig. 11.11) is almost certainly explained by herd compositions unique to *Dama*. Kersten (1987:121) noted the similarity between Epipaleolithic exploitation of Mesopotamian fallow deer at Ksar 'Akil and the age structure of one recent European fallow deer reference population (*Dama dama*, Ueckerman and Hansen 1983, cited in Kersten 1987). Similar distributions are seen for fallow deer archaeofaunal samples from Mediterranean Turkey, Israel, and Italy. Again, there are no clear signs of predator pressure until the end of the Epipaleolithic.

DISCUSSION OF RESULTS FROM THE DENTAL REMAINS

The ungulate assemblages discussed in this section were associated with good preservation conditions in limestone shelters where sediment chemistries were dominated by carbonate and phosphate minerals and

Table 11.3

Three-cohort system results for ungulates from Upper Paleolithic and Epipaleolithic Mediterranean sites in Israel, Italy, Turkey, and Lebanon, based on the dP4–P4 or dP3–M3 dental eruption and wear sequence

Site (Country)	Layer(s)	Prey	Total MNE	% Juveniles	% Prime Adults	% Old Adults
Early Upper Paleolithic						
Kebara Cave (Israel)	All cuts	Gazelle	85	46	46	8
	All cuts	Fallow deer	95	45	46	8
Üçağızlı Cave (Turkey)	C–I	Roe deer	15	27	73	0
	C–I	Wild goat	42	22	64	14
	C–I	Fallow deer	18	44	50	6
	B–B3	Roe deer	10	20	60	20
	B–B3	Wild goat	9	33	56	11
	B–B3	Fallow deer	16	31	63	6
Riparo Mochi (Italy)	F–G	Red deer	8	13	75	12
	F–G	Ibex	5	20	80	0
Upper Paleolithic						
Riparo Mochi (Italy)	C–D	Red deer	22	27	64	9
	C–D	Ibex	11	36	64	0
	C–D	Roe deer	7	29	71	0
Epipaleolithic						
Grotta Palidoro (Italy)	33–34	Red deer	10	20	70	10
	33–34	Aurochs	7	29	71	0
Ksar 'Akil (Lebanon)	I–V	Wild goat	43	14	56	30
	II	Fallow deer	21	48	38	14
	III	Fallow deer	20	55	40	5
	IV	Fallow deer	17	35	41	23
	V	Fallow deer	33	48	42	9
Terminal Epipaleolithic						
Grotta Polesini (Italy)	1–12	Red deer	784	33	51	16
	1–12	Aurochs	19	0	95	5
	1–12	Wild ass	111	25	59	16
	1–12	Horse	120	22	36	42
	1–12	Roe deer	202	21	66	13
	1–12	Chamois	79	48	27	25
	1–12	Ibex	13	0	85	15
Ksar 'Akil (Lebanon)	I	Fallow deer	20	80	15	5
Hayonim Cave (Israel)	B	Gazelle	18	50	39	11
Hayonim Terrace (Israel)	All cuts	Gazelle	31	42	39	19

SOURCES: For Kebara Cave, Speth and Tchernov 1998:231; for the Epipaleolithic at Ksar 'Akil, Kersten 1987; for the Natufian at Hayonim Cave and Hayonim Terrace, Munro 2001:294.

NOTE: Total MNE refers only to the total number of dental specimens that could be age-scored. The equid dental sequence uses P2 rather than P4.

where carnivore gnawing damage was rare or absent (but see Speth and Tchernov 1998 on Kebara Cave). Assuming that the chances of bone preservation generally improve as the time between deposition and archaeological discovery decreases, the consistent presence of prime-dominant mortality patterns contradicts what would be expected if this mortality pattern were due to differential preservation of juvenile and adult ungulate teeth. In fact, the emphasis on prime-adult prey was geographically and temporally widespread from Middle Paleolithic through Holocene cultures (e.g., Enloe 1997; Hoffecker et al. 1991; Jaubert et al.

Table 11.4

Three-cohort system results for artiodactyls from Late Pleistocene Mousterian Mediterranean sites in Israel and Italy, based on the dP4–P4 or dP3–M3 dental eruption and wear sequence

Site (Country)	Layer(s)	Prey	Total MNE	% Juveniles	% Prime Adults	% Old Adults
Mousterian (Late Pleistocene)						
Grotta di Sant'Agostino (Italy)	S0–3	Red deer	46	41	44	15
	S0–3	Fallow deer	13	35	50	15
	S0–3	Roe deer	9	44	39	17
Grotta Breuil (Italy)	Br	Aurochs	48	1	87	12
	Br	Red deer	42	0	71	29
	Br	Ibex	16	37	26	37
	B3/4	Red deer	18	55	44	1
	B3/4	Ibex	30	0	87	13
Grotta dei Moscerini (Italy)	M2–4,6	Red deer	39	10	38	52
Grotta Guattari (Italy)	G4–5	Red deer	25	24	44	32
	G4–5	Aurochs	13	15	62	23
Kebara Cave (Israel)	All cuts	Gazelle	316	30	62	8
	All cuts	Fallow deer	114	38	47	16
	All cuts	Red deer	40	18	55	27
	All cuts	Aurochs	46	20	67	13

SOURCES: For Kebara Cave, Speth and Tchernov 1998:231; for Italian sites, Stiner 1990, 1994.

NOTE: Total MNE refers only to the total number of dental specimens that could be age-scored.

1990; Pike-Tay et al. 1999; Rick and Moore 2001; Smith 1974; Speth and Tchernov 1998; Stiner 1990). The Wadi Meged data merely push the known onset of prime-focused ungulate hunting back to at least the late Middle Pleistocene.

Prime-focused ungulate hunting can be a relatively fragile predator-prey relationship, because human hunters disproportionately seek adults. An emphasis on adult male prey can alleviate some of the pressure this harvesting strategy might place on ungulate populations but cannot eliminate it entirely. Though seemingly antithetical to "prudent predation" models (Pianka 1978:210), prime-age-focused hunting is

Figure 11.8

Mortality patterns in artiodactyl species from Mediterranean assemblages dating to the terminal Epipaleolithic: Natufian in Hayonim Cave and Hayonim Terrace (gray infill), late Epipaleolithic Grotta Polesini (Italy), and Ksar 'Akil (Lebanon). Symbols are as in figure 11.7.

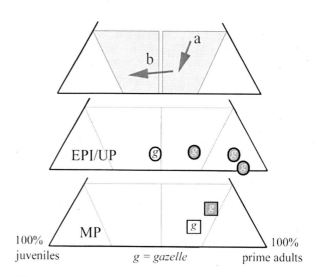

Figure 11.9

Mortality patterns in mountain gazelles from Epipaleolithic–Upper Paleolithic and Mousterian sites in the Galilee. Data are a subset of the points shown in figure 11.7.

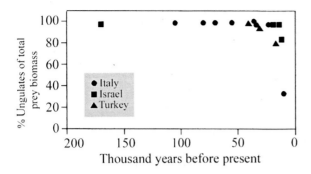

Figure 11.10
Changes in the amount of ungulate biomass in total prey spectra of Mediterranean Paleolithic faunal series from Israel, Italy, and Turkey.

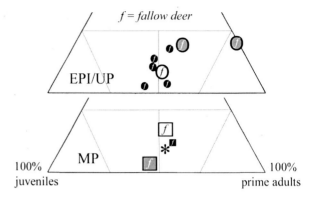

Figure 11.11
Mortality patterns in two species of fallow deer (*D. mesopotamica* and *D. dama*) from Mediterranean Epipaleolithic, Upper Paleolithic, and Mousterian sites in Israel, Turkey (Üçağızlı Cave), Lebanon (Ksar 'Akil), and Italy (Latium and Liguria). The asterisk denotes the living-structure of one recent and apparently stable European fallow deer population. Data are a subset of points presented in figure 11.7.

ecologically complementary to the patterns of prey age selection by cursorial predators and, to a lesser extent, to those of nonhuman ambush predators. Habitual prime-focused ungulate hunting may be feasible, however, for omnivorous predators, who can switch to other foods when the densities of favored prey decline (Stiner 1994; Winterhalder and Lu 1997). The Wadi Meged data, in combination with data from other Mediterranean faunal series, may suggest that greater pressure on ungulate resources began to manifest itself in the Upper Paleolithic, but clear signs of it in the three-cohort approach appear only in the late Epipaleolithic.

MORTALITY PATTERNS FROM BONE FUSION DATA

Comparisons of bone fusion states across limb elements offer detailed information on juvenile age structures (e.g., Horwitz and Goring-Morris 2001; Kersten 1987; Zeder and Hesse 2000), because bone fusion follows a fixed order during development (table 11.5). Barring radical differences in nutrition among individuals, the timing of fusion for each type of element is predictable relative to the month of birth. Birthing schedules vary with species, and the span of a species' birthing season may vary across environments, but some general assumptions about age of fusion are justified by the wildlife literature. Only gazelles are abundant throughout the Wadi Meged faunal series, and so they are the focus of analysis here.

Mountain gazelles of the Levant today are born in late spring, generally in May and June, although a second, minor birthing peak may occur in the autumn (Ayal and Baharav 1983; Baharav 1974, 1983a, 1983b;

Frankenburg 1992; Mendelssohn 1974; Nowak 1991; Shy et al. 1998; see reviews by Dean 1997; Munro 2001). Variation in the birthing season is affected by the availability of surface water from year to year, and the birthing season is less restricted as conditions become wetter. Gazelles, including mountain gazelles,

Table 11.5
Fusion schedules in months for selected appendicular elements of mountain gazelles and European fallow deer

Species and element	Proximal end	Distal end
Gazelle (Gazella gazella)		
First phalanx	5–8	—
Tibia	12–18	8–10
Calcaneum (tuber calcanei)	—	10–16
Femur	10–16	10–16
Metapodials	—	10–16
Radius	~2	12–18
Fallow deer (Dama dama)		
First phalanx	12–14	—
Tibia	27–28	18
Calcaneum (tuber calcanei)	—	23–24
Femur	22–24	22–24
Metapodials	—	22–24
Radius	6–7	24

SOURCES: Davis 1980 for mountain gazelles; Kersten 1987 for European fallow deer.

are relatively productive ungulates, mainly because of their rapid maturation rate (Munro 2001). Gestation lasts six months, and although normally each doe bears only one fawn per year, females bear their first fawn between one and two years of age and in each year thereafter if conditions are reasonably good. Environments were different during the Pleistocene, but the peak month of birth probably varied less in the past than did the duration of the birthing season.

METHODS AND SAMPLES

Although four fusion states were initially recorded per element (Appendix 2), unfused through nearly fused elements were grouped as "immature" for this analysis, and fully fused bones formed a second (adult) cohort. Percentages of unfused and fused specimens were calculated for the Wadi Meged samples on the basis of MNE in order to guard against double-counting an element. This procedure differed from the more common practice of making NISP-based counts of fused and unfused specimens (compare Bar-Oz et al. 1999; Davis 1980a, among others), in which there is a significant risk of double-counting elements. Both the NISP- and MNE-based methods will duplicate counts of individuals, however, since paired elements are nearly always combined, as they were in this study.

Six major appendicular elements of gazelles were compared. These were the proximal epiphysis of the first phalanx; the distal tibia; the distal calcaneum (tuber calcanei); the proximal head of the femur; the distal metapodials; and the distal radius. The fusion schedule for mountain gazelles *(G. gazella)* is taken from Davis 1980a. All units of Layer E of Hayonim Cave were combined to obtain a Mousterian sample large enough for comparison with those of later periods, as was done for the dental-based mortality analysis described earlier.

RESULTS FOR GAZELLE BONE FUSION AT HAYONIM CAVE

Some important differences existed between the Mousterian, Kebaran, and Natufian periods at Hayonim Cave in terms of the exploitation of juvenile mountain gazelles (tables 11.6 and 11.7). Figure 11.12 compares the Kebaran pattern of juvenile exploitation with that of the whole of the early Mousterian. The white area in the graph highlights the fact that these differences are concentrated in the second year of life in young gazelles, or yearlings; for immature gazelles under one year of age, exploitation patterns are nearly

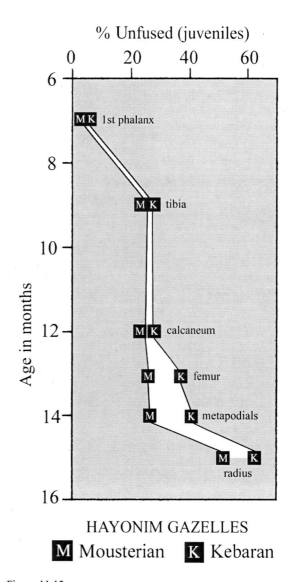

Figure 11.12
Percentages of unfused (immature) specimens for six limb elements of gazelles from the Kebaran and Mousterian layers of Hayonim Cave. The elements fuse over the course of 6 to 16 months, beginning with the proximal first phalanx, which is followed by the distal tibia, the distal calcaneum, the proximal femur head, the distal metapodials, and finally the distal radius.

identical. Very few fetuses or fawns occur in the Mousterian or Kebaran assemblages.

The situation is quite different for the early Natufian of Hayonim Cave and Terrace (fig. 11.13), due mainly to hunters' remarkable emphasis on very young gazelles. Many of the first phalanges of gazelles from the Natufian are unfused, composed of porous immature bone, or both, indicating that a third or more of these individuals were fawns or even near-term fetuses

Table 11.6
Percentages of fused and unfused gazelle and fallow deer limb bones in the Wadi Meged Kebaran and Mousterian faunal assemblages

Element-End	Age (mo.) at Fusion	Hayonim Kebaran			Meged Kebaran UP			Hayonim Mousterian		
		MNE	% Unfused	% Fused	MNE	% Unfused	% Fused	MNE	% Unfused	% Fused
Gazelle										
First phalanx-P	5–8	70	4	96	28	4	96	68	2	98
Tibia-D	8–10	17	24	76	8	25	75	25	24	76
Calcaneum-D	10–16	15	20	80	11	18	82	29	21	79
Femur-P (or -D)	10–16	28	36	64	8	12	87	27	26	74
Metapodials-D	10–16	69	38	62	38	27	72	55	25	75
Radius-D	12–18	10	60	40	3	*	*	12	50	50
Fallow deer										
First phalanx-P	12–14	11	9	91	0	—	—	39	5	95
Tibia-D	18	3	*	*	0	—	—	14	7	93
Calcaneum-D	23–24	6	*	*	0	—	—	12	58	42
Femur-P (or -D)	22–24	6	*	*	0	—	—	3	*	*
Metapodials-D	22–24	7	29	71	0	—	—	26	38	61
Radius-D	24	0	—	—	0	—	—	8	12	87

NOTE: Total MNE refers only to those elements for which fusion state could be recorded; all estimates of unfused versus fused bones are MNE-corrected counts. In archaeological specimens, fused elements are only those for which the process was complete at the time of death; unfused elements include those in which the epiphysis is unattached or partly fused.

KEY: (P) Proximal end; (D) distal end. (*) Remains were too few to warrant calculation.

Table 11.7
Percentages of fused and unfused gazelle limb bones in the Natufian faunal assemblage from Hayonim Cave and Terrace

Element-end	Age (mo.) at Fusion	Early Natufian (HC)			Late Natufian (HT)		
		MNE	% Unfused	% Fused	MNE	% Unfused	% Fused
First phalanx-P	5–8	99	38	62	72	43	57
Tibia-D	8–10	30	17	83	11	27	73
Calcaneum-D	10–16	23	4	96	15	20	80
Femur-P (or -D)	10–16	14	21	79	8	25	75
Metapodials-D	10–16	54	28	72	46	41	59
Radius-D	12–18	9	44	56	6	50	50

SOURCE: Munro 2004.

NOTE: Total MNE refers only to those elements for which fusion state could be recorded; all estimates of unfused versus fused bones are MNE-corrected counts. In archaeological specimens, fused elements are only those for which the process was complete at the time of death; unfused elements include those in which the epiphysis is unattached or partly fused.

KEY: (P) proximal end; (D) distal end. (*) Remains were too few to warrant calculation.

(unpublished data from N. D. Munro). The early and late phases of the Natufian from Hayonim Cave and Hayonim Terrace are similar to each other (table 11.7), but the Natufian as a whole differs greatly from both the Kebaran and the Mousterian in terms of gazelle assemblages. Because individuals just one year older than fawns are present but not abundant in the Natufian gazelle assemblages, the emphasis on fawns can-

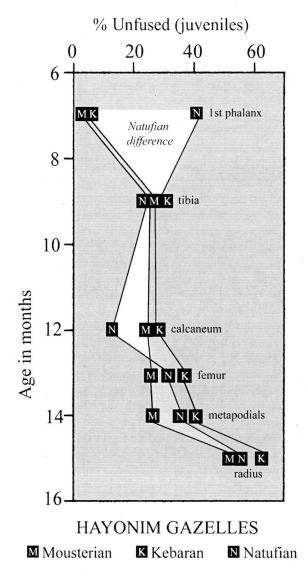

% Unfused (juveniles)

HAYONIM GAZELLES

Ⓜ Mousterian   Ⓚ Kebaran   Ⓝ Natufian

Figure 11.13
Percentages of unfused (immature) specimens for six limb
elements of gazelles from the Natufian (all phases) of
Hayonim Cave and Terrace, contrasted with earlier periods.
Element fusion sets are arranged as in figure 11.12. (Data from
Munro 2004)

not be explained simply by seasonal differences in site
occupations. Horwitz and Goring-Morris (2001:119)
obtained a similar result for a small sample from the
generally coeval Ramonian site of Upper Besor 6.

There seems, then, to have been a net increase in
juvenile off-take with time, consistent with the observations of other authors working in the Galilee (Davis
et al. 1994; Munro 2001; but see Bar-Oz et al. 1999).
These changes may have begun with slightly increased
exploitation of yearlings during the Kebaran. By the
Natufian, a conspicuous number of fawns or very

young juveniles had been added to the death assemblages, and perhaps at the expense of yearlings. The
trend toward greater exploitation of young animals
could be the result of increased harvesting intensity.
Certainly the early Mousterian pattern represents the
lightest effect on mountain gazelle populations, and
the Natufian, the severest. In addition, gazelles became
progressively more important than other ungulates in
the Wadi Meged series, whether quantified in terms
of NISP or prey biomass (chapter 7).

Figure 11.14 compares trends in the ratios of juvenile to adult gazelles as determined on the basis of two
criteria, metapodial fusion (which occurs between 10
and 16 months of age) and dental eruption and wear,
for cultural periods represented in the Wadi Meged
and other Galilee sites (table 11.8); the latter are documented by Davis (1983), Davis et al. (1994), Munro
(2001), Rabinovich (1998), and Speth and Tchernov
(1998). Although the comparison obscures variation
within the younger juvenile cohorts, and the added
cases involve NISP-based counts, I believe the comparison is useful because fusion of the distal metapodials corresponds best to the timing of the dP4–P4/M3
dental replacement boundary. Davis et al. (1994) noted
from these types of data a gradual increase in the proportion of juvenile animals hunted with time. Davis
(1981, 1983) has also argued that increased culling of
juveniles began early in the Natufian period because of
the coincidence of the human occupations with the
spring birthing peak, apparently the result of greater
overall sedentism, new hunting methods, or both. Bar-
Oz et al. (1999:78–79), also working with NISP-based
fusion counts, found no significant differences in their
data set in juvenile off-take overall between the Natufian and earlier Epipaleolithic periods. The results for
the Wadi Meged series—a faunal series from a fixed
location—are in general agreement with Davis's
results provided that dental and early-fusing bone elements are considered. The trend seems less gradual
than previously supposed, and the proportion of juveniles is somewhat higher in the dental data than in
the bone fusion data (perhaps because milk teeth preserve better than immature bone).

GAZELLE SEX RATIOS

Considerable size overlap is expected to exist between
the sexes in ungulates, particularly in small-bodied
taxa such as gazelles (e.g., Davis et al. 1994). The rarity of discrete size distributions in plots of linear osteometric traits makes the analysis of sex ratios in

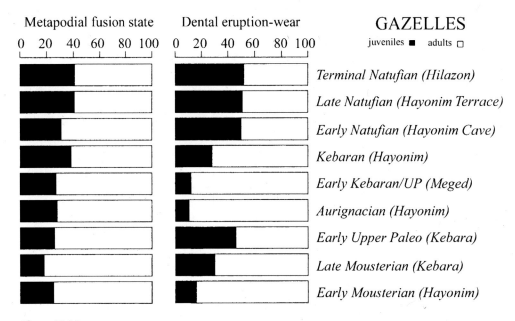

Figure 11.14
Trends in ratios of juvenile to adult gazelles for the early Mousterian through terminal Natufian periods at various Galilee sites, based on metapodial fusion and dental eruption and wear data.

Table 11.8
Ratios of juvenile and adult mountain gazelles and Mesopotamian fallow deer in Galilee Paleolithic sites, based on metapodial fusion and dental evidence

Period	Cave Site	Metapodials			Teeth			Source
		MNE	Juveniles	Adults	MNE	Juveniles	Adults	
Gazelles								
Terminal Natufian	Hilazon	—	.41	.59	—	.52	.48	Munro 2001
Late Natufian (HT)	Hayonim	—	.41	.59	—	.50	.50	Munro 2001
Early Natufian (HC)	Hayonim	—	.28	.72	—	.50	.50	Munro 2001
Kebaran	Hayonim	69	.38	.62	18	.28	.72	Stiner
Kebaran	Hayonim	—	.25	.75	—	—	—	Davis 1983
Early/pre-Kebaran	Meged	38	.27	.72	8	.12	.88	Stiner
Aurignacian	Hayonim	—	.28	.72	—	—	—	Davis 1983
Aurignacian	Hayonim	—	—	—	128	.11	.89	Davis et al. 1994
Early UP	Kebara	—	.26	.74	—	—	—	Davis 1983
Early UP	Kebara	—	—	—	85	.46	.54	Speth and Tchernov 1998
Late Mousterian	Kebara	—	.18	.82	—	—	—	Davis 1983
Late Mousterian	Kebara	—	—	—	316	.30	.70	Speth and Tchernov 1998
Early Mousterian	Hayonim	—	.21	.79	—	—	—	Davis 1983
Early Mousterian	Hayonim	55	.25	.75	36	.16	.84	Stiner
Fallow deer								
Kebaran	Hayonim	7	.29	.71	4	.25	.75	Stiner
Early Mousterian	Hayonim	26	.38	.61	16	.56	.44	Stiner

NOTE: Dental eruption and wear data are based on the dP4–P4 or dP4–M3 sequence. MNE refers to the total number of specimens that could be age-scored.

ungulate remains challenging but not impossible. As explained earlier, only gazelles were sufficiently common in the Wadi Meged series to permit comparisons across culture periods, and in making them, I used astragalus measurements because of their abundance and measurable state. The astragalus is not a fusing element but rather grows by accretion. Juvenile astragali are distinguished from those of adults by their rounded, porous contours. Specimens in the final stages of juvenile development, however, are not altogether different in appearance from fully mature specimens, a point I took into consideration for the analysis described in what follows.

Figure 11.15 shows the ranges of astragalus sizes in gazelles, based on the greatest length of the lateral face and the distal breadth, for the Wadi Meged Kebaran/pre-Kebaran and early Mousterian assemblages. The shaded area of the graph highlights the difference in maximum astragalus length between the two chronological groups, showing some of the Mousterian gazelles (the three specimens in the unshaded area) to be larger than any of the individuals in the later set of assemblages. Osteometric samples for distal humeri, distal tibias, and mandibular condyles of the gazelles (chapter 6, table 6.2) confirm a greater maximum size during the Mousterian. Davis et al.'s (1994) measurements of astragalus distal breadth (15.5–19.5 mm) indicate that the Aurignacian gazelles of Hayonim Cave were similar in size range to the Mousterian sample.

Three possible explanations for the observed differences in gazelle body (astragalus) sizes between the Mousterian and the Upper Paleolithic–Epipaleolithic death assemblages from the Wadi Meged are sex ratio differences between the periods, population-level differences in the potential for somatic growth, and age structure differences. The mortality evidence indicates that the Mousterian sample contains a greater number of mature adults than juveniles, but not many more. This fact and the deliberate omission of immature and burned astragali from the measured sample eliminates the third hypothesis. These results suggest but do not prove that mountain gazelles experienced body size diminution by the Kebaran period. The sex structure is examined next in an attempt to eliminate the first hypothesis.

Two studies, one by Horwitz et al. (1990) and a second by Davis et al. (1994), used osteometric measurements of astragali and other skeletal elements of recent, known-sex, adult mountain gazelles from northern and central Israel to infer gazelle sex ratios in archaeofaunas of the same region. Their results for

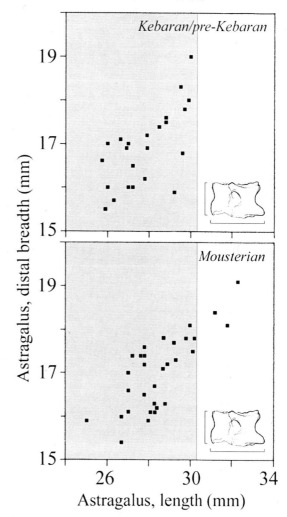

Figure 11.15
Mountain gazelle astragalus size distributions in the Kebaran/pre-Kebaran and early Mousterian assemblages from the Wadi Meged, based on total length and distal breadth. Shading highlights the difference in maximum length between periods.

astragalus size ranges in control populations differed, with Davis et al. obtaining a somewhat greater overall size range and greater overlap between the sexes in a larger sample. The two sets of researchers also disagreed over the feasibility of inferring gazelle sex ratios. Davis et al. (1994) noted size decreases, or diminution, in Galilee mountain gazelles during the Late Pleistocene, as well as an absence of visible bimodality in element size distributions. They interpreted the diminution trend as a consequence of climate warming and related changes in habitat structure (see also Bar-Oz et al. 1999). Horwitz et al. (1990) and Cope (1992) also recognized the likely imprint of climate effects on the Galilee gazelles but argued that the effect was pressed further by a new pattern of human predation at the end of the Epipaleolithic, or Natufian, period.

Figure 11.15 suggests a possibly bimodal distribution for the distal breadth of gazelle astragali, although the size ranges for males and females overlap appreciably. The chance that a fraction of the male population is mixed with counts of females, and vice versa, is an acceptable risk in mixture analysis (Josephson et al. 1996; Stiner et al. 1998), an informal version of which is applied here. This approach compares two peak positions and, by extension, the distance between them, if the normal distributions of the two peaks overlap. On the side of caution, the more inclusive astragalus size ranges of Davis et al. (1994:94) are used for the sex-specific distributions of recent gazelles. The recent adult female distal breadth range is 15.0–17.5 mm, and the male range 16.0–18.5 mm, with an imaginary division at roughly 16.5 mm (fig. 11.16, arrow *a*), based on the assumption that a crevice or plateau must exist midway between the peaks of two underlying normal distributions. From this perspective, only about 38 percent of the adult gazelles taken during the Kebaran/pre-Kebaran in the Wadi Meged were females. In other words, there was a mild bias toward adult males at 62 percent. A cutoff of 16.5 mm or less suggests that 65 percent of adult astragali were from males in the Natufian layer of Hayonim Cave, based on Munro's (2001) fine stratigraphic divisions within the Natufian samples. This is less than the figure of about 80 percent males obtained by Horwitz et al. (1990) for Natufian and prepottery Neolithic (phase A) gazelle assemblages.

Cope (1992) proposed that greater harvesting of male gazelles took place during the Natufian, and she cited evidence of diminution and greater morphological variation in gazelles in an argument for their "proto-domestication." Dayan and Simberloff (1995) reanalyzed Cope's morphometric data and found no evidence for greater morphological variation during the Natufian period. Munro (2001:175–177), in a separate study, found that despite the existence of large caches of male gazelle horn cores in the Natufian layer of Hayonim Cave, the postcranial data (principally distal humerus measurements) indicated only a mild bias toward adult males.

If gazelles were larger during the Mousterian, the midpoint between the sex-specific distributions must also have been somewhat greater. Shifting the division

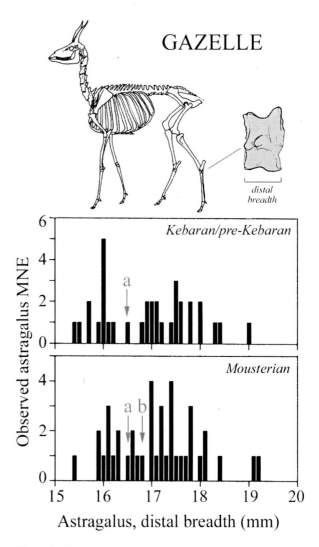

Figure 11.16
Distribution by distal breadth of gazelle astragali from the Kebaran/pre-Kebaran and early Mousterian assemblages of the Wadi Meged. The label *a* denotes the 16.5 mm cutoff between the sexes based on recent populations; the label *b* marks the adjusted cutoff at 16.8 mm based on the assumption that the larger sizes of Mousterian gazelle astragali represent larger body sizes attained.

between the adult sexes slightly upward, to about 16.8 mm (fig. 11.16, prediction *b*), on the basis of the maximum size range for the early Mousterian sample indicates that roughly 35 percent of the adults taken were females and 65 percent males. This ratio is quite similar to that for the Kebaran/pre-Kebaran and Natufian gazelles from Hayonim Cave. Thus, in all periods there was a bias toward adult male prey of roughly 60–65 percent. Some aspect of human hunting methods seems to have produced the adult male bias, but this aspect of culling existed throughout the Paleolithic in the Wadi Meged. The sex bias in gazelle hunting seems not to have been a late development, and it did not increase during the Natufian period. No doubt the debate over gazelle sex ratios from osteometric measurements will continue, particularly with regard to the later periods and as more data become available. What is most important for this study is the consistency of the sex ratio over time in a single, long faunal series in a fixed location.

It would be desirable to know about fallow deer sex ratios as well, but measurable specimens from the Mousterian layers of Hayonim Cave are few. There are fewer still from the Kebaran/pre-Kebaran and almost none from the Natufian in the Wadi Meged sites. The maximum dimensions of available samples, such as those of the astragalus and mandibular condyle, appear similar for the Mousterian and Kebaran periods (chapter 6, table 6.2), but nothing can be said about fallow deer sex ratios from these data.

CONCLUDING DISCUSSION

The data on artiodactyl age structures speak to questions about the hunting capabilities of early humans and the antiquity of prime-age-focused hunting, a uniquely human predator-prey relationship. The results for skeletal decomposition presented in earlier chapters, as well as the quantitative methods employed in this one, ensure that the mortality patterns were not substantially biased by poor preservation or analytical oversights. Dental and bone fusion samples from the Wadi Meged sites, though relatively small, are sufficient to infer ungulate mortality patterns and, in gazelles, sex structures as well.

The ungulate remains from the early Mousterian of Hayonim Cave indicate procurement primarily or exclusively by hunting. The mortality patterns for mountain gazelles, fallow deer, and aurochs in the early Mousterian assemblages either mimic the age structures of living ungulate populations (nonselective hunting) or are biased toward prime adult animals. In this regard, the results for the early Mousterian are consistent with results for the late Mousterian and Upper Paleolithic in the Galilee (Speth and Tchernov 1998) and for Middle and Upper Paleolithic sites of the Mediterranean Rim more generally. The early Mousterian assemblages from Hayonim Cave extend the antiquity of this niche characteristic backward into the late Middle Pleistocene, to at least 200–250 KYA. Humans' place in large predator guilds seems to have been shaped long ago by periodic competition with large carnivore genera such as *Panthera, Lupus, Lycaon,* and *Crocuta,* probably well before the early Mousterian. Hominids' emphasis on prime adult prey in artiodactyl species is ecologically unique among ungulate predators and persists in human behavioral repertoires worldwide, throughout Upper Paleolithic, Holocene, and recent cultures. Weapons technology must have been an important element of Paleolithic foraging adaptations, yet the previously noted shift in the hominid niche evolved long before stone- or bone-tipped hunting weapons came into regular or widespread use in the Old World (chapter 12).

Opportunistic exploitation of ungulates, including scavenging, is also apparent in some Paleolithic foraging regimens, but it played a small role and was just one facet of a more complex array of foraging behaviors (Stiner 1994; Stiner and Kuhn 1992). Scavenging is also suggested by some late Upper Paleolithic cases, but it involved species that were less important in the local diet (such as horses at Grotta Polesini, Stiner 1994). No evidence of scavenging behavior was recognized in the Wadi Meged faunal series, although the gathering of sessile small game was widespread.

One conclusion from the mortality analyses of the Galilee faunas is that the Middle and Upper Paleolithic culture periods were quite similar in terms of hominids' impact on herbivore populations. Increases in human population densities and greater pressure on ungulate resources are indicated only for the Epipaleolithic. The Mousterian samples are the most scattered with respect to the old adult age axis, suggesting that artiodactyl populations had ample recovery times, not unlike the situation for the Wadi Meged tortoises of this period (chapter 7). Prime-age-focused hunting might have begun reducing the mean age at first reproduction in prey populations sometime in the Upper Paleolithic, but this is less clear. However, the effect is apparent during the Natufian to a point that suggests that heavy exploitation of juvenile animals could not be avoided. Similar trends may be evidenced in other Mediterranean regions around the time of the Pleistocene-Holocene boundary, although background data on these cases are less complete.

The possible, subtle age compression of ungulate populations during the Upper Paleolithic does not indicate a high risk of extinction. It merely suggests that human hunters were a major source of mortality for these ungulate populations and that hunting was sufficiently heavy to push populations into growth mode for greater lengths of time. The strong juvenile biases in the late Epipaleolithic Mediterranean cases are exceptional by contrast, and they occurred around the same time as a precipitous drop in total ungulate bio-mass acquired by hunters (fig. 11.10). The biomass data lend credence to the idea that radical shifts in human ecology took place around this time, perhaps resulting ultimately in the demise of Paleolithic hunt-ing and gathering ways of life in some areas.

The ungulate mortality data are clear about inter-specific differences in predatory niches and the unique character of predation by the genus *Homo* by some time in the Middle Pleistocene. And the effects of hunting pressure on artiodactyl populations were pro-nounced near the end of the Epipaleolithic period in at least two areas of the eastern Mediterranean Basin. What remains in doubt for the Wadi Meged series is whether hunting pressure on large game resources was felt prior to the Epipaleolithic. There seems to be good evidence of this between the Middle and Upper Pale-olithic in the Italian sequence, but it is less clear-cut in the Levantine sequence. Therefore it is interesting that adult gazelle body sizes decreased with the Kebaran period in the Wadi Meged. Body size reduction in mountain gazelles is not explained by changes in the proportions of the two sexes hunted; adult male gazelles are represented consistently at a level of 60 to 65 percent in the Natufian, Kebaran/pre-Kebaran, and early Mousterian periods. A mildly male-biased sex ratio in gazelle archaeofaunas makes ecological sense in a predator-prey relationship that focused on the most reproductively active portion of a mammal pop-ulation. On the other hand, the Kebaran and earlier gazelle assemblages from Hayonim Cave pose a stark contrast to the juvenile-biased Natufian assemblage, signaling the development of a truly marginal preda-tor-prey relationship in the latter period.

CHAPTER TWELVE

The Changing Shape of Paleolithic Society

METAPHORICALLY SPEAKING, RESEARCH ON PALEOLITHIC societies is not unlike the now abandoned discipline of phrenology, the practice of feeling for bulges on the surface of the human cranium to learn about a person's character. The methods and information are different, of course, but paleoanthropologists are after secrets locked within an entity whose surface belies a remarkable organization. Archaeology has a better track record than phrenology for producing replicable results, but interpretations about the nature of extinct societies still vary from compelling to vague or baldly unbelievable. Frequently it is only the general contours of extinct social systems that are palpable to us, yet inferences about their internal order are made in spite of this.

Zooarchaeology might not seem the most direct way to investigate changes in human social adaptations, since the bulk of the information gained by its methods concerns food acquisition. Indeed, the faunal record of the Mediterranean Basin testifies foremost to shifts in human dietary breadth, population densities, and technological investment in the food quest.

There is at least one area, however, in which zooarchaeological data may be central to learning about the changing shape and organization of ancient forager societies. This is the division of labor between the sexes and between humans of different ages—specifically, shifts in the personnel who had direct access to animal protein. The Middle Paleolithic is striking for its narrow adherence to large game, almost irrespective of latitude (Kuhn and Stiner 1998b, 2001). Small game animals were widely exploited at lower latitudes, but always in small quantities relative to the total prey biomass consumed and usually in the form of taxa that could be collected easily. Biomass comparisons relegate small game to the status of what we might call "backup" resources, yet this was where some of the most dramatic changes took place in hunting agendas and costs during the Upper Paleolithic and Epipaleolithic. Patterns of small game exploitation diversified with time, encompassing shifts in the types of animals routinely sought, the substrates and habitats commonly searched, the means of capture, and the depth of technological investment. One gets the sense that a wider array of human adults and children became more directly involved in protein acquisition from the Upper Paleolithic onward, especially in the Epipaleolithic. How did this come about? As in any other major zooarchaeological study, a series of taphonomic and behavioral questions had to be answered in order to arrive at general conclusions. These were presented in

detail in earlier chapters as a means toward a larger end. Only the main outcomes of the research are summarized in what follows, and their evolutionary implications are explored.

BEHAVIORAL VARIATION IN THE MIDDLE PALEOLITHIC

The late Middle Pleistocene must have been a volatile time in human evolutionary history. The human fossil record indicates considerable encephalization, culminating in a brain size essentially the same as our own. Archaeofaunal assemblages became more clearly centered on ungulates by this period (Stiner 2002a). A great many of the archaeological sites occur in natural caves and shelters, where, in some cases, cultural strata intercalate or are mixed with the debris of large carnivore dens (Brugal and Jaubert 1991; Gamble 1986, 1999; Stiner 1991b, 2002a; Stiner et al. 1996; Straus 1982; Villa 1991, among others). Gamble (1983, 1986) was the first to demonstrate the prevalence of alternating occupations by hominids and hyenas, bears, or wolves in caves across the Mediterranean Basin. This situation was common, for example, in Middle and Late Pleistocene caves in Italy (Stiner 1991b, 1994; Tozzi 1974), Spain (Lindly 1988; Straus 1982), southern France (Brugal and Jaubert 1991; Villa and Soressi 2000), European Turkey (Stiner et al. 1996), and some Levantine sites (Speth and Tchernov 1998). Where hominid components are thin, carnivore components often are thick and easily recognized. Carnivores also visited the middens made by humans, as is evidenced by extensive secondary modification of the bones of prey animals (e.g., Grotta di Sant' Agostino, Stiner 1994), implying a wide range of contact situations among these predators in Pleistocene environments.

Yet other Mousterian sites are dominated nearly exclusively by hominid-generated debris that is thick with artifacts and the remains of large, medium, and some small ungulates. One of the earliest examples of this pattern—from the late Middle Pleistocene—was found in Hayonim Cave. The quantities of faunal material there were not tremendous, but all of the high-quality body parts we expect to be valued by hunters were represented, and there was evidence of human modification of carnivore remains as well. Later Middle Paleolithic (or Late Pleistocene) examples of ungulate hunting are much more common, and

the cultural layers often are richer in faunal material. Examples in Israel include Kebara Cave (Bar-Yosef et al. 1992; Speth and Tchernov 1998), Amud Cave (Hovers 1998), and Qafzeh Cave (Rabinovich and Tchernov 1995). Comparable situations have been reported for late Middle Paleolithic cave sites in Italy (Stiner 1990, 1991a, 1994; Tozzi 1970), France (Chase 1986; David and Poulain 1990; Jaubert et al. 1990), Germany (Gaudzinski 1995), and Russia (Hoffecker et al. 1991). Indeed, many of the European Middle Paleolithic sites that preserve faunal remains contain impressive quantities of both ungulate bones and stone artifacts. The relative frequencies of these materials are positively and strongly correlated in the richer cave sites (Stiner 1994:362–368; Stiner and Kuhn 1992), implying that tool manufacture and maintenance were intimately connected to large game exploitation.

The abundance of human-modified bones, prime-dominated ungulate age structures, and relatively complete patterns of body part representation in prey indicate that Mousterian humans enjoyed uninhibited, early access to large game and that they hunted large animals regularly. Without concluding that Mousterian foraging habits were wholly equivalent to the adaptations of recent foragers, it is fair to say that the Middle Paleolithic faunal record includes some straightforward evidence of large game hunting. But the Mousterian faunal record also presents facets that are less easy to interpret, such as a consistently narrow diet of high-ranked prey types. Such properties of the Mousterian record diverge from what we expect based on knowledge of recent hunter-gatherers and many Upper Paleolithic peoples as well (Gamble 1999; Kuhn and Stiner 2001).

Though the trends within the Mousterian period are subtle, and no match for the demographic increases of the Upper Paleolithic, the Mousterian does seem to embody some change. There was an increase in Middle Paleolithic site numbers after about 70 KYA (e.g., Mirazón Lahr and Foley 2003; van Andel et al. 2003). Moreover, some of these sites appear to have been occupied more intensively than others, in the sense of more frequent visits or visits by somewhat larger social groups. The Wadi Meged study offers no information on site numbers in the Levant, but the roughly consecutive Mousterian sequences from Hayonim and Kebara Caves do provide a good example of differences in occupation intensity based on the character of bone and lithic debris and wood ash deposits.

COMPARISON OF THE MOUSTERIAN OCCUPATIONS AT HAYONIM AND KEBARA CAVES

Numerous reoccupations of Hayonim Cave and Kebara Cave took place during the Middle and Late Pleistocene. Micromorphological, mineralogical, and phytolith analyses show that the deposits in both sites were mainly anthropogenic, with much trampling of ashy hearth areas (Albert et al. 2000, 2003; Goldberg and Bar-Yosef 1998; Weiner et al. 2002). However, the intensity of occupation as indicated by debris generation and hearth building relative to sedimentation rates seems to have been substantially greater during the later Mousterian at Kebara Cave than in the early Mousterian of Hayonim Cave. These observations raise questions about qualitative and quantitative differences between the human occupations of the two sites. Are we seeing evidence that more people were present in the area during the later part of the Mousterian, differences in the ways the sites were being used, or both? In what follows, I summarize the insights offered by Meignen et al. (n.d.; see also Bar-Yosef et al. 1988, 1992) on the differences between these consecutive Mousterian sites in the Galilee, along with faunal observations specific to the Wadi Meged faunal study.

Hayonim Cave

The Early Mousterian occupations in Hayonim Cave were repetitive but largely ephemeral. A low density of artifacts might in other circumstances be explained simply by a rapid rate of sedimentation, but the prevalence of owl pellets and generally thin hearth lenses in Hayonim Cave cannot be accounted for in this way. Thermoluminescense dates suggest a sediment accumulation rate of about one meter per 10,000–15,000 years, whereas in Kebara Cave each meter of depth accumulated in roughly 3,000 years (Bar-Yosef 1998).

Although the intensity of human occupation at Hayonim Cave was generally low, all stages of stone tool production were in evidence there. Its users collected nodules of various flint raw materials mainly within 10–15 km of the cave, and numerous Eocene and Cenomanian outcrops of good-quality flint were available within 7 km. A few artifacts testify to the use of raw material sources some 20 km to the south and to others even 30–40 km away (Delage et al. 2000). Exotic materials constituted more than 10 percent of the assemblages in some Mousterian units, but the imported blanks did not always arrive in the form of finished products; Levallois and laminar blanks were brought in, along with debitage by-products. The non-local flint component indicates exploitation of a territory larger than the immediate surroundings of Hayonim Cave, but it does not appear to have been the result of a specific curation strategy.

A laminar technology aimed at the production of elongated blanks was recognized throughout the lower part of Layer E and in Layer F in the Mousterian sequence, together with Levallois core reduction geared toward producing both short and elongated products (Meignen 1998). The former technology was more developed in the lower Mousterian units, and numerous diversified, retouched blanks—characteristic elongated retouched points, retouched blades, side scrapers and inversely retouched scrapers on Levallois blanks, and typical burins—were present in Mousterian (MP) units 7 through 10 of Layer F. In unit 7, these artifacts were more abundant at the entrance of the cave than in the central (interior) area, possibly suggesting some spatial partitioning of activities in this large shelter.

The faunal data from Hayonim Cave indicate repeated exploitation of ungulates and tortoises, but never in great quantities. These and other observations point to low human population densities and narrow diets dominated by high-yield types of game. The small number of individual ungulates (MNI) in Layer E is striking, particularly if the large areas excavated and the great time span they represent are taken into account, and this situation contrasts with the dense faunal accumulations in the late Mousterian of Kebara Cave. Evidence of carnivore activity is virtually absent in the Mousterian assemblages of Hayonim Cave, despite the existence of hyenas and canids in Middle Pleistocene ecosystems of the region. The lack of gnawing damage may simply be another indication that refuse accumulation in the site was scant per occupation event—perhaps insufficient to attract large carnivores with any regularity.

Lithic artifacts were fairly common throughout Mousterian Layers E and F of Hayonim Cave, but their densities varied through the sediment column. This vertical variation in artifact densities is not explained by sediment compaction from mineral dissolution or variable rates of sedimentation (chapter 4). It therefore must reflect the frequency and duration of human occupation of the cave, indicating several pulses in site use over time. However, the densities of lithics seldom, if ever, approach those observed in Kebara Cave.

Oddly, there was little variation over time in the transport of high meat-utility or marrow-utility animal parts to Hayonim Cave, other than an apparent reluctance on the part of foragers to transport much of the vertebral column. Almost every meat- or marrow-yielding part of deer and gazelles was carried to the site for processing. Body part transport was only slightly more selective for the largest prey, aurochs, if different at all (chapter 10). Extraction of medullary marrow was thorough but always stopped short of grease rendering of any sort (chapter 5), a pattern typical for the Mousterian of Eurasia in general. Tortoises were roasted and then cracked open with the aid of a hammer and anvil. There is also much evidence of fire-aided food preparation.

Few horizontal divisions, if any, were discerned in the use of space in the central areas of Hayonim Cave in which Mousterian bone was preserved (ca. 8–10 square meters). The area excavated, however, was considerably less than that in Kebara Cave. Postdepositional burning of bone and lithic material was widespread in the Mousterian deposits, and the sediments were rich in wood ash products. Micromorphological evidence indicates considerable trampling around the hearth areas. These observations testify to much reuse of certain locations in the cave. Rather than yielding evidence for spatial partitioning of human activities, Hayonim Cave offers evidence of humans' attraction to certain spots within the cave time and again, an attraction that led to extensive postdepositional modification of material laid down previously, in spite of low rates or intensity of reoccupation. Diverse activities are apparent from the lithic technology, faunal remains, and evidence of fire building in the early Mousterian. Together the data suggest repeated, brief uses of the cave as a residential camp, probably by small numbers of people at any one time, within a strategy of high mobility (Meignen et al. n.d.). Complete core reduction on-site, together with a diversified tool kit, rules out the possibility of a task-specific or special-use camp.

Kebara Cave

The Late Mousterian occupations in Kebara Cave (units XI–IX) were much more frequent or "intense" over the stratigraphic sequence, which spans only about 3,000 years (ca. 60–57 KYA). The paucity of small rodent bones from owl pellets in the deposits is striking, given the abundance of small solution cavities and crevices in the walls and ceiling of this large cave. A relatively high density of human occupations is evident from the great number of stone artifacts, usually 1,000–1,200 per meter of depth. The same is true for the animal bones: the remarkably concentrated bone midden along the north wall in units XI–IX indicates high-volume game acquisition and the processing of multiple carcasses in a comparatively short time frame (Speth and Tchernov 1998, n.d.). In recent forager camps, the deposition of bone trash along the peripheries of habitation areas normally increases as the duration of occupation increases (Binford 1978, 1991, 1998; Brooks 1984; Galanidou 2000; Gorecki 1991; see also Schiffer 1983 for a general discussion). This phenomenon may explain the peculiar concentrations of bone debris in Kebara Cave. The hearths were also well developed and deeply stacked, creating the thick, zebra-striped deposits for which this site is famous (Bar-Yosef et al. 1992; Meignen et al. 1989; Weiner et al. 1995).

The large faunal samples from Kebara Cave permitted detailed analyses of prey body part representation and processing techniques (Speth and Tchernov 2001, n.d.). The largest game animals were represented principally by body parts of high nutritive value, whereas all of the body parts of medium and small prey generally were transported to the site. Foragers' food transport decisions appear to have been conditioned by the bulk and food utility of the parts, not unlike the situation at Hayonim Cave. In contrast to Hayonim, however, there is evidence that carcass processing and consumption took place at spatially discrete locations in Kebara Cave.

No major changes in lithic reduction strategies were detected through units XI–IX in Kebara Cave (Meignen and Bar-Yosef 1992), nor were any changes observed in the spatial organization of the central excavation area. A full suite of lithic production activities is represented in the Mousterian series, judging from the proportions of cortical and noncortical knapping products, ordinary flakes, Levallois products, and cores. Blocks of flint were imported from a maximum distance of 10–15 km (Shea 1991). Highly exhausted cores remained at the site; they were not included in the transportable tool kit, apparently because of an inefficient ratio of utility per unit mass (*sensu* Kuhn 1995; see also Hovers 1997:172). Few tools exhibit wear from prolonged use (Shea 1991). The small percentage of retouched tools, the low intensity of retouching on each piece, and observed patterns of use-wear all confirm a pattern of generally low-level utilization of stone implements. The tools, retouched and unretouched, were used for diverse tasks including butchery, woodworking, cutting and scraping of hard and medium materials, graving, and wedging (Plisson and Beyries 1998; Shea 1991), but each tool was soon abandoned for another. Even the retouched component of the lithic assemblage demonstrates no prolonged cycles of use and reuse.

The absence of imported, exotic raw materials and the complete reduction sequence of cores on-site indicate large, well-equipped, and relatively long-term encampments in Kebara Cave. Numerous hearths testify to successive episodes of burning, and human burials are present (Meignen et al. n.d.). In units VIII–VII, however, some of the ungulate carcasses were processed only to the extent necessary to prepare the higher-utility parts for further transport; mostly lower-utility skeletal parts were left behind in the cave (Speth and Tchernov 2001, n.d.). Interpretations of seasonal variation in site occupations at Kebara Cave have been advanced from this and other faunal evidence. Lithic and faunal data suggest the same general kinds of dense occupations for units VIII–VII as those seen in units XI–IX, but with rather different hunting agendas among the units.

NUMBERS OF PEOPLE AND OCCUPATION INTENSITY

One may argue from the foregoing observations that the most readily visible differences between the early and late Mousterian in Hayonim and Kebara Caves have to do with numbers of people on a landscape—rates of visitation and perhaps the sizes of the social groups present. Both caves appear to have served as residential encampments throughout most or all of their histories of occupation, judging from the diverse ranges of activities represented. The occupations at Kebara Cave, however, lasted longer or were in some sense more intense. The basic treatment of prey was pretty much the same at the two sites. The larger size of the faunal sample from Kebara Cave allowed for a more detailed investigation of seasonality and game processing techniques; we know less from the much older, smaller samples from Hayonim Cave. Yet to the extent that the samples are truly comparable, their contents are not terribly different, except with respect to quantities of bones and stone tools per meter of accumulated sediment. Mousterian foragers were avid hunters of large game, and they maintained remarkably narrow diets despite the great diversity of plants and animals in Mediterranean biotic communities. Foragers were thorough and judicious in the transport and use of meat and marrow throughout the Middle Paleolithic period in the Levant.

From the perspective of site structure we see greater, substantive contrasts between the two sites and the successive phases of the Mousterian that they represent. Hayonim Cave seems to have been characterized by redundant, spot-specific use of domestic space, whereas Kebara Cave displays a more rigidly partitioned spatial pattern, probably in response to faster rates of debris generation. This difference does not necessarily represent a cognitive shift in the hominids. It likely represents a natural response within a preexisting adaptation to slightly greater crowding in the most immediate sense—the point when debris begins to interfere with the continuation of specific activities in a limited space, as may occur in natural shelters because of wall configurations, drainage, illumination, and ventilation. Finer variations in resource scheduling may also be apparent in the late Mousterian, but this is less certain, owing to the limitations of the Hayonim sample sizes.

More than anything, these observations raise questions about variation in human population densities during the Mousterian period. There are few indications, if any, of significant increases in predator pressure on traditional resources during the Mousterian of the Levant, except perhaps on tortoises in the Wadi Meged sequence (the later Mousterian sample represented by MP units 1–2). If indications of increases in human numbers exist for the late Mousterian, they are subtle. The ungulate mortality evidence is suggestive but as yet unclear with respect to variation within the Mousterian (but see Speth and Tchernov n.d.; Stiner 1994). It is interesting that the most convincing indications of more people in the area are confined mainly to two spatial aspects of the archaeological record: internal differentiation in site structure during the later Mousterian and, on a geographic scale, greater numbers of sites that may also be richer in material. Differences in preservation and sampling attenuate these contrasts, but not to the extent that we may discount them.

TRENDS IN PALEOLITHIC HUNTING, POPULATION ECOLOGY, AND TECHNOLOGY

The debate over whether Middle Paleolithic people were hunters or scavengers is by now a rather old horse, and one ridden harder than most. Plenty of large game hunting took place during the Middle Paleolithic, but more to the point, it was only one of several important dimensions of the hominid predatory niche. Three trends are evidenced by the zooarchaeological and related data for the last 200,000 years: the early development of humans' niche as hunters of prime-age ungulates, a shift toward greater dietary breadth, including greater use of fast-moving, fast-reproducing small game, and technological changes toward greater efficiency in weapons and processing of animal carcasses. Recognition of these trends is based on observations about ungulate exploitation, about small game use and its demographic correlates, and, in a seemingly counterintuitive way, about the nature of hunting technology. The Mousterian is not noted for rapid change or pronounced regional variations in technology and subsistence. As mentioned earlier, however, there was a mild increase in Middle Paleolithic site numbers and occupation intensity after about 60,000 years ago, and there may be hints of new pressures on human populations and their traditional food supplies toward the close of the Middle Paleolithic period. Yet only the rapid radiations in Upper Paleolithic and Epipaleolithic technology were accompanied by notable expansions in dietary breadth.

TREND 1: HUMANS' HUNTING OF PRIME-AGE UNGULATES

Considerable overlap in the foraging interests of Pleistocene hominids and carnivores is apparent from their exploitation of large hoofed animals. Assuming that the niches of ungulate predators were shaped in part by the risks of interference competition for large resource packages, interspecific competition should have fostered ecological differentiation. Hominids appear always to have been ambush hunters of one sort or another. Hominids' tactics for hunting ungulates nonetheless grew increasingly distinct from the ways of coeval large cats, spotted hyenas, and large canids (chapter 11). Specifically, humans became the only predators that frequently targeted the reproductive core—prime adults—of ungulate populations. Spotted hyenas and large canids generally focus on juveniles and old adults in the same prey species, and most cats apart from cheetahs tend to take prey more randomly. Humans' focus on prime adult prey therefore is ecologically unprecedented and partly complementary to the niches of longer-established nonhuman predators—strong evidence for niche separation.

The emphasis on prime adult prey has been geographically and temporally widespread in human cultures from the Middle Paleolithic through the Holocene. The possibility that this pattern, when observed in archaeofaunas, can be explained by the differential decomposition of young ungulate teeth is refuted by a variety of taphonomic observations (chapters 5 and 11). Though it seems somewhat counterintuitive, prime-age-biased ungulate hunting is largely insensitive to regional and temporal variations in weapons technology. What is more, prime-biased

ungulate hunting can be a relatively fragile predator-prey relationship, because reproductive-age adults—including adult females—are disproportionately sought. Such a relationship may be most sustainable for omnivorous predators that can switch to other foods when the densities of favored prey decline (chapter 11). Neandertal populations nonetheless appear to have been quite carnivorous and uniformly dependent on highly ranked prey; later peoples appear to have been somewhat less so.

That prime-adult harvesting of bovids and cervids was well established by the Late Pleistocene raises the question of when it first evolved. Current evidence places this behavior in the Middle Pleistocene and perhaps earlier. It is clearly in evidence for the early Middle Paleolithic of Hayonim Cave. The expression of prime-dominant ungulate hunting by the earliest Middle Paleolithic suggests a deeper history for more generalized forms of large game hunting. It seems likely on theoretical grounds and from limited empirical evidence that a more basic adaptation for ungulate hunting had evolved in hominids by at least 500,000 years ago, if not earlier. For the moment at least, relatively few cases from this time range are available for comparison, and most of them are subject to many questions about site formation history. Nonconfrontational scavenging is also evidenced in some Middle Paleolithic cases in Italy, but it was just one facet of a more complex array of foraging behaviors and never a major source of meat. Scavenged meat seems to have been a backup resource in some situations, just as gatherable small game animals were in others.

Prime-age-focused hunting was a robust feature of the human niche from the Mousterian through the Upper Paleolithic. Some pressure on resources may be indicated by mild distortion of ungulate age structures in the Late Pleistocene, but this predator-prey system seems to have broken down only around the time of the Pleistocene-Holocene boundary, as is signaled by severe compression of ungulate age structures in some parts of the Mediterranean Basin.

The subject of ungulate mortality patterns merits continued investigation, and interpretations can no doubt be improved upon with good data on prey sex ratios and seasonality and related information on food quality from a variety of sites and environments.

TREND 2: CHANGING EMPHASES IN SMALL GAME FORAGING

The breadth of foragers' diets can vary a great deal, depending on the availability of high-quality foods.

Subsistence diversification is expected if foragers put excessive pressure on key resources, forcing them into decline. Indication of increasing dietary breadth during the Paleolithic in the Mediterranean Basin takes the form of greater proportional evenness between high-ranked and low-ranked prey types. Narrow diets, in which lower-quality prey is usually ignored, are sustainable only if the chances of finding superior types remain high. If rates of encounters with preferred prey decline, predators must broaden their diets by taking more low-yield types.

Archaeological applications of this reasoning assume that resources can be ranked in the energetic terms of the predator and that the relative values (payoffs) of prey can be evaluated at least partly from the adaptations of those that are exploited. One means for ranking prey of similar body sizes is in terms of handling costs, irrespective of the phylogenetic histories of the species. Such criteria place Mediterranean tortoises and rock-dwelling marine shellfish in a single category of "sessile" game—that which is immobile or sluggish. Quick-running and fast-flying animals, such as hares and partridges, have body weights similar to those of tortoises or an armload of shellfish, but they are far more difficult to catch by hand.

The relative emphasis that Pleistocene humans placed on three general types of small animals—slow-moving or sessile types, fast-running hares and rabbits, and quick-flying game birds—changed dramatically over time (chapter 7). Mousterian foragers seldom bothered with small prey unless it could be gathered easily. The volume of meat biomass obtained in this way was quite limited in an absolute sense but apparently was always of some importance to Middle Paleolithic populations. The situation changed with the beginning of the Upper Paleolithic and more still with the Epipaleolithic. A simple measure of evenness in the prey types eaten reveals significant expansion in human dietary breadth as early as 40,000–50,000 years ago in the eastern end of the Mediterranean Basin (chapter 9). Though beginning in oxygen isotope stage 3, most of the expansion throughout the basin took place during a phase of climate cooling (OIS 2, following Martinson et al. 1987). Had it occurred only in conjunction with global warming, the trend might not be distinguishable from natural shifts in animal community diversity and structure with effective latitude (sensu Keeley 1995). Instead, the evidence indicates a categorical change in the way humans interacted with small animal populations, beginning around the time of the Middle to Upper Paleolithic cultural transition.

The large-to-small body size contrast in the three Mediterranean series from Israel, Turkey, and Italy appeared trendless in NISP-based comparisons, but the relative productivity of the prey types emphasized increased with time across the entire prey spectrum. Biomass-corrected prey counts indicate greater use of faster-reproducing species such as gazelles and roe deer among ungulates, especially after about 15 KYA. Before that time, the proportional contribution of small game biomass to Paleolithic diets was constant at about 3 percent. Much earlier signals of dietary change (40–50 KYA) are found in the small game fraction of the Wadi Meged and other Mediterranean faunal series.

The differential productivity of small animal populations is a key to understanding the tenacity and ecological significance of the trend in small game exploitation. In the Mediterranean Basin, a simple distinction in the "catchability" of small animals happens to correspond to great differences in prey population resilience, which is governed mainly by the rates at which individual prey animals mature (chapter 8). Slow-moving tortoises and certain shellfish are especially susceptible to overharvesting, because it takes them several years to reach reproductive age. It is striking, therefore, that to the extent that Mousterian foragers exploited small animals at all, they focused so consistently on slow-growing types. Overharvesting of tortoises caused a reduction in the mean size of individuals available to foragers by either the earliest Upper Paleolithic (approximately 44 KYA) or the late Middle Paleolithic in Israel, and this effect was sustained throughout multiple climate oscillations thereafter. The maximum body size attained by long-lived individual tortoises remained about the same from the early Mousterian through the Natufian in the Wadi Meged, with size-skewing toward juveniles in the later part of the series. The timing and duration of diminution in the tortoises were largely independent of global climate trends—the other potential cause of diminution—and so the data point largely to a human cause.

The small game trend in the Wadi Meged series therefore is not unique. Similar phenomena have been observed on the Hatay coast of Turkey, on the northern Mediterranean Rim, and in certain other areas, although cases to the west generally postdate those of the Galilee. Bird exploitation became widespread with the early Upper Paleolithic. The burgeoning importance of lagomorphs in human diets was also common by the Epipaleolithic, certainly including the northern interior of Europe and arid lands to the south, in open and cave sites alike. Lagomorphs were common prey at many Solutrean and (especially) Magdalenian and Epigravettian sites in southern Europe (Clark 1987; Davidson 1983; Hockett and Bicho 2000; Stiner et al. 1999; Straus 1990; Zilhão 1990; but early in Portugal, Hockett and Bicho 1999; Stiner 2003), northwestern Europe (Albrecht and Berke 1982–1983; Barton 1999; Berke 1984; Bratlund 1996, 1999), southwestern Germany (Jochim 1998), Moravia (Svoboda 1990), the Dnestr region (Kosoutsky Layer 4; Borziyak 1993), and even at Norgorod-Severskii on the central Russian Plain (Soffer 1990). The surge in lagomorph exploitation occurred somewhat earlier in western Asia (compare Byrd and Garrard 1990; Henry and Garrard 1988; Klein 1995; Munro 1999; Tchernov 1992d), yet quite late in North Africa (Smith 1998). Environmental changes brought on by global warming (e.g., Madeyska 1999) may have expanded the habitats favored by lagomorphs and thus their numbers in Eurasia. However, paleontological evidence indicates that lagomorphs existed in most or all of these regions in earlier times but were largely ignored by humans (Stiner 1994; Tchernov 1994).

An important quality of small prey populations that rebound quickly is their greater potential reliability as a food source. Any forager population that can grow faster on low-value but more resilient foods will have a demographic advantage over competing populations. Warm-blooded small animals, mainly partridges, hares, and rabbits, mature in one year or less, and their populations rebound easily from heavy hunting by humans. The predator-prey computer simulations presented in chapter 8 indicate that lagomorph and partridge populations can support 7 to 10 times the annual off-take that tortoise populations can support. Thus, the greater dependence on slow-growing animals during the Middle Paleolithic, and on larger individual prey, implies that these early human populations were very small and dispersed. The first detectable human demographic pulse occurred more or less at the threshold of the Middle to Upper Paleolithic cultural transition, which falls quite early in the eastern Mediterranean Basin. Some late Mousterian populations must also have been affected by this process, perhaps with the expansion of early Upper Paleolithic populations east to west. The zooarchaeological evidence testifies to additional demographic pulses over the course of the Late Pleistocene, the intensity of which increased after the Last Glacial Maximum.

Although small game animals were ubiquitous, biomass estimates show that they were backup resources, means for adjusting to variation in the availability of large game to individual consumers, male and female, young and old. Because the exploitation of small animals was highly conditional, these data can be exceptionally revealing about changes in foraging adaptations and other aspects of human ecology. Not everything that predator-prey relationships can tell us concerns food volume alone.

TREND 3: LATE DEVELOPMENT OF MORE EFFICIENT TECHNOLOGIES

There is much evidence for large game hunting in the Middle Paleolithic, but very few Middle Paleolithic tools can reasonably be called hunting weapons. Stone-tipped weapons appeared late in the Eurasian Middle Paleolithic (e.g., Shea 1989, 1993), but they were not widespread (Kuhn and Stiner 1998b). Bone working and bone-tipped weapons designed specifically for hunting were generally confined to Upper Paleolithic and later cultures, when rapid turnover in weapon designs and other aspects of material culture was typical. However, some inventiveness with respect to weapon heads cannot be denied for certain later Middle Paleolithic cases in Eurasia, where bifacial (*blatspitzen*, Müller-Beck 1988) and hafted Levallois-type stone points occur (Shea 1989), nor for bifacial stone and bone points in the Middle Stone Age of Africa (e.g., Brooks et al. 1995; d'Errico et al. 2001; McBrearty and Brooks 2000; Yellen et al. 1995).

It is paradoxical nonetheless that humans routinely hunted medium-size and large ungulates long before the undisputed or regular appearance of stone-tipped and bone-tipped weapons in Paleolithic records anywhere. Prime-age-focused ungulate hunting is evidenced by at least 200–250 KYA, about the time the earliest Middle Paleolithic technologies first evolved and well before every indisputable innovation in weapons technology save the simple wooden spear (e.g., Lehringen, ca. 200 KYA, Jacob-Friesen 1956; Schöningen, ca. 400 KYA, Thieme 1997). The elaborate weapons traditions of the Eurasian Upper Paleolithic and African Later Stone Age were separated from the emergence of prime-focused ungulate hunting by more than 100,000 years. Even the remarkable and apparently precocious examples from African Middle Stone Age sites (e.g., Brooks et al. 1995; McBrearty and Brooks 2000; Yellen et al. 1995) are much too young to close this temporal gap. This lag in technological change suggests that cooperation among early hunters was essential for the capture of large game animals. It also indicates that later acceleration in the rates of change in hunting-weapon designs was largely independent of the evolution of humans' basic capacity to bring down large, dangerous prey.

Many of the changes in weapon designs of the later Paleolithic certainly were connected to humans' dietary interest in meat, but less directly than one might imagine. Improvements in weapon designs toward greater efficiency did not necessarily raise the number of large prey animals available to Paleolithic hunters over the long run. Instead, improvements in weapons efficiency were more likely to have reduced the time it took for an individual to procure an animal, the risk per foray, and possibly also the minimum size of a hunting party needed to capture large animals (Kuhn and Stiner 2001; Stiner 2002a). This implies a change in the value of foragers' time—time that could be allocated to other tasks, thus avoiding undue opportunity costs (*sensu* Hames 1992; Pianka 1978:258). Large-scale resource pooling also could have favored greater individual task specialization. Thus, weapons innovations might have been driven partly by a need for greater mechanical efficiency, but the incentives for doing so might have originated in the pressures of time allocation for diverse social or foraging concerns. Such behavioral and social changes were confined mainly to the Upper Paleolithic period and afterward. Few changes in hunting agendas, if any, are apparent for the Middle Paleolithic.

Many weapons innovations of the Upper Paleolithic and Epipaleolithic relate to the exploitation of small prey (e.g., Knecht 1997). Although some small animal populations are very productive and can serve as reliable resources as human population density increases, humans can tap the productivity of these animals only if the work of capturing them is reduced with new technology such as tended or untended traps (Oswalt 1976). There is good reason to think that trap technology developed rapidly from the late Upper Paleolithic through the Mesolithic (e.g., Adovasio et al. 1996; Gamble 1986; Gramsch and Kloss 1989; Hayden 1981; Jochim 1998; Kuhn and Stiner 2001; Mordant and Mordant 1992; Nadel et al. 1994). The tools for overcoming the quick flight strategies of birds and small mammals no doubt also permanently altered foragers' systems for ranking prey. The pressure to develop such tools began, however, with predator-prey relationships already gone sour, a process that seems to have begun earlier in western Asia than in western and northern Europe. The differences in timing across Eurasian regions may be explained primarily by the geographic origins of early Upper Paleolithic populations (Asia-Africa) and only secondarily by latitudinal variations in animal community structure.

A different side of the technological record concerns carcass processing, which also grew more complex with time. Major shifts in processing efficiency included grease rendering via stone boiling, as evidenced by the thick litter of fire-cracked stones in some later Upper Paleolithic and Epipaleolithic sites (chapter 5). Such heat-in-liquid techniques are labor intensive but can raise protein and fat (mainly grease) yields per carcass well beyond what is possible through "cold" extraction techniques (Binford 1978; Brink 1997; Lupo and Schmitt 1997). During the Middle Paleolithic, only cold extraction techniques that were focused on the concentrated marrow reserves in large medullary cavities were employed. Marrow processing during this period was efficient within the limits of cold extraction methods, but it never intensified beyond that simple technique. Concerted harvesting of quick small animals antedated or accompanied the changes in marrow processing of large mammals in Mediterranean Paleolithic sequences, specifically the addition of grease rendering methods. Increases in carcass processing efficiency are forms of resource intensification.

AN UNSHAKEABLE MIDDLE PALEOLITHIC?

The assertion that Middle Paleolithic humans were large game hunters is almost certainly true, but it reveals little about human subsistence organization, predator-prey interactions, or demography. Middle Paleolithic humans in the Mediterranean region made considerable use of small animals but maintained remarkably narrow diets across a wide range of latitudes. Few subsistence trends are apparent within the Levantine Middle Paleolithic, with the exceptions of possible mild harvesting pressure on slow-turnover prey populations and accelerated debris buildup in late Middle Paleolithic sites. The evidence indicates that human populations were exceptionally small throughout the Mediterranean Middle Paleolithic. A categorical shift in human predator-prey dynamics accompanied by demographic expansion seems to demarcate the Middle to Upper Paleolithic cultural boundary. It is not difficult, therefore, to imagine how Mousterian populations might have been swamped toward the end of the period by the influx of faster-growing populations such as those of anatomically modern humans possessing Upper Paleolithic culture.

The Middle Paleolithic persisted longer in some areas than in others, but its internal features seem remarkably consistent—and very stable—if its long duration and wide geographic distribution are taken into account. The apparent inflexibility of Middle Paleolithic adaptations has been equated with a sort of human stupidity. Yet human brain volume—which corresponded to metabolic costs about equal to our own, irrespective of internal organization—reached its maximum in this period. The persistence of Mousterian populations in time and space seems enviable in light of the brief existence of "modern" cultures.

The ecological data on these early humans indicate that the "inflexibility" we see in Middle Paleolithic culture was a product of the success and stability of the adaptation, and not a matter of basic intelligence. There seems to have been a lack of pressure or economic incentive for these clever, mobile hunters to squeeze more out of their traditional food supplies—that is, there was little selection, if any, for greater foraging efficiency. One may marvel that so little change took place during the Middle Paleolithic, because this is quite different from recent human experience. Yet this sort of stability and the vast geographic range of the adaptation are very much what one would expect to see in a successful generalist adaptation—longevity across diverse environments and latitudes.

More difficult to explain are the downward shifts in trophic level so characteristic of later humans. These shifts took the form of subsistence diversification via the inclusion of lower-ranked foodstuffs associated with greater processing costs, and they coincided more than once with increases in human population densities. Larger networks for spreading risk might also have appeared in conjunction with expanding diets, possibly setting some Upper Paleolithic populations at an advantage (*sensu* Gamble 1986 on archaeological cases; Cashden 1985 on ethnographic cases).

The strongest signals or themes of change in the Mediterranean faunal series are demographic, and the conditions of natural selection for humans' prey species must have shifted accordingly. For the bulk of prehistory, mobility has been humans' primary solution to local resource scarcity. With increasing population, humans in some regions seem to have had fewer options for solving problems of resource availability through mobility, beginning sometime in the Upper Paleolithic, and the situation deteriorated further in the Epipaleolithic (*sensu* Bar-Yosef 1981; L. Binford 1968, 1999; Cohen 1977, 1985; Flannery 1969; Tchernov 1993b, 1998b). Increasing dependence on more biologically "productive" or resilient prey populations over time might have allowed people to obtain a greater volume of meat per unit of habitat area. Perhaps more important is that prey population resilience could have substantially increased the reliability (i.e., reduced the variance) and diversity of meat sources to which a population had access, especially as the costs of acquisition or processing were controlled through technology. A more reliable supply of animal protein and fats has significant implications for child survivorship.

POPULATION GROWTH, NUTRITION, AND SURVIVORSHIP

The Wadi Meged and other Mediterranean faunal series expose the likely effects of increasing predator density in the evolutionary process, but not necessarily its cause. Why did higher human population densities become a permanent condition? It is possible to draw some speculative conclusions, if only as fodder for future research.

Hominids' tendencies to manipulate and restructure their environment are manifested in a variety of sophisticated tactics for insulating human groups from the unpredictability of their food supplies. Among these, small-scale storage of consolidated animal tissues and/or seeds and nuts may have been pivotal in the Upper Paleolithic and Epipaleolithic. Storage buffers human groups against low points in annual resource abundance, especially in situations where residential mobility, exchange, or sharing cannot solve the problem (see discussions by Soffer 1985, 1989b; Testart 1982).

To humans, animal protein represents something more than a source of food energy. Meat is one of the few sources of complete protein in nature. Moreover, the human body cannot store undedicated protein as it does the nutrients that yield food energy, nor can it assimilate protein effectively in the absence of energy supplements (Speth and Spielmann 1983). Ideally, humans should consume usable protein according to a schedule determined directly by what is needed for bodily maintenance and growth. Daily requirements for complete dietary protein are modest but constant, especially for growing children and the women who produce children. Children are at particular risk in lean times, not least because their immune systems decline sharply under conditions of poor nutrition. Thus, there may be a direct reproductive advantage to lessening the effects of seasonal and annual oscillations in the availability of the critical nutrients, protein and fats, and a related advantage of resilience with respect to population size. The trends in small game use along the Mediterranean Rim, which increasingly came to include very productive animals, might inadvertently have stabilized humans' access to meat as the abundance of highly ranked but relatively unproductive prey declined (for a related argument, see Winterhalder and Goland 1993).

The rarity and value of meat resources, along with their discontinuous availability, sharpen humans' interest in obtaining meat in large packages. Because the optimal metabolic schedule of meat consumption is small but regular doses, small game animals, though some are available only seasonally, can represent reliable and relatively continuous sources for appreciable intervals. Large game animals are available more sporadically—although sharing remedies this unevenness in supply to some extent. For all of these reasons, small game animals, as complete protein supplements, are more essential to health and survival than standard foraging models applied to humans may recognize. Variations in the fat content of small animals may be significant—lagomorphs tend to be lean, for example, and birds and reptiles are fat only seasonally—but the changes in small game exploitation in the Paleolithic generally coincided with increases in grease rendering in some areas and plant seed processing in others. All of these are added sources of food energy that may form the needed complement to protein intake from large game (*sensu* Speth and Spielmann 1983).

Subsistence behaviors that enhance the predictability of supplies of critical nutrients can improve childhood survivorship and thereby help a population grow without a change in birthrate. Recent research by Hawkes, O'Connell, and Blurton Jones (1997), for example, suggests that the most consistent sources of protein, and in some cases fat, for hunter-gatherer children in arid environments are the small animals and certain nuts and roots that children procure for themselves or that are provided by female kin. If this is so, then small game use, along with intensive exploitation of plant mast, is relevant to human population growth in prehistoric foraging societies. Such changes in the character and regularity of meat acquisition therefore hold social implications. Opportunities to obtain small animals are considerably more diverse and widespread than are opportunities to obtain large game. Not everyone can hunt large mammals, because of the physical and reproductive demands of human existence, nor can everyone be first in line for a big piece of meat. But nearly any child who can walk efficiently on its own can collect berries, insects, and certain mollusks if the risks of moving about in an environment are relatively low.

The small-large dichotomy in prey body size and the slow-quick dichotomy in small prey may correspond to socially significant divisions in labor networks among modern hunter-gatherer cultures. Immobile or sluggish small animals are essentially gatherable resources and thus are directly accessible to both sexes and all age groups in human societies. Fresh meat from large game animals generally must be obtained by hunting, normally the job of grown men. Quick small animals present other challenges: they are most efficiently caught with special tools and, in some cases, a substantial measure of vigilance. Although access to small quick game is limited by technical skills, these often are learned in late childhood.

The development of capture devices such as snares, deadfalls, and nets might have led to even more reliable access to small protein packages from formerly elusive small animals. The price of these activities was higher labor investment in tool preparation and maintenance or direct inputs of cooperative labor to capture small animals in quantity. It is doubtful that all evolution in tool design can be explained by superior mechanical performance and efficiency. Yet some of the technical evolution seen during the Upper Paleolithic and Mesolithic might have been spurred by dwindling supplies of high-quality meat resources. In western Asia, demographic pressure preceded rather than followed the earliest technological innovations of the Upper Paleolithic and Epipaleolithic periods (chapters 7–9). We do not know who the inventors were in Paleolithic societies, but innovations in trap, snare, and net technologies for hunting small prey could have been the province of women, older children, and the elderly (see also L. Binford 1968). Unfortunately, few elements of trap technology preserve well in the archaeological record, because they are normally made from sinew, cordage, wood, and other biodegradable materials. We see only rare hints of these complex tools in the form of possible bone triggers and cord imprints in mud and, less directly, in art, a topic that merits continued study.

TRANSITIONS IN EIGHT NICHE DIMENSIONS

The findings of the Wadi Meged study and comparisons with other Mediterranean faunal series indicate some broad changes in human ecology. These probably were not matters of simple gains or losses of fundamental behaviors, such as hunting and scavenging, or basic responses to fluctuations in prey abundance. Perhaps most enlightening with respect to human carnivory are "niche boundary shifts"—here summarized

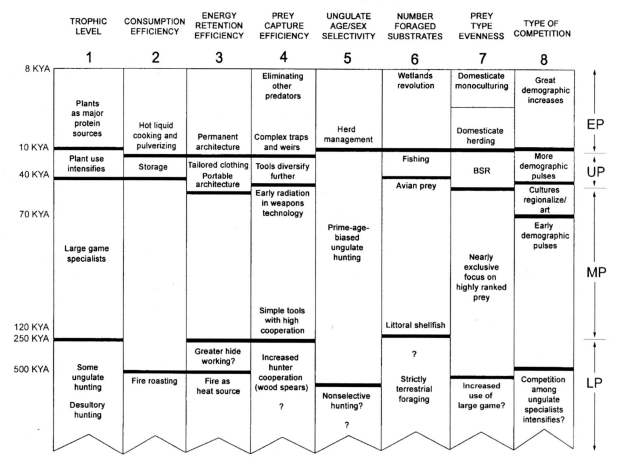

Figure 12.1

Niche threshold shifts in hominids on a logged time scale from 500,000 to 8,000 years ago. Annotations refer to particular behavioral and trophic developments associated with each transition. (EP) Epipaleolithic period; (UP) Upper Paleolithic; (MP) Middle Paleolithic; (LP) Lower Paleolithic.

in eight dimensions, following concepts synthesized by Pianka (1978) and other ecologists—and the way they clustered in time. These shifts may represent ways in which competition was reduced in coevolutionary relations with other predators and, eventually, with neighboring human societies. A logged time scale is used in figure 12.1 to detail the accelerated pace of behavioral change in the later periods.

A strict reliance on meat is the least direct or efficient way of accessing energy from primary production in an ecosystem (dimension 1). High levels of carnivory therefore support proportionally fewer consumers than do lower trophic levels. It was for this reason, as much as any other, that the most carnivorous of Eurasian hominids, including Middle Paleolithic populations, were probably also the most highly dispersed on their landscapes. Humans' position in the food chain dropped significantly only with the Upper Paleolithic and especially the Epipaleolithic–

Mesolithic, allowing humans to exist at higher densities (L. Binford 1968, 1999; Flannery 1969; Harpending and Bertram 1975; Keeley 1988). In fact, Middle Paleolithic humans maintained the most concerted focus on large game observed for any time period. It is possible that they existed at the highest of all human positions in the food chain.

In hominids, the efficiency of food consumption (dimension 2) is mediated by technology, particularly processing equipment that enhances the nutritional yield of a given food unit. Fire is one early innovation in this regard, appearing by perhaps 400–500 KYA, if not earlier, and it was very much a part of early Middle Paleolithic technology. Upper Paleolithic and Epipaleolithic developments in storage, container, and grinding technologies ultimately revolutionized humans' ability to maximize food value from both plant and animal sources (see, for example, Keeley 1988 and Testart 1982 on food storage; Lupo and

Schmitt 1997 on bone grease rendering; and Keeley 1995, Miller 1992, and Wright 1994 on seed processing). Extracting more nutrition from any available food package or food patch is one of the few ways in which a forager can improve upon natural patterns of food abundance without actually manipulating the reproduction of food species. In Eurasia these innovations took the forms of hot-liquid cooking, pulverizing, and rendering techniques for processing animal carcasses and plant seeds, behaviors that evolved most rapidly during and after the terminal Pleistocene.

Efficiency of energy retention (dimension 3) is an elusive property of archaeological records, but a few obvious increases in energy retention are apparent. Hearths lessen the total energy demands on the human metabolism. As evidence for fire increases in the archaeological record, so does that for hide working. Both are in evidence for the Mousterian (Meignen et al. 1989; Villa et al. 2001). Insulation mechanisms may have reached a new threshold in the later Upper Paleolithic with portable architecture evidenced by "tent rings," or rock alignments (e.g., Kozlowski 1999), and abandoned framing materials (e.g., Pidoplichko 1998; Soffer 1990), elaborations in hearth construction, and tailored clothing in the northern Epipaleolithic, as indicated by abundant needles and awls.

Increasing investments of time and labor in tool manufacture are associated with the bone technology of the Eurasian Upper Paleolithic, as well as with the sophisticated woodwork and cord craft of the Epipaleolithic and Mesolithic. As effective as Paleolithic hunters of large game might have been, foragers cannot raise prey densities above environmental carrying capacity. There generally is room, however, for changes in the amount of time and the associated risks that people incur while hunting large prey; these are essentially matters of the efficiency of prey capture (dimension 4). Modest diversification in weapons might have occurred in the late Middle Paleolithic, but these innovations were soon eclipsed by the appearance and rapid radiation of Upper Paleolithic technological systems, culminating in a profusion of trapping gear in particular (nets, snares, weirs). During the Middle Paleolithic, only close cooperation within groups overcame the limitations of simple wooden spears.

Ungulate age and sex selectivity (dimension 5) is an important indicator of change in human predator-prey dynamics. This niche dimension concerns humans' effects on prey animals' population structure. A shift is indicated by the rise of prime-age-biased ungulate hunting by the early Middle Paleolithic; the practice then remained quite stable for the duration of the Middle and Upper Paleolithic periods. This predator-prey relationship collapsed in some regions only around the time of the Pleistocene-Holocene boundary, followed by the domestication of certain ungulates and the rise of food-production economies. Because of sample sizes and poor temporal resolution, we generally know less about sex selectivity and seasonality on the part of ungulate hunters, but these are potentially related to the mortality phenomena explored by this study.

Diversification in the foraging substrates typically exploited by Eurasian hominids (dimension 6) began from a terrestrial precedent. The addition of marine shellfish exploitation by at least 120,000 years ago seems almost trivial in the sense that it was practiced on such a small scale and seems to have been confined to collecting along shore margins. Indeed, use of shellfish seems to have been a natural extension of the collecting of slow-moving small animals on land. Marine exploitation nonetheless is significant in that very different food chains and energy production systems are involved. After becoming a part of human foraging repertoires, marine littoral exploitation appeared again and again in prehistory, and these resources grew to monumental importance in some late Mediterranean cases.

The major introduction of avian prey by about 45 KYA in western Asia represents another novel foraging substrate and contributed to greater evenness in the small prey spectrum (dimension 7). Birds more than most other small game resources signal the earliest onset of the broad spectrum revolution (BSR) in the early Upper Paleolithic. Another new foraging substrate is reflected by free-swimming fish harvested from marine, brackish, and freshwater habitats. This change took place after roughly 20,000 years ago in subtropical Eurasia (e.g., the Nile Valley of Africa: Peters 1991; Stewart 1989; Van Neer 1986) and —especially after 12 KYA—culminated in the so-called wetlands revolution by 8,000 years ago (e.g., Coles 1992; Enghoff 1991). Fish were not a major source of food in the Levantine sequences examined for this study, but they appear in later parts of the Wadi Meged sequence (Natufian, Munro 2001), at the 19,000-year-old forager occupations at Ohalo II on the margin of Lake Kinneret (Sea of Galilee, Nadel et al. 1994), and at Üçağızlı Cave in Levantine Turkey (Stiner et al. 2002).

Interspecific competition (dimension 8) substitutes here for the more traditional contrast between *r*-selected and *K*-selected species discussed in the early days by population ecologists (e.g., MacArthur and Pianka 1966; Pianka 1978:246–247). All hominids were essentially *K*-selected, or "low-turnover," organisms that existed close to environmental carrying capacity, a condition that is thought to select for competitive efficiency over reproductive efficiency. The rapid spread of Upper Paleolithic cultures at the expense of Neandertals and, much later, the spread of Neolithic lifeways across Eurasia represent only minor variations on this theme, although their consequences were remarkable. In these cases, demic expansion, apparently founded on humans' new ability to reproduce more efficiently and maintain good population size on a diet of lower-ranked foodstuffs, seems to have outstripped the growth rates of other, coeval societies (*sensu* Ammerman and Cavalli-Sforza 1984; Harpending and Bertram 1975; Pianka 1978:208).

A related point concerns the broadest types of competitive forces that operated on hominid populations—namely, the declining influence of competition with other species of large predators over time and a corresponding increase in intraspecific competition as human population densities increased in some regions. Hominids' competition with members of the order Carnivora was almost certainly responsible for the early emergence of prime-adult-biased ungulate hunting. Only after 20,000 years ago were large carnivores conspicuously excluded from shelters that attracted human occupants (Gamble 1986), roughly coincident with the evolution of "complex" hunter-gatherers in many areas (Bar-Yosef 1995; Belfer-Cohen 1991; L. Binford 1968; Henry 1989; Keeley 1988). The evolutionary significance of competition is in principle greatest in high-density situations and situations where food is highly concentrated (e.g., Kruuk 1972; Sinclair and Norton-Griffiths 1979; Stanford 2001; Winterhalder 1997, 2001). Increases in population density heighten the rate of intraspecific contact and, by extension, the potential scales at which social ties can be manipulated by humans. Such conduits for managing resource risks are unlikely to evolve below a certain density threshold, but once evolved, they may persist even among later populations living at very low densities (e.g., Gould 1980).

It is significant that many archaeological records are suddenly enriched with ornaments at the beginning of the Upper Paleolithic (e.g., d'Errico et al. 1998; Hahn 1972; Kuhn et al. 1999, 2001; Stiner 1999; Taborin 1993; White 1993; Zilhão and d'Errico 1999). The circumstances under which geographically extensive human partnerships and resource transfers—hunter-gatherer insurance systems, in effect—might have evolved are not easily modeled (but see Conkey 1985; Gamble 1986; Soffer 1985; Wiessner 1983; Wobst 1974, 1977). Human population increase in the Upper Paleolithic and Epipaleolithic, however, might have created feedback situations in which social networks and

institutions could be exploited more efficiently to spread foraging risk over larger areas in bad times. Though provocative, such possibilities are among the most challenging to test with archaeological data. The marked regionalization of artifact styles during the Upper Paleolithic and the shrinking territories of the Epipaleolithic and Mesolithic, by the same kinds of criteria (e.g., Newell et al. 1990; Price 1991), tell us nonetheless that the challenges to human survival came from the lay of cultural landscapes as much as from natural ones. Although resource sharing and cooperative foraging must have been components of Middle Paleolithic lifeways, and early people certainly had social relationships to maintain, the demographic conditions that make large, open-ended networks numerically possible seem to have evolved considerably later. Whatever the causes of demographic increase, the conditions of selection on human societies and foraging behavior shifted to a more profoundly intraspecific forum.

The earliest niche shifts outlined in figure 12.1 seem to cluster at about 500,000 years ago. Though it is conjectural, this was when fire might have become a standard component of Paleolithic technology, and we know from many published sources that site and hominid fossil numbers in Eurasia generally increase to form an unambiguous record of occupation around this time (reviewed in Stiner 2002a). A second and more certain set of shifts centers on 250,000 years ago. This boundary shift might include the early appearance of prime-aged-biased ungulate hunting, although 250 KYA may simply represent a minimum age of the phenomenon. Middle Paleolithic technology appeared around 250 KYA, likely including greater hide working and tanning, and a narrower range of prey animals was exploited than before, nearly all of them ungulates and easily gathered small animals. High levels of carnivory no doubt kept hominid population densities low throughout the Middle Paleolithic. Human adaptations of this period display great stability.

Early Upper Paleolithic cultures appeared between 50,000 and 45,000 years ago in western Asia and soon spread across Eurasia, replacing all Mousterian cultures by about 30–28 KYA. Early population pulses accompanied this cultural transition. From the early Upper Paleolithic the pace of change in material culture accelerated on multiple fronts, with the novel additions of body ornaments, decorated bone and antler tools, and art. The final series of shifts corresponded to extreme climate changes, beginning with the Last Glacial Maximum 20,000–19,000 years ago and followed by rapid global warming.

PARTING THOUGHTS

The conclusion that human population densities were rising during the later Paleolithic will surprise few archaeologists. Yet the unique potential of small game data for examining when and where early demographic increases took place may surprise many. A close look at the Middle Paleolithic record relative to the records of later periods revealed an appropriate way to test the hypothesis of expanding dietary breadth in response to human population pressure. The Wadi Meged faunas indicate much diversity in Paleolithic prey species, including many small game animals. The Mediterranean spur-thighed tortoise, a species long ignored in zooarchaeological research in the Levant, provided new insights into the nature of Mousterian subsistence and demography. The kinds of phenomena demonstrated by this study in the Levant also occurred elsewhere, although the details and timing of subsistence changes varied. Much work remains to be done on Paleolithic small game exploitation. If we are to make sense of interregional comparisons of Paleolithic game use, it will also be necessary to hold geography and environment as constant as possible while assembling local faunal series.

Of course, not all of the small animal remains found in archaeological sites are attributable to humans, but those that can be linked to Paleolithic human activities hold the unique power to clarify the timing and geographic centers of rapid population growth suggested by research on human molecular phylogenetics, as well as to clarify some of the factors that contributed to the earliest forager-producer transitions. To date, most information on human population history has been obtained from studies of modern human genetic diversity, which on the whole suggest several demographic pulses originating from western Asia, Africa, or both that ultimately affected peripheral populations of Europe and elsewhere (e.g., Ammerman and Cavalli-Sforza 1984; Barbujani and Bertorelle 2001; Reich and Goldstein 1998; Relethford 1998, among others). Time is the most difficult variable to control for in these studies: biological clocks inferred from gene mutation rates are notoriously inaccurate, and so there is dismaying variety in their interpretation. Demographic pulses evidenced by the zooarchaeological record can be dated by radiometric techniques over geographical gradients, permitting independent tests of prehistoric human population dynamics and population history.

Demography is a substrate condition that often participates in the process of natural selection by narrowing the range of final outcomes. Other factors, including history, also play crucial roles in this process, but this is not a good reason to discount the effects of variation in population densities. Suffice it to say that demography is back on the table for archaeological research on the social evolution of early humans.

Finally, it is well worth noting that the success of this zooarchaeological study owes much to two basic research conditions. The first was the very long time frame available for examining changes in the Wadi Meged and nearby areas, thanks to a combination of good preservation, good luck, and a wealth of earlier research in the area. Equally important was the condition of efficient information exchange, which can be attributed only to the organization of the projects at Hayonim and Kebara Caves under the leadership of Ofer Bar-Yosef, Liliane Meignen, and Bernard Vandermeersch. The prompt timing and rapid circulation of project results facilitated the discovery process in important and novel ways. The Hayonim project has been particularly helpful for solving problems that no single expert could have solved by working alone. Researchers are keenly aware of the benefits of intellectual exchange but too often are disadvantaged by the slow pace of information synthesis and dissemination. To enhance the exchange of observations and ideas in the Hayonim project, many preliminary analyses were conducted on-site by a broad range of specialists. Discussion led to prearrangements to complete specific analyses for dissemination within the following year. Nowhere in the projects was this more evident than with regard to our understanding of site formation processes. Steve Weiner's on-site FTIR analysis of sediments revolutionized the taphonomic studies at Hayonim and Kebara Caves. The insights gained into bone and wood ash preservation in Hayonim Cave allowed us to overcome a wide range of proximate taphonomic and chronological obstacles. Collaboration in "real time" (on-site) also helped us make the conceptual and methodological transitions needed to link the consequences of microscale site formation processes to the consequences of macroscale ones. None of this would have been possible without prompt analyses and information sharing.

Appendixes

■

In this study, teeth were quantified separately from bone regardless of whether they remained within the bony armature of the jaw. Bones and teeth have different material properties that greatly affect their chances of preservation, including differing microstructures, mineral densities, sizes, and shapes. Tooth enamel generally resists decomposition factors better than all kinds of bone. Any anatomical connection of tooth to bone in archaeological specimens was recorded, as were anatomical connections (articulations) of any pair or group of bone elements. This procedure enables one to consider the potential effects of fragmentation and disarticulation on the counting units.

DEFINITIONS OF COUNTING UNITS

NSP is the number of faunal specimens (a primary counting variable); it is the most fundamental inventory measure employed in this study. Specimens need not be identifiable to taxon or body part. Most of the taphonomic inquiries of burning and fragment size distributions, weathering, and cut marks begin here.

NUSP, or number of unidentified specimens (a primary counting variable and subset of NSP), refers to those specimens that could not be identified to taxon or body part. This variable is employed in some of the taphonomic comparisons, such as the analyses of burning damage.

NISP, or number of identified specimens (a primary counting variable and subset of NSP), refers to items that could be identified to the level of species, genus, or a more general taxonomic group and to a body part or anatomical region. Analysts' criteria for specimen identifiability vary with experience and research circumstances, but because NISP is the building block of most quantitative analyses, some standard conventions have been developed and are widely used (Grayson 1984). NISP usually serves as the independent variable in an analysis.

Portion-of-element, or PE (a primary counting variable and subset of NISP), follows basic principles of anatomy but may be defined and emphasized somewhat differently by different analysts. If an assemblage is highly fragmented, then portion types must be unique, small, and localized (discrete). Ideally, several portions are considered per element type and are widely distributed over its surface. Portion counts are used to compare the differential representation of

spongy, compact, and stony skeletal tissues. They are also essential to constructing body part profiles and estimating MNE and MNI.

MNE (a derived variable) is an estimate of the minimum number of skeletal elements from which the identified fragments could have originated. A skeletal element is a whole bone such as a femur, tibia, mandible, or fourth upper premolar. Paired (lateral) elements are classified as left, right, or unknown. Bone fusion and tooth replacement were used in some analyses to refine estimates of MNE.

MNI (a derived variable) represents the minimum number of whole living animals from which the faunal fragments could have originated. MNI is usually determined from the most common portion of the most common skeletal element within a taxonomic category, but it (or Binford's [1978] simpler calculation of minimum animal units, MAU) can also be derived by element or portion-of-element as a way to investigate differential destruction and transport phenomena. MNI estimates never promise that every fragment from the whole skeleton is present in the assemblage, because many fates may befall fragments or portions of whole bones.

Taxon may refer to species-level or more general animal types, following the hierarchy of biological nomenclature (e.g., genus and order) or, alternatively, following body size groups such as small mammal, medium ungulate, and so on. Because taxonomic terms tend to be hierarchical, each zooarchaeological analysis must specify the taxonomic "plane" of consideration and employ nonoverlapping categories.

Properties of the Quantitative Variables

Primary counting units employ mutually exclusive categories, meaning that one category cannot substitute for another in the recording phase of a study, although the categories can, and often do, embody hierarchical relations among terms. NISP, for example, requires that a specimen be identifiable to taxon and body part, but it can represent any fraction of a whole skeletal element. In addition, that specimen might be attributed to the species, *Dama mesopotamica,* implying that it also falls into the more general category "medium-size ungulate." Analysts normally push taxonomic identifications to the most specific level possible, but not all specimens retain enough diagnostic features to permit species-level identification. For this reason, specific and certain generic taxonomic categories must be

treated as distinct during data collection, although some taxonomic categories will be subsumed by certain others for the purpose of higher-order analysis.

PE is also a primary, irreducible counting unit whose categories tend to overlap hierarchically. Some PE categories are especially important for estimating MNE because they are unique features that occur only in certain areas of the skeleton. Less fragmented specimens may preserve many skeletally unique features, whereas small fragments of the same kind of element might at best preserve only one or two. Thus, some PE classifications are more inclusive than others (e.g., proximal one-half of femur versus femoral head alone), following a hierarchical arrangement that depends partly on fragmentation and varies from perfect completeness to small but recognizable, unique fractions of the whole (see appendix 2).

Primary counting variables (NSP, NISP, PE) are considered recombinable, and they generally are not subject to aggregation errors caused by "rounding up" to an idealized, complete anatomical model such as a whole element or a whole skeleton. Although primary counting units do not qualify as continuous variables, they often are treated as such in zooarchaeological analyses (see Grayson 1984 for a complete review).

Derived variables (MNE, MNI) are estimated from primary counting variables. Because a rounding-up procedure normally is involved, they are subject to aggregation error. Most errors introduced by rounding in zooarchaeological studies are directional (always upward), and the intensity of their effects depends on how many stages of calculation (derivation) are involved. Because research on faunal assemblages normally begins with fragmented and disarticulated remains, with a goal of understanding more coherent anatomical and population patterns, MNI is the more subject of the two to aggregation error; MNE is somewhat less so because it is focused on elements rather than whole skeletons. Some variables can be indexed to create a continuous ratio scale (see Lyman 1994), provided one knows the proper order of units, the interval between units, and the location of true zero. MAU (minimum animal unit, from Binford 1978, 1981), for example, is a standardization of observed MNE against expected MNE in the complete skeleton model; it is less subject to aggregation error than MNI. (Standardized MNE, used in this study, is merely a variation of the MAU concept.) Most or all derived counting units should be treated as ordinal variables (Grayson 1984; Lyman 1994).

APPENDIX 2: FAUNAL CODING KEYS

The following coding keys, employed in the present study, were developed by Stiner for use in research on Mediterranean faunas. Portions-of-elements can occur singly or in combination with others, and thus the coding system is hierarchical. Note also that foramen positions on elements vary by species, but positions tend to be fixed in the wild species to which this coding system has been applied. Foramen positions vary more in domesticated animals.

Coding Keys for Skeletal Elements and Portions-of-Elements

Element	Portion-of-Element
Horn/antler (10s)	*Horn/antler*
11 Horn core	10 Rosette (base)
12 Antler	11 Pedicle-braincase
	12 Shaft-rosette-pedicle-braincase
Skull (20s)	13 Tip/tine (2 = shaft fragment; 80 = diaphysis section)
21 Half cranium, l or r	
22 Half mandible, l or r	*Cranium*
	19 Hyoid
Neck (30s)	20 Premaxilla (also used for mandible)
31 Atlas	21 Nasal
32 Axis	22 Zygomatic (jugal-squamous)
33 Cervical vertebra	23 Maxilla (~complete half)
	24 Maxilla fragment (241 ant; 242 post)
Main axial column (40s)	25 Petrous
40 Vertebra, type unknown	26 Auditory bulla
41 Thoracic vertebra	27 Braincase fragment
42 Rib	28 Occipital
43 Lumbar vertebra	29 Occipital condyle (l or r)
44 Sacral vertebra	30 Frontal foramen (or ant foramen on mandible)
45 Innominate (1/2 pelvis)	31 Orbit lower rim (or gonial angle on mandible)
46 Caudal vertebra	32 Lacrimal (foramen)
47 Sternal segment	
	Mandible, base missing
Front limb (50s and 60s)	33 Middle horizontal ramus
51 Scapula	34 Mid-anterior horizontal ramus
52 Humerus	35 Anterior horizontal ramus (incl. P2 socket)
53 Coracoid (e.g., birds)	36 Mid-posterior horizontal ramus
61 Radius	37 Posterior horizontal ramus (incl. dorsal ridge behind M3)
62 Ulna	38 "Dip" between condyle-coronoid
63 Carpal (type unknown)	39 Base of horizontal ramus
64 Metacarpal (bird = carpometacarpus)	40 Condyloid process
65 Cuneiform	41 Coronoid process
66 Magnum	42 Condyle and coronoid
67 Lunate	43 Ascending ramus with lingual foramen
68 Scaphoid	
69 Unciform	*Mandible, base intact*
	44 Horizontal ramus (whole)
Hind limb (70s and 80s)	45 Middle horizontal ramus
71 Femur	46 Anterior horizontal ramus
81 Tibia	47 Posterior horizontal ramus
82 Patella	48 Mid-anterior horizontal ramus
83 Astragalus	49 Mid-posterior horizontal ramus
84 Calcaneum	
85 Tarsal (type unknown)	*Innominate*
86 Metatarsal (bird = tarsometatarsus)	57 Acetabulum fragment
87 Naviculo-cuboid	58 Acetabulum section—pubic body
88 External and middle cuneiform	59 Acetabulum, complete
89 Lateral malleolus	

Continued on the next page

CODING KEYS FOR SKELETAL ELEMENTS AND PORTIONS-OF-ELEMENTS, *CONTINUED*

Element	Portion-of-Element
	60 Acetabulum and ilium (~complete)
Feet (90s)	61 Acetabulum section—iliac body fragment
90 Sesamoid	62 Acetabulum-ischium (~complete)
91 First phalanx	63 Acetabulum section—ischial body fragment
92 Second phalanx	64 Iliac body (diaphysis)
93 Third/terminal phalanx	65 Iliac blade
	66 Ilium
General element categories	67 Ischial body
1 Metapodial (type unknown)	68 Ischial blade
2 Long bone shaft (type unknown)	69 Ischium
3 Flat bone (skull or scapula fragment)	
4 Carpal or tarsal (type unknown)	*Vertebrae*
5 Spongy element (axial)	50 Epiphysis
6 Auxiliary third phalanx	51 Centrum (body)
7 Auxiliary second phalanx	52 Transverse process
8 Auxiliary first phalanx	53 Pre-zygopophysis (5353 = intact pair)
9 Auxiliary metapodial	54 Post-zygopophysis (5454 = intact pair)
10 Eggshell (bird)	55 Dorsal spine
	56 Half
Teeth (100s, mammals only)	57 Anterior-ventral articulation
9 _ _ _Deciduous tooth*	58 Zygopophysis (type unknown)
_ 100 From upper jaw	
_ 200 From lower jaw	*Limb ("long") bones*
_ 300 Dental position unknown	70 Proximal (P) epiphysis
_ _10 Incisor (type unknown)	71 P epiphysis fragment (see also 91–94)
_ _11 First incisor	72 P < 1/2
_ _12 Second incisor	73 P 1/2
_ _13 Third incisor	74 P > 1/2
_ _20 Canine	75 Distal (D) > 1/2
_ _30 Premolar (type unknown)	76 D 1/2
_ _31 First premolar	77 D < 1/2
_ _32 Second premolar	78 D epiphysis fragment (see also 81–84)
_ _33 Third premolar	79 D epiphysis
_ _34 Fourth premolar	
_ _40 Molar (type unknown)	*Long bone epiphysis subportions*
_ _41 First molar	81 Medial distal (D) epiphysis
_ _42 Second molar	82 Lateral D epiphysis
_ _43 Third molar	83 Anterior D epiphysis
	84 Posterior D epiphysis
Tortoise shell	91 Anterior proximal (p) epiphysis
53 Carapace (tortoise only)	92 Posterior P epiphysis
54 Plastron (tortoise only)	93 Medial P epiphysis
	94 Lateral P epiphysis
	Long bone shaft features
	990 W/ foramen present
	991 W/ attachment scar (proximal end of radius feature)
	994 Anterior "angle" or keel (tibia)
	995 Muscle insertion scar
	996 Posterior rugosities (tibia)
	997 Interior diagonal lattice (humerus)
	998 Anterior groove or sulcus (metapodials)
	999 Posterior groove (metapodials)
	General portion codes
	1 Complete
	2 Nearly complete
	56 Half (lateral dimension)
	80 Short diaphysis (tube)

Continued on the next page

CODING KEYS FOR SKELETAL ELEMENTS AND PORTIONS-OF-ELEMENTS, *CONTINUED*

Element	Portion-of-Element
	85 Long diaphysis (tube)
	86 Diaphysis w/ foramen
	90 Shaft fragment
	93 Epiphysis fragment
	95 Spongy bone fragment
	97 Flat bone fragment
	50 Epiphysis
	Tortoise bony shell features
	993 Tortoise shell bridge (anterior or posterior edge)
	952 Tortoise shell edge fragment
	953 Tortoise nuccal carapace plate
	954 Tortoise anal carapace plate

* Underscores denote placement of codes that follow for upper or lower position in jaws, tooth functional type, and specific tooth element, respectively.

CODING KEYS FOR AGEING CRITERIA

Bone fusion and development	Tooth eruption and wear
0 Not applicable or no information	0 Not applicable or no information
1 Unfused	1 Alvaeolus just opening
2 Partly fused	2 Alv. open, tooth crown partly formed
3 Nearly fused (line visible)	3 Tooth crown beginning to erupt
4 Fully fused	4 Tooth crown 1/2 erupted
50 Woven bone tissue (immature)	5 Tooth crown >1/2 erupted
81 Tetus or neonate	10 No occlusal wear
88 Probably juvenile	11 Slight wear
90 Antler base, shed	12 Light wear
99 Antler base, unshed	13 Light–medium wear
	14 Medium wear
Side	15 Medium–advanced wear
1 Left	16 Advanced wear
2 Right	17 Crown completely destroyed by wear
0 Not applicable or unknown	41 Pathology, infection/resorption, age unclear

CODING KEYS FOR BONE MODIFICATION

Weathering	Burning (location in comments)
0 None	0 Unburned
1 Fine linear cracks only	1 <1/2 carbonized (<1/2 blackened)
2 Fine cracks, some "open"	2 >1/2 carbonized
3 Many cracks, most "open"	3 Fully carbonized
4 Some exfoliation	4 <1/2 calcined (<1/2 whitened)
5 Advanced exfoliation	5 >1/2 calcined
6 Chemical weathering	6 Fully calcined
	88 Darkened but could be organic or mineral staining

Tool mark types	Tool mark orientations
C Cutmark	P Perpendicular or transverse (to main axis)
H Hack mark (sharp)	A Axial (to main axis)
I Blunt impact w/ local crushing	D Oblique or diagonal (to main axis)
^ Cone fracture	*(Also record mark length, number, and position)*
S Sawing	
Z Scraping, scratching	*General abrasion damage*
A Abrasion by deliberate grinding method	0 None
G Girdled incision	1 Mild, 2 medium, 3 severe

Continued on the next page

CODING KEYS FOR BONE MODIFICATION, *CONTINUED*

Fracture types	*Gnawing types*
TR Transverse to main axis	Carnivore (record presence and comment on
SR Spiral fracture	types, abundance, position)
SL Split relative to main axis	Damage types include
CR Crushing	Punctures
RI Ripped	Salivary rounding
RR Very ragged	Tooth drag or shear marks
HG Hinge fracture	Crenelation
CONE Cone fracture *(measure diameter)*	Rodent (record presence and comment on
OPP Opposing cone fractures	types, abundance, position)
I Impact damage (bruise, dent)	

APPENDIX 3: UNGULATE TOOTH SAMPLES FROM HAYONIM CAVE SUBMITTED TO J. RINK AND H. P. SCHWARCZ (MCMASTER UNIVERSITY) FOR ESR DATING

Recovery Date	Square	Layer	X	Y	Z	Catalog No.	Taxon and Element(s)
Submitted to Schwarcz and Rink, 1994							
1992	J28c	E?	—	—	—	1	*Bos,* fragment
28-7-92	H23d	E	92	61	341	45	*Bos,* fragment
29-7-92	H23b	E	30	89	338	46	*Bos,* fragment
29-7-92	H23-	E	47	89	344	49	*Bos,* fragment
14-7-93	G27c	E	64	24	431	315	*Bos,* upper molar
17-7-93	K24c	E	72	45	373	176	*Cervus,* UM3?
1-7-93	K24d	E	86	70	325	168	*Bos,* UM3
6-7-93	J21a	E	3	48	328	7	*Bos,* UP3
1-7-93	F26a	C/E	38	32	385	8	*Bos,* UM3
2-7-93	F26a	C/E	10	7	395	19	*Dama,* UP2–4
1994	I20d	E	—	—	—	268	*Cervus,* upper molar
Submitted to Rink, 1995							
5-7-95	I23a	E	30	30	417	667	*Bos,* UP4
3-7-95	I23c	E	56	43	403	668	*Bos,* LP4
11-7-95	I23c	E	58	28	435	669	*Equus,* LM3
10-7-95	I24b	E	45	66	435	341	*Bos,* UM3?
13-7-95	D28b	E	9	54	538	29	*Dama,* lower molar
24-6-95	K21c	E	98	36	427	387	*Bos,* UM1
Submitted to Rink, 1997							
15-7-96	K24b	E	15	105	~467	239	*Dama,* LM3
18-7-96	J20c	E	94	33	467	311	*Bos,* lower molar
25-7-96	J22b	E	40	70	474	182	*Bos,* lower molar
17-7-96	J20c	E	92	8	465	321	*Bos,* fragment
28-6-97	J21c	E	75	10	485	332	*Bos,* UM3
2-7-97	I23a	E	3	2	460	897	*Bos,* upper molar
25-6-97	I20a	E	39	7	483	576	*Bos,* UM3
25-6-97	J21d	E	85	93	477	321	*Bos,* molar fragment
25-7-96	K21c	E	60	90	460	544	*Dama,* LM2–3
25-7-96	K21c	E	55	5	470	543	*Bos,* lower molar
Submitted to Rink, 1998							
19-7-97	J23c	E	98	40	498	101–2	*Bos,* lower molar
26-6-98	I22b	E	3	99	506	None	*Dama,* LM2
1-7-98	J22d	E	—	—	510-15	None	*Bos,* UM2?
27-6-98	I23b	E	22	79	509	None	*Cervus,* UM2?
29-6-98	I23b	E	35	77	510	None	*Cervus,* UM1?

NOTE: The column heads X, Y, and Z refer to northing, easting, and depth locations, respectively, in the excavation grid of the central trench.

KEY: (M) Molar; (P) premolar; (U) upper dentition; (L) lower dentition; (—) no data.

APPENDIX 4: BONE-BASED AND TOOTH-BASED MNE ESTIMATES FOR UNGULATE CRANIA AND MANDIBLES FROM HAYONIM CAVE

SEVEN MIDDLE PALEOLITHIC ASSEMBLAGES (MP UNITS)

Taxon	Element	1	2	3	4	5	6	7	All
Gazelle	Cranial bone	1	2	2	16	8	3	1	33
	Cranial teeth	0	4	6	16	11	4	0	41
	Mandibular bone	2	1	5	18	7	5	1	39
	Mandibular teeth	1	2	12	21	8	4	1	49
Fallow deer	Cranial bone	1	0	2	12	5	3	1	24
	Cranial teeth	0	0	4	12	3	1	0	20
	Mandibular bone	1	1	6	14	4	2	1	29
	Mandibular teeth	1	2	5	15	3	3	1	30
Aurochs	Cranial bone	0	1	1	4	1	3	0	10
	Cranial teeth	0	2	1	5	2	1	1	12
	Mandibular bone	1	1	0	1	1	1	0	5
	Mandibular teeth	1	1	1	1	2	2	0	8

NOTE: Dental counts include the most common deciduous element, provided that its developmental or wear condition did not indicate a high risk of double-counting individuals.

EPIPALEOLITHIC AND UPPER PALEOLITHIC ASSEMBLAGES

Taxon	Element	Hayonim Natufian	Hayonim Kebaran	Meged Early Kebaran	Meged Pre-Kebaran UP
Gazelle	Cranial bone	28	9	5	2
	Cranial teeth	28	12	8	2
	Mandibular bone	n.d.	8	7	4
	Mandibular teeth	n.d.	14	6	3
Fallow deer	Cranial bone	n.d.	3	3	1
	Cranial teeth	n.d.	3	4	0
	Mandibular bone	n.d.	4	2	2
	Mandibular teeth	n.d.	6	1	1
Aurochs	Cranial bone	n.d.	1	0	0
	Cranial teeth	n.d.	1	0	0
	Mandibular bone	n.d.	1	0	0
	Mandibular teeth	n.d.	3	0	0

NOTE: Dental counts include the most common deciduous element, provided that its developmental or wear condition did not indicate a high risk of double-counting individuals. The entry *n.d.* indicates that the element was absent or too rare to consider.

APPENDIX 5: PORTION-BASED MNE COUNTS FOR SMALL UNGULATES (GAZELLES) IN THE MOUSTERIAN (MP) UNITS OF HAYONIM CAVE AND PHOTON DENSITOMETRY CONTROL DATA FOR PRONGHORN

Element	Portion	Scan Site	Avg. Density	MP Unit (MNE)						
				1	2	3	4	5	6	7
Mandible	Ant hor ramus	DN1	n.d.	2	–	4	17	7	5	–
	Ant-mid hor ramus	DN3	n.d.	2	–	5	17	7	5	–
	Mid hor ramus[a]	DN4	n.d.	1	1	4	16	5	4	1
	Post hor ramus	DN5	n.d.	1	1	3	15	6	1	–
	Mand condyle	DN7	n.d.	1	–	1	18	6	3	–
Atlas	Complete or nearly so	AT2	0.13	–	–	1	1	–	–	–
	Ant artic[a]	AT3	0.32	–	–	1	5	1	1	–
Axis	Ant-vent artic[a]	AX1	0.13	1	–	3	11	1	1	1
Other cerv v	Pre-/post-zyg[a]	CE1/2	0.12	1	3	8	22	6	7	1
Thor vert	Body (centrum)	TH1	n.d.	–	~1	9	19	8	9	1
	Dorsal spine[a]	TH2	n.d.	2	–	7	23	1	7	–
Lumb vert	Pre-zyg[a]	LU1	0.15	3	2	1	16	9	7	1
	Post-zyg[a]	LU2	0.11	3	2	1	16	9	7	1
Rib	Head (prox)	RI2	n.d.	2	4	11	39	9	4	1
	Prox diaph[a]	RI3	n.d.	2	4	11	40	9	4	2
Sacrum	Ant body[a]	SC1	0.11	–	–	1	3	2	2	–
Innom	Acetabulum	AC1	0.14	2	2	4	10	6	4	1
	Iliac body[a]	IL2	0.33	1	–	3	8	5	2	–
	Ischial blade	IS1	0.28	–	–	3	8	6	1	–
Scapula	Dist end[a]	SP1	0.27	1	2	5	19	6	6	–
	Dist diaph[a]	SP2	0.10	–	~1	4	13	7	3	–
	Prox rim	SP5	0.21	–	–	2	6	1	–	–
Humerus	Prox end	HU1	0.06	1	1	1	15	1	2	–
	Diaph or foramen	HU3	0.25	–	–	1	3	3	1	~1
	Dist end[a]	HU5	0.33	1	2	4	16	5	5	–
Radius	Prox end[a]	RA1	0.26	2	2	3	14	3	4	1
	Diaph or attach[a]	RA3	0.57	–	1	3	10	4	3	–
	Dist end	RA5	0.34	1	–	2	2	2	2	1
Ulna	Prox end[a]	UL2	0.26	–	~1	4	8	1	5	–
Femur	Prox end[a]	FE1	0.16	–	2	5	13	2	1	–
	Diaph or foramen[a]	FE4	0.33	–	–	3	5	1	1	–
	Dist end	FE6	0.27	–	2	2	6	3	3	–
Tibia	Prox end	TI1	0.18	–	–	–	4	1	3	2
	Prox diaph	TI2	0.26	–	–	1	3	3	1	–
	Diaph or foramen[a]	TI3	0.48	1	–	3	5	4	4	1
	Dist end[a]	TI5	0.29	4	2	6	15	4	5	–
Calcaneum	Dist end	CA1	0.29	–	2	3	12	13	2	2
	Diaph	CA2	0.55	–	2	3	11	13	2	2
	Prox end[a]	CA3	0.50	–	3	2	11	7	3	1
Astragalus	Complete	AS1	0.39	–	1	3	12	15	10	1
	Middle	AS2	0.48	–	1	3	15	18	10	1
	Dist end[a]	AS3	0.57	–	1	3	15	18	10	1

Continued on the next page

APPENDIX 5 (CONTINUED)

| Element | Portion | Scan Site | Avg. Density | MP Unit (MNE) | | | | | | |
				1	2	3	4	5	6	7
Metacarpal	Prox end[a]	MC1	0.33	–	1	–	13	2	5	–
	Diaph (long)	MC3	0.57	–	–	1	3	1	1	–
Metatarsal	Prox end[a]	MR1	0.47	–	–	1	8	9	7	1
	Diaph (long)	MR3	0.57	–	~1	~2	6	2	2	–
All metapod	Dist end[a]	MC6/MR6	0.44	2	2	11	18	16	10	1
Phal 1	Prox end	P11	0.24	5	2	8	19	23	11	2
	Dist end[a]	P13	0.45	1	2	8	25	17	8	–
Phal 2	Prox end	P21	0.23	2	3	11	25	25	10	–
	Dist end[a]	P23	0.30	2	5	9	32	29	8	–
Phal 3	Prox end[a]	P31	0.25	1	1	12	31	21	16	1

SOURCE: Photon densitometry control data for pronghorn are from Lyman 1982, 1984.

NOTE: Scan sites and codes follow Lyman. Average density is a composite value based on multiple photon densitometry scans for measuring relative differences in structural density of bone (see Lyman 1994). "N.d." indicates that no density data were available. The "similar to" symbol (~) indicates that the nature of the morphologic feature did not permit a very accurate estimate of MNE.

[a] Densest portion for this kind of element.

APPENDIX 6: PORTION-BASED MNE COUNTS FOR MEDIUM UNGULATES (FALLOW DEER) FROM THE MOUSTERIAN (MP) UNITS OF HAYONIM CAVE AND PHOTON DENSITOMETRY CONTROL DATA FOR AMERICAN DEER

Element	Portion	Scan Site	Avg. Density	MP Unit (MNE)						
				1	2	3	4	5	6	7
Mandible	Ant hor ramus	DN1	0.55	–	–	4	13	4	2	–
	Ant-mid hor ramus	DN3	0.55	–	1	4	14	4	2	–
	Mid hor ramus[a]	DN4	0.57	1	1	5	12	4	2	–
	Post hor ramus	DN5	0.57	1	~1	3	10	4	1	–
	Mand condyle	DN7	0.36	1	–	1	9	3	–	~1
Atlas	Complete or nearly	AT2	0.15	–	–	–	–	–	–	–
	Ant artic[a]	AT3	0.26	–	–	2	6	1	–	–
Axis	Ant-vent artic[a]	AX1	0.16	–	–	2	6	~1	~1	–
Other cerv v	Pre-/post-zyg[a]	CE1/2	0.19/0.15	–	–	3	18	6	5	2
Thor vert	Body (centrum)	TH1	0.24	–	1	1	13	3	2	1
	Dorsal spine[a]	TH2	0.27	–	–	2	29	4	4	1
Lumb vert	Pre-zyg[a]	LU1	0.29	~1	–	2	16	2	3	2
	Post-zyg[a]	LU2	0.30	~1	–	2	16	2	3	2
Rib	Head (prox)	RI2	0.25	–	1	6	34	4	7	3
	Prox diaph[a]	RI3	0.40	–	1	6	34	4	7	3
Sacrum	Ant body[a]	SC1	0.19	–	1	1	2	1	–	–
Innom	Acetabulum	AC1	0.27	–	–	2	8	–	1	–
	Iliac body[a]	IL2	0.49	–	–	2	8	1	1	–
	Ischial blade	IS1	0.41	–	–	–	7	–	1	–
Scapula	Dist end[a]	SP1	0.36	–	–	1	14	4	2	–
	Dist diaph[a]	SP2	0.49	–	–	1	15	5	3	1
	Prox rim	SP5	0.28	–	–	–	4	1	1	–
Humerus	Prox end	HU1	0.24	1	–	2	4	–	1	–
	Diaph or foramen	HU3	0.53	–	–	2	8	2	1	~1
	Dist end[a]	HU5	0.39	–	~1	1	7	2	2	–
Radius	Prox end[a]	RA1	0.42	–	–	1	10	2	1	–
	Diaph or attach[a]	RA3	0.68	1	~1	3	11	2	2	~1
	Dist end	RA5	0.43	–	–	3	5	1	–	–
Ulna	Prox end[a]	UL2	0.45	~1	–	3	7	1	2	–
Femur	Prox end[a]	FE1	0.41	~1	1	1	4	2	–	–
	Diaph or foramen[a]	FE4	0.57	–	~1	4	4	2	1	2
	Dist end	FE6	0.28	~1	~1	4	5	3	1	–
Tibia	Prox end	TI1	0.30	–	–	3	8	~3	1	–
	Prox diaph	TI2	0.32	–	–	5	14	1	2	–
	Diaph or foramen[a]	TI3	0.74	–	–	1	14	4	1	~1
	Dist end[a]	TI5	0.50	~1	–	5	13	–	1	–
Calcaneum	Dist end	CA1	0.41	1	–	4	9	4	–	1
	Diaph	CA2	0.64	1	–	4	7	4	–	1
	Prox end[a]	CA3	0.57	1	1	4	5	2	2	–
Astragalus	Complete	AS1	0.47	–	–	–	12	1	1	1
	Middle	AS2	0.59	–	–	2	15	2	1	1
	Dist end[a]	AS3	0.61	–	–	2	15	2	1	–

Continued on the next page

APPENDIX 6 (CONTINUED)

Element	Portion	Scan Site	Avg. Density	MP Unit (MNE)						
				1	2	3	4	5	6	7
Metacarpal	Prox end[a]	MC1	0.56	–	–	5	7	1	1	–
	Diaph (long)	MC3	0.72	–	–	5	7	~1	1	~1
Metatarsal	Prox end[a]	MR1	0.55	–	2	3	7	2	2	1
	Diaph (long)	MR3	0.74	–	2	3	8	~5	2	1
All metapod	Dist end[a]	MC6/MR6	0.51	–	1	9	17	6	3	1
Phal 1	Prox end	P11	0.36	–	–	8	23	4	4	–
	Dist end[a]	P13	0.57	–	2	7	33	8	3	~1
Phal 2	Prox end	P21	0.28	–	–	3	30	4	2	–
	Dist end[a]	P23	0.35	–	–	5	25	3	3	2
Phal 3	Prox end[a]	P31	0.25	–	1	7	37	2	9	3

SOURCE: Photon densitometry control data for *Odocoileus* deer are from Lyman 1982, 1984.

NOTE: Scan sites and codes follow Lyman. Average density is a composite value based on multiple photon densitometry scans for measuring relative differences in structural density of bone (see Lyman 1994). The "similar to" symbol (~) indicates that the nature of the morphologic feature does not permit a very accurate estimate of MNE.

[a] Densest portion for this kind of element.

APPENDIX 7: PORTION-BASED MNE COUNTS FOR LARGE UNGULATES (AUROCHS) FROM THE MOUSTERIAN (MP) UNITS OF HAYONIM CAVE AND PHOTON DENSITOMETRY CONTROL DATA FOR AMERICAN BISON

Element	Portion	Scan Site	Avg. Density	MP Unit (MNE) 1	2	3	4	5	6	7
Mandible	Ant hor ramus	DN1	0.53	–	–	–	–	1	–	–
	Ant-mid hor ramus	DN3	0.62	1	–	–	1	1	1	–
	Mid hor ramus[a]	DN4	0.53	1	~1	–	1	1	1	–
	Post hor ramus	DN5	0.53	–	–	–	–	–	–	–
	Mand condyle	DN7	0.49	–	–	–	~1	–	~1	–
Atlas	Complete or nearly	AT2	0.91	–	–	–	–	–	–	–
	Ant artic[a]	AT3	0.34	–	–	–	2	1	–	–
Axis	Ant-vent artic[a]	AX1	0.65	–	–	–	~1	–	1	–
Other cerv v	Pre-/post-zyg[a]	CE1/2	0.37/0.62	–	1	–	2	2	–	–
Thor vert	Body (centrum)	TH1	0.42	–	–	–	2	–	–	–
	Dorsal spine[a]	TH2	0.38	–	–	–	3	–	–	–
Lumb vert	Pre-zyg[a]	LU1	0.31	–	?	1	1	1	–	–
	Post-zyg[a]	LU2	0.11	–	?	–	–	1	1	–
Rib	Head (prox)	RI2	0.35	–	–	2	3	3	1	–
	Prox diaph[a]	RI3	0.57	–	–	1	3	1	1	–
Sacrum	Ant body[a]	SC1	0.27	–	–	–	–	–	–	–
Innom	Acetabulum	AC1	0.53	–	–	–	1	1	1	–
	Iliac body[a]	IL2	0.52	–	–	–	1	–	1	–
	Ischial blade	IS1	0.50	–	–	–	–	1	–	–
Scapula	Dist end[a]	SP1	0.50	–	–	1	1	2	1	–
	Dist diaph[a]	SP2	0.48	~1	–	~1	2	–	–	–
	Prox rim	SP5	0.17	–	–	–	–	–	–	–
Humerus	Prox end	HU1	0.24	–	–	2	1	–	1	–
	Diaph or foramen	HU3	0.45	–	–	~1	2	3	1	–
	Dist end[a]	HU5	0.38	–	–	–	1	1	–	–
Radius	Prox end[a]	RA1	0.48	–	–	–	1	–	–	–
	Diaph or attach[a]	RA3	0.62	–	1	~1	2	–	1	–
	Dist end	RA5	0.35	–	–	–	1	–	–	–
Ulna	Prox end[a]	UL2	0.69	–	–	1	2	2	–	1
Femur	Prox end[a]	FE1	0.31	–	–	–	–	–	–	–
	Diaph or foramen[a]	FE4	0.45	–	–	–	1	1	–	–
	Dist end	FE6	0.26	–	–	–	–	–	–	–
Tibia	Prox end	TI1	0.41	–	–	–	2	1	–	–
	Prox diaph	TI2	0.58	–	–	–	2	~1	–	–
	Diaph or foramen[a]	TI3	0.76	–	–	~1	5	–	–	–
	Dist end[a]	TI5	0.41	–	–	–	2	2	1	–
Calcaneum	Dist end	CA1	0.46	–	–	–	–	–	–	–
	Diaph	CA2	0.80	–	–	–	~1	–	–	–
	Prox end[a]	CA3	0.49	–	–	–	1	–	–	–
Astragalus	Complete	AS1	0.72	–	–	–	1	–	–	–
	Middle	AS2	0.62	–	–	–	2	–	–	–
	Dist end[a]	AS3	0.60	–	–	–	2	–	–	–

Continued on the next page

APPENDIX 7 (CONTINUED)

Element	Portion	Scan Site	Avg. Density	MP Unit (MNE)						
				1	2	3	4	5	6	7
Metacarpal	Prox end[a]	MC1	0.59	–	–	–	1	?	1	–
	Diaph (long)	MC3	0.69	–	–	–	~1	–	~1	–
Metatarsal	Prox end[a]	MR1	0.52	–	1	–	1	?	1	–
	Diaph (long)	MR3	0.67	–	–	~1	~1	–	~1	–
All metapod	Dist end[a]	MC6/MR6	0.53/0.48	–	–	1	3	–	–	–
Phal 1	Prox end	P11	0.48	–	–	1	2	1	–	–
	Dist end[a]	P13	0.48	–	–	–	7	2	2	–
Phal 2	Prox end	P21	0.41	–	–	1	1	–	–	–
	Dist end[a]	P23	0.46	–	–	1	2	1	1	–
Phal 3	Prox end[a]	P31	0.32	–	1	–	3	~1	–	–

SOURCE: Photon densitometry control data for American bison are from Kreutzer 1992.

NOTE: Scan sites and codes follow Lyman. Average density is a composite value based on multiple photon densitometry scans for measuring relative differences in structural density of bone (see Lyman 1994). The "similar to" symbol (~) indicates that the nature of the morphologic feature did not permit a very accurate estimate of MNE.

[a] Densest portion for this kind of element.

APPENDIX 8: PORTION-BASED MNE COUNTS FOR UNGULATES FROM KEBARAN AND PRE-KEBARAN UNITS OF HAYONIM CAVE AND MEGED ROCKSHELTER, WITH PHOTON DENSITOMETRY SCAN SITE CODES

Element	Portion	Scan Site	Gazelles			Fallow Deer			Aurochs
			Hay Keb	Meg E. Keb	Meg Pre-Keb	Hay Keb	Meg E. Keb	Meg Pre-Keb	Meg E. Keb
Mandible	Ant hor ramus	DN1	8	5	3	3	1	2	1
	Ant-mid hor ramus	DN3	8	7	3	4	3	2	1
	Mid hor ramus[a]	DN4	8	7	3	4	3	3	1
	Post hor ramus	DN5	6	5	3	3	2	2	1
	Mand condyle	DN7	4	4	2	3	2	0	0
Atlas	Complete or nearly	AT2	0	0	0	1	0	0	0
	Ant artic[a]	AT3	2	1	0	1	1	0	0
Axis	Ant-vent artic[a]	AX1	4	1	2	1	3	0	0
Other cerv v	Pre-/post-zyg[a]	CE1/2	8	4	4	3	3	2	1
Thor vert	Body (centrum)	TH1	16	1	1	3	1	1	0
	Dorsal spine[a]	TH2	15	5	3	4	1	0	0
Lumb vert	Pre-zyg[a]	LU1	5	3	2	2	3	2	0
	Post-zyg[a]	LU2	5	3	2	2	3	2	0
Rib	Head (prox)	RI2	34	10	2	12	4	4	1
	Prox diaph[a]	RI3	~34	~10	~2	12	~4	~4	1
Sacrum	Ant body[a]	SC1	1	0	0	1	0	0	0
Innom	Acetabulum	AC1	7	6	3	2	2	2	0
	Iliac body[a]	IL2	5	3	4	2	0	1	1
	Ischial blade	IS1	5	6	4	1	1	1	0
Scapula	Dist end[a]	SP1	11	2	5	1	1	0	1
	Dist diaph[a]	SP2	11	3	1	1	~1	~1	1
	Prox rim	SP5	0	1	1	0	0	0	0
Humerus	Prox end	HU1	4	3	2	0	0	0	0
	Diaph or foramen	HU3	4	4?	1?	3	~1	~1	1
	Dist end[a]	HU5	21	10	5	3	2	2	0
Radius	Prox end[a]	RA1	8	9	2	3	0	0	3
	Diaph or attach[a]	RA3	7	~1	4	4	1	1	2
	Dist end	RA5	6	2	1	3	1	1	0
Ulna	Prox end[a]	UL2	10	11	3	2	~1	2	1
Femur	Prox end[a]	FE1	20	5	2	4	3	0	2
	Diaph or foramen[a]	FE4	8	~2	~1	4	4	~1	1
	Dist end	FE6	4	2	2	1	0	1	0
Tibia	Prox end	TI1	3	3	1	3	0	1	1
	Prox diaph	TI2	7	4	~1	3	3	2	1
	Diaph or foramen[a]	TI3	4	0	0	4	1	1	0
	Dist end[a]	TI5	12	9	2	2	0	0	0
Calcaneum	Dist end	CA1	15	4	5	3	1	4	0
	Diaph	CA2	11	9	5	2	1	3	0
	Prox end[a]	CA3	9	9	5	1	1	2	0
Astragalus	Complete	AS1	28	26	5	3	2	1	0
	Middle	AS2	27	27	6	3	3	1	0
	Dist end[a]	AS3	28	27	6	4	3	1	2

Continued on the next page

APPENDIX 8 (CONTINUED)

| Element | Portion | Scan Site | Gazelles | | | Fallow Deer | | | Aurochs |
			Hay Keb	Meg E. Keb	Meg Pre-Keb	Hay Keb	Meg E. Keb	Meg Pre-Keb	Meg E. Keb
Metacarpal	Prox end[a]	MC1	4	2	2	5	1	0	0
	Diaph (long)	MC3	8	1	2	4	2	0	1
Metatarsal	Prox end[a]	MR1	7	5	3	4	1	1	1
	Diaph (long)	MR3	6	?	?	~5	2	2	1
All metapod	Dist end[a]	MC6/MR6	42	32	18	7	9	1	1
Phal 1	Prox end	P11	54	24	10	13	5	1	3
	Dist end[a]	P13	68	27	14	19	15	5	5
Phal 2	Prox end	P21	62	27	15	7	6	3	3
	Dist end[a]	P23	58	29	13	16	8	3	2
Phal 3	Prox end[a]	P31	64	26	10	6	1	1	0

SOURCE: Photon densitometry scan sites are from Lyman 1982, 1984; Kreutzer 1992.

NOTE: Scan sites and codes follow Lyman. Average density is a composite value based on multiple photon densitometry scans for measuring relative differences in structural density of bone (see Lyman 1994). The "similar to" symbol (~) indicates that the nature of the morphologic feature did not permit a very accurate estimate of MNE.

[a] Densest portion for this kind of element.

APPENDIX 9: END-BASED AND SHAFT-BASED MNE ESTIMATES FOR MAJOR LONG BONES OF UNGULATES FROM THE MOUSTERIAN (MP) UNITS OF HAYONIM CAVE

Element (Exp)[a]	Portion	MP Unit (MNE)							
		1	2	3	4	5	6	7	All
Small ungulates (gazelles)									
Scapula (2)	Epiphysis	1	2	5	19	6	6	0	39
	Diaphysis	0	1	4	13	7	3	0	28
Humerus (2)	Epiphysis	1	2	4	16	5	5	0	33
	Diaphysis	0	0	1	3	3	1	~1	9
Radius (2)	Epiphysis	2	2	3	14	3	4	1	29
	Diaphysis	0	1	3	10	4	3	0	21
Femur (2)	Epiphysis	0	2	5	13	3	3	0	26
	Diaphysis	0	0	3	5	1	1	0	10
Tibia (2)	Epiphysis	4	2	6	15	4	5	2	38
	Diaphysis	1	0	3	5	4	4	1	18
Calcaneum (2)	Epiphysis	0	3	3	12	13	3	2	36
	Diaphysis	0	2	3	11	13	2	2	33
Metapodial (4)	Epiphysis	2	2	11	21	16	12	1	65
	Diaphysis	0	1	3	9	3	3	0	19
Medium ungulates (fallow deer)									
Scapula (2)	Epiphysis	0	0	1	14	4	2	0	21
	Diaphysis	0	0	1	15	5	3	1	25
Humerus (2)	Epiphysis	1	1	2	7	2	2	0	15
	Diaphysis	0	0	2	8	2	1	~1	14
Radius (2)	Epiphysis	0	0	3	10	2	1	0	16
	Diaphysis	1	~1	3	11	2	2	~1	21
Femur (2)	Epiphysis	~1	~1	4	5	3	1	0	15
	Diaphysis	0	~1	4	4	2	1	2	14
Tibia (2)	Epiphysis	~1	0	5	13	~3	1	0	23
	Diaphysis	0	0	5	14	4	2	~1	26
Calcaneum (2)	Epiphysis	1	1	4	9	4	2	1	22
	Diaphysis	1	0	4	7	4	0	1	17
Metapodial (4)	Epiphysis	0	2	9	17	6	3	1	38
	Diaphysis	0	2	8	15	6	3	2	36
Large ungulates (aurochs)									
Scapula (2)	Epiphysis	0	0	1	1	2	1	0	5
	Diaphysis	~1	0	~1	2	0	0	0	4
Humerus (2)	Epiphysis	0	0	2	1	1	1	0	5
	Diaphysis	0	0	~1	2	3	1	0	7
Radius (2)	Epiphysis	0	0	0	1	0	0	0	1
	Diaphysis	0	1	~1	2	0	1	0	5
Femur (2)	Epiphysis	0	0	0	0	0	0	0	0
	Diaphysis	0	0	0	1	1	0	0	2
Tibia (2)	Epiphysis	0	0	0	2	2	1	0	5
	Diaphysis	0	0	~1	5	~1	0	0	7

Continued on the next page

APPENDIX 9 (CONTINUED)

Element (Exp)[a]	Portion	MP Unit (MNE)							
		1	2	3	4	5	6	7	All
Calcaneum (2)	Epiphysis	0	0	0	0	1	0	0	1
	Diaphysis	0	0	0	0	1	0	0	1
Metapodial (4)	Epiphysis	0	1	1	1	1	1	1	6
	Diaphysis	0	0	1	1	1	1	1	5

NOTE: Epiphysis MNE counts represent the more common of two ends of a long bone element. The counts are based on unique morphological features such as the medial face of the distal tibia and the proximal femoral head. MNE counts for diaphyses are also based on unique morphological features, principally foraminae, muscle insertions, and shaft contact points between long bone elements (ulna, radius).

[a] Numbers in parentheses represent the expected MNE for one whole carcass.

APPENDIX 10: FRAGMENT LENGTH STATISTICS FOR IDENTIFIED FAUNAL SPECIMENS (NISP) FOR THE NATUFIAN, KEBARAN, AND MOUSTERIAN ASSEMBLAGES FROM HAYONIM CAVE

Taxon	Size Rank	Natufian (Layer B)			Kebaran (Layer C)			Mousterian (Layer E, All Units)		
		NISP	Mean (cm)	SD	NISP	Mean (cm)	SD	NISP	Mean (cm)	SD
Agama	1	215	1.49	0.7	—	—	—	4*	1.10	0.2
Ophisaurus	2	65	1.81	0.8	21	1.39	0.5	226	1.40	0.4
Small bird	3	28	1.61	1.3	14	2.20	0.6	20	2.13	0.8
Medium bird	4	635	1.55	1.0	28	2.02	0.8	37	2.54	1.0
Large bird	5	115	2.34	1.6	14	2.56	0.9	22	2.88	1.5
Huge bird	6	28	3.23	2.7	11	3.04	1.0	15	4.89	3.6
Testudo	6	6,331	1.43	0.8	277	2.03	0.6	4,675	2.27	1.0
Lepus	7	2,314	1.75	1.3	7	2.03	1.2	5*	2.60	1.4
Sciurus	7	28	1.21	0.9	—	—	—	9*	2.64	1.0
Small mammal	7	680	1.67	1.1	21	1.89	0.5	75	2.43	1.0
Ostrich *(Struthio)* egg	8	—	—	—	7	2.27	0.6	91	2.15	0.7
Gazella	11	2,976	2.66	2.0	842	2.47	1.2	1,587	2.97	1.8
Small ungulate	12	1,727	2.54	2.1	559	2.74	1.3	1,800	3.41	1.6
Capra	13	26	3.06	1.4	30	2.99	2.0	4*	3.02	0.8
Sus	14	113	2.61	1.4	46	2.64	1.6	153	3.82	2.2
Dama	15	88	3.04	1.5	136	3.10	1.6	952	3.91	2.1
Equus (small)	15	—	—	—	—	—	—	8*	3.87	1.2
Medium ungulate	16	147	3.75	2.1	369	4.11	2.0	2,407	4.35	1.9
Cervus	17	42	3.83	2.0	55	3.9	1.9	136	4.53	2.3
Equus (large)	18	—	—	—	—	—	—	8*	6.99	3.8
Bos	19	19	4.55	1.7	42	3.92	2.0	234	4.82	2.6
Large ungulate	20	34	5.25	2.2	78	6.94	3.2	419	5.86	2.6
All taxa	—	18,904	1.97	1.6	2,810	2.88	1.8	13,857	3.22	1.9

NOTE: Only animals representing human food are considered here. Body size ranks are ordered from smallest to largest. NISP counts are for measured specimens only. An asterisk (*) highlights small sample size.

APPENDIX 11: TAXONOMIC ABUNDANCES (NISP) FOR MOUSTERIAN (MP) UNITS OF THE CENTRAL TRENCH IN HAYONIM CAVE

Taxon	MP Unit								
	1	2	3	4a	4b	5	6	7	All
Ungulates									
Gazella gazella	30	71	244	422	370	468	278	29	1,912
Capreolus capreolus	—	1	7	2	—	—	1	—	11
Small ungulate	46	59	268	664	342	310	279	28	1,996
Sus scrofa	1	4	20	45	39	50	17	—	176
Capra aegagrus	—	1	4	1	—	1	—	1	8
Equus cf. hemionus	—	—	3	3	—	5	1	—	12
Dama mesopotamica	14	20	156	438	300	159	75	16	1,178
Medium ungulate	21	33	335	959	735	358	192	48	2,681
Large cervid	4	3	44	94	89	46	37	4	321
Cervus elaphus	4	7	24	55	37	9	17	2	155
Bos primigenius	5	12	29	60	67	55	50	3	281
Equus caballus	—	—	2	5	3	—	—	—	10
Large ungulate	6	22	69	95	97	103	63	5	460
Dicerorhinus hemitoechus	—	—	1	1	—	—	—	—	2
Reptiles									
Testudo graeca	14	48	426	1,469	1,686	1,417	1,184	113	6,357
Coluber sp.	—	10	81	262	101	38	53	2	547
Ophisaurus apodus	—	2	22	112	49	27	12	3	227
Indet. lizard	—	1	—	3	—	—	—	—	4
Aves									
Indet. bird	—	1	—	5	—	—	—	—	6
Small bird (songbird)	—	—	4	5	11	8	1	—	29
Medium bird (partridge/dove)	—	4	15	13	8	14	3	2	59
Large bird (predator)	1	1	9	4	3	3	5	1	27
Huge bird (predator)	—	2	9	—	7	—	—	—	18
Struthio camelus eggshell	—	2	1	22	21	28	16	1	91
Small herbivorous game mammals									
Lepus capensis	1	1	3	—	—	—	—	—	5
Sciurus anomalous	—	—	1	2	9	—	—	—	12
Indet. small mammal	1	7	20	37	14	11	3	1	94
Carnivores									
Hyaenidae	—	1	—	—	—	—	—	2[a]	3
Vulpes vulpes	2	1	7	4	4	—	—	—	18
Canis sp.	—	—	—	4	—	—	1	—	5
Lycaon sp.	—	—	—	—	—	4	—	—	4
Panthera pardus	—	—	—	5	3	—	1	—	9
Ursus arctos	—	1	3	3	—	1	13	—	21
Felis cf. sylvestris	1	1	—	1	—	4	1	—	8
Felis chaus?	—	—	—	—	1	—	—	—	1
Martes foina	—	—	1	—	2	—	—	—	3

Continued on the next page

APPENDIX 11 (CONTINUED)

Taxon	MP Unit								
	1	2	3	4a	4b	5	6	7	All
Small carnivore	1	—	—	—	5	—	1	2	9
Large carnivore	—	—	3	1	1	—	3	—	8
Indet. carnivore	—	—	3	2	—	—	—	—	5
Other									
Erinaceus sp.	—	—	—	—	—	2	—	—	2
Indet. large mammal	9	36	54	56	50	54	44	7	310
Total	161	352	1,868	4,854	4,054	3,175	2,351	270	17,085

NOTE: All taxonomic categories listed in this table are mutually exclusive. Microfauna (small rodents, bats, amphibians) are not considered.

[a] Hyena coprolite.

APPENDIX 12: TAXONOMIC ABUNDANCES (NISP) FOR EPIPALEOLITHIC AND UPPER PALEOLITHIC FAUNAL ASSEMBLAGES FROM HAYONIM CAVE

Taxon	Hayonim B Natufian	Hayonim C Kebaran	Meged E. Kebaran <200 cm bd	Meged UP 200–214 cm bd	Meged UP 215+ cm bd	Hayonim D Aurignacian
Ungulates						
Gazella gazella	2,602	1,039	453	42	118	6,253
Capreolus capreolus	8	16	1	—	—	77
Small ungulate	1,716	722	427	27	104	n.d.
Sus scrofa	98	63	7	1	3	52
Capra aegagrus	26	36	19	5	2	103
Equus cf. *hemionus*	—	3	1	—	1	?
Dama mesopotamica	80	198	58	5	11	798
Medium ungulate	151	537	210	22	67	n.d.
Large cervid	81	93	14	1	4	51
Cervus elaphus	39	72	10	1	5	236
Alcelaphus buselaphus	—	—	—	—	—	12
Bos primigenius	19	53	1	—	5	77
Equus caballus	1	—	—	—	—	3
Large ungulate	33	110	14	1	5	n.d.
Dicerorhinus hemitoechus	—	2	—	—	—	1
Reptiles						
Testudo graeca	4,793	453	462	35	89	1,118
Coluber sp.	n.d.	164	4	—	1	371
Ophisaurus apodus	62	34	4	—	—	18
Agama stellio	212	—	3	—	—	87
Indet. lizard	—	2	—	—	—	331
Aves						
Indet. bird	—	91	35	2	1	—
Small bird	28	17	31	1	1	—
Medium bird	620	29	62	4	10	—
Large bird	112	16	—	2	—	—
Huge bird	26	11	—	—	—	—
Anseriformes	4	—	—	—	—	—
Falconiformes	431	—	64	3	5	—
Phasinidea (mostly chukar)	1,174	—	—	—	—	—
Gruiformes	40	—	—	—	—	—
Columbiformes	14	—	—	—	—	—
Strigiformes (owls)	6	—	—	—	—	—
Passeriformes	19	—	—	—	—	—
Large predatory bird	—	—	—	—	—	572
Other birds	—	—	—	—	—	306
Struthio camelus eggshell	—	11	—	—	—	?

Continued on the next page

APPENDIX 12 (CONTINUED)

Taxon	Hayonim B Natufian	Hayonim C Kebaran	Meged E. Kebaran <200 cm bd	Meged UP 200–214 cm bd	Meged UP 215+ cm bd	Hayonim D Aurignacian
Small mammals						
Lepus capensis	2,219	12	25	—	—	~72
Sciurus anomalous	28	4	—	—	1	—
Erinaceus sp.	21	1	—	—	—	—
Procavia sp.	2	—	—	—	—	—
Indet. small mammal	677	32	65	2	7	~75
Carnivores						
Vulpes vulpes	294	9	6	1	1	147
Canis aureus	4	1	—	—	—	1
Canis sp.	6	—	—	—	—	—
Panthera pardus	5	3	—	—	—	—
Ursus arctos	—	2	—	—	—	—
Hyaenidae	—	—	—	—	—	1
Felis sylvestris	—	6	1	—	2	67
Felis chaus?	115	—	—	—	—	—
Meles meles	38	—	—	—	—	—
Martes foina	42	—	4	—	—	5
Mustelidae	18	—	—	—	—	—
Vormela peregusna	24	—	—	—	—	—
Indet. carnivore	152	2	1	—	—	—
Other						
Indet. medium mammal	742	57	37	5	5	n.d.
Fish	32	—	—	—	—	—
Total	16,814	3,901	2,019	160	448	(≥ 10,834)

SOURCES: Data on the Hayonim Natufian fauna are from Munro 2001. Data on the Hayonim Aurignacian are based on Stiner's NISP counts of small game taxa in the old collection and Rabinovich's (1998) counts of carnivores and ungulates.

NOTE: Taxonomic categories are mutually exclusive. Chukar *(Alectoris chukar)* dominates among the nonpredatory bird remains that could be identified to species. Birds from Meged Rockshelter are raptors. The Levantine Aurignacian is rich in human-modified vulture bones. The entry *n.d.* indicates that no comparable data are available.

APPENDIX 13: "BONE-BASED" MNI COUNTS FOR COMMON PREY ANIMALS IN THE MOUSTERIAN (MP) UNITS OF THE CENTRAL TRENCH IN HAYONIM CAVE

Taxon	MP Unit							
	1	2	3	4	5	6	7	All Units
Ungulates								
Gazella gazella	1	2	3	9	4	3	1	23
Sus scrofa	<	<	<	1	1	<	0	2
Dama mesopotamica	1	1	3	8	3	2	1	19
Cervus elaphus	<	<	<	2	<	<	<	2
Bos primigenius	<	1	1	2	2	2	<	7
Reptiles								
Testudo graeca	1	2	7	67	16	9	2	104
Ophisaurus apodus	0	1	1	2	1	1	1	7
Aves								
Small-medium bird	0	<	2	2	2	<	<	6
Large bird	<	<	1	1	<	<	<	2
Struthio camelus eggshell	0	<	<	2	1	1	<	4
Small mammals								
Lepus capensis	<	<	1	0	0	0	0	1
Sciurus anomalous	0	0	1	2	0	0	0	3
Erinaceus sp.	0	0	0	0	1	0	0	1

NOTE: MNI counts are based exclusively on bone, except for ostrich eggshell fragments. A less-than symbol (<) means that one individual is represented by a small number of specimens.

APPENDIX 14: "BONE-BASED" MNI COUNTS FOR COMMON PREY ANIMALS IN THE EARLY NATUFIAN, KEBARAN, AND PRE-KEBARAN ASSEMBLAGES FROM HAYONIM CAVE AND MEGED ROCKSHELTER

Taxon	Hayonim Natufian	Hayonim Kebaran	Meged Early Kebaran	Meged Pre-Kebaran UP
Ungulates				
Gazella gazella	17	14	11	5
Sus scrofa	<	<	<	<
Capra aegagrus	<	<	<	<
Dama mesopotamica	2	2	3	2
Cervus elaphus	<	<	<	<
Bos primigenius	<	<	<	<
Reptiles				
Testudo graeca	14	8	6	2
Ophisaurus apodus	1	1	1	0
Aves				
Small-medium bird	65	3	<	<
Large bird	10	1	4?	4?
Struthio camelus eggshell	0	1	0	0
Small Mammals				
Lepus capensis	26	1	3	0
Sciurus anomalous	<	<	0	<
Erinaceus sp.	<	<	0	0

SOURCE: Natufian data (early phase) are from Munro 2004.

NOTE: MNI counts are based exclusively on bone, except for ostrich eggshell fragments. A less-than symbol (<) means that one individual is represented by a small number of specimens.

APPENDIX 15: GENUS-SPECIFIC COUNTS FOR PLEISTOCENE FAUNAL ASSEMBLAGES FROM HAYONIM CAVE AND MEGED ROCKSHELTER

Taxon	Hay B	Hay C	Meged Keb	Meged Pre-Keb UP	Hay D	Hay E Unit 1	Hay E Unit 2	Hay E Unit 3	Hay E Unit 4
Large game									
Gazella	1,483	1,039	451	160	6,253	30	185	721	674
Capreolus	5	16	1	—	77	—	6	1	—
Sus	50	63	7	4	52	—	17	66	61
Capra	17	36	21	7	103	—	4	1	—
Equus	—	3	1	1	3	—	5	10	6
Dama	47	198	58	16	798	12	140	587	199
Cervus	23	72	7	6	236	2	19	98	21
Alcelaphus	—	—	—	—	12	—	—	—	—
Bos	11	53	1	5	77	8	27	139	81
Dicerorhinus	—	2	—	—	1	—	1	1	—
Slow small game									
Testudo	1,777	453	462	124	1,118	8	367	2,326	2,268
Ophisaurus	9	34	4	—	18	—	21	170	29
Agama	113	—	—	—	87	—	—	—	—
Struthio, egg only	—	11	0	0	?	—	1	48	32
Quick small game									
Fliers									
Anseriformes	2	—	—	—	—	—	—	—	—
Galliformes or medium bird	823	~25	—	—	~306	3	12	39	12
Gruiformes	25	—	—	—	—	—	—	—	—
Columbiformes	5	—	—	—	—	—	—	—	—
Passeriformes	—	—	—	—	—	—	—	—	—
Falconiformes (human coll.)	—	—	~25	~10	—	—	—	—	—
Runners									
Lepus	1,559	12	25	—	72	—	3	—	—
Sciurus	27	4	—	1	—	—	1	7	—
Erinaceus	11	1	—	—	—	—	—	—	2
Swimmers									
Fish (one indet. genus)	23	—	—	—	—	—	—	—	—
Total	6,010	2,022	1,063	334	9,213	63	809	4,214	3,385

NOTE: Counts are number of specimens identified to genus (NISP). Vertebrate specimens that could be identified to anatomical element but not to genus or finer-level taxonomic affiliations are omitted from the counts. Taphonomic evidence indicates that all of these animals were consumed by Paleolithic humans. Data are subsets of total samples from the two sites, but the samples are representative.

APPENDIX 16: GENUS-SPECIFIC COUNTS FOR PLEISTOCENE FAUNAL ASSEMBLAGES FROM COASTAL SITES IN ITALY

LATE EPIPALEOLITHIC THROUGH LATE MOUSTERIAN ASSEMBLAGES FROM RIPARO MOCHI (LIGURIA) AND GROTTA BREUIL (LAZIO)

Taxon	Mochi A	Mochi C	Mochi D	Mochi F	Mochi G	Breuil B3/4	Breuil Br	Mochi I
Large game								
Rupicapra	—	7	—	1	—	1	—	—
Capreolus	25	117	189	26	35	14	35	14
Sus	—	29	16	36	44	—	1	89
Capra	32	360	229	74	29	184	22	22
Equus	1	18	1	3	5	9	14	42
Dama	—	3	12	1	—	9	2	4
Cervus	45	238	542	150	124	112	168	111
Megaceros	—	21	1	—	8	—	—	3
Bos	1	4	42	30	35	22	40	48
Dicerorhinus	—	—	—	—	—	—	6	—
Elaphas	—	—	—	—	—	—	1	—
Slow small game								
Patella	498	43	43	164	98	—	—	—
Monodonta	87	18	3	9	12	—	—	—
Glycymeris	3	5	12	5	23	—	—	—
Mytilus	191	88	57	122	327	—	—	—
Pecten	9	15	21	9	35	—	—	—
Acanthocardia/Cerastoderma	3	3	3	8	8	—	—	—
Callista	4	5	6	2	4	—	—	—
Quick small game								
Fliers								
Anseriformes	—	8	11	2	8	—	—	—
Galliformes	—	41	74	5	9	—	—	—
Gruiformes	—	5	11	4	5	—	—	—
Columbiformes	—	110	22	1	3	—	—	—
Passeriformes	—	180	51	18	13	—	—	—
Runners								
Lepus	—	7	2	5	3	—	1	—
Oryctolagus	2	529	173	6	28	—	—	2
Marmota	—	6	28	10	4	—	—	—
Total	901	1,860	1,549	691	860	351	290	335

SOURCE: Stiner 1994 for Grotta Breuil.

NOTE: Counts are number of specimens identified to genus in the case of vertebrates (NISP), but MNI is used for mollusks to control for much higher fragmentation in the latter group. Vertebrate specimens that could be identified to anatomical element but not to genus or finer-level taxonomic affiliations are omitted from the counts. Taphonomic evidence indicates that all of these animals were consumed by Paleolithic humans.

Continued on the next page

APPENDIX 16 (CONTINUED)

Mousterian Assemblages from Grotta di Sant'Agostino (Lazio) and Grotta dei Moscerini (Lazio)

Taxon	S0	S1	S2	S3	M1–2	M3	M6	M4
Large game								
Capreolus	146	106	52	28	8	65	16	18
Sus	71	68	30	13	—	4	—	—
Capra	74	73	33	17	—	15	5	—
Equus	9	22	7	3	—	1	—	—
Dama	138	118	47	23	4	13	8	28
Cervus	246	283	162	69	25	240	123	120
Bos	51	72	16	7	2	4	—	8
Hippopotamus	—	—	—	—	—	—	2	—
Dicerorhinus	9	10	2	—	2	2	1	—
Slow small game								
Testudo	—	—	—	—	—	—	—	10
Emys	—	—	—	—	—	—	—	29
Patella	—	—	—	—	5	6	6	—
Monodonta	—	—	—	—	2	3	1	1
Glycymeris	—	—	—	—	2	10	5	13
Mytilus	—	—	—	—	131	30	11	—
Pecten	—	—	—	—	1	—	—	1
Acanthocardia/Cerastoderma	—	—	—	—	1	2	1	1
Callista	—	—	—	—	1	69	4	5
Ostrea	—	—	—	—	1	4	—	1
Conch (one indet. genus)	—	—	—	—	2	3	2	1
Quick small game (runners)								
Lepus	17	13	4	3	6	—	—	1
Oryctolagus	10	6	2	1	—	—	—	—
Total	771	771	355	164	193	471	185	237

Source: Stiner 1994.

Note: Counts are number of specimens identified to genus in the case of vertebrates (NISP), but MNI is used for mollusks to control for much higher fragmentation in the latter group. Vertebrate specimens that could be identified to anatomical element but not to genus or finer-level taxonomic affiliations are omitted from the counts. Taphonomic evidence indicates that all of these animals were consumed by Paleolithic humans.

Key: (S) Grotta di Sant'Agostino; (M) Grotta dei Moscerini.

APPENDIX 17: GENUS-SPECIFIC COUNTS BY STRATIGRAPHIC LAYER FOR EPIPALEOLITHIC THROUGH INITIAL UPPER PALEOLITHIC FAUNAL ASSEMBLAGES FROM ÜÇAĞIZLI CAVE (HATAY, TURKEY)

Taxon	Epi	B	B1–3	C–D	E–E2	F–F2	G–I
Large game							
Capreolus	35	131	276	14	55	37	55
Sus	3	11	9	2	8	24	31
Capra	35	74	64	12	66	59	133
Dama	8	106	178	5	20	1	13
Cervus	—	1	1	—	18	7	1
Bos	—	1	2	1	3	12	26
Slow small game							
Testudo	16	8	15	1	—	4	1
Patella	33	80	165	1	—	—	1
Monodonta	72	342	471	4	1	—	1
Quick small game							
Fliers							
Galliformes or medium bird	41	39	58	—	3	1	3
Runners							
Lepus	71	1	2	—	1	1	—
Mustelids (human-predated)	15	1	3	—	—	—	6
Swimmers							
Rockfish (one indet. genus)	7	1	6	—	—	—	—
Total	336	796	1,250	40	176	146	271

NOTE: Counts are number of specimens identified to genus in the case of vertebrates (NISP), but MNI is used for mollusks to control for much higher fragmentation in the latter group. Vertebrate specimens that could be identified to anatomical element but not to genus or finer-level taxonomic affiliation are omitted from the counts. Taphonomic evidence indicates that all of these animals were consumed by Paleolithic humans. Data are subsets (1999 excavations only) of the current total sample from this site, because studies are ongoing, but the samples are representative.

APPENDIX 18: MNE COUNTS FOR UNGULATE TAXA IN THE KEBARAN (LAYER C) OF HAYONIM CAVE

Element	SU and Gazelle	MU	Fallow Deer	Pig	LU	Aurochs
Horn core/antler	4	—	1	—	—	—
Half cranium	9	2	1	1	1	—
Hemimandible	8	2	2	—	—	1
Atlas	2	1	1	—	—	—
Axis	4	1	—	—	—	—
Other cervical vertebra	8	3	~1	—	1	—
Thoracic vertebra	16	4	—	—	—	—
Rib	~34	~12	—	—	1	—
Lumbar vertebra	5	2	—	—	—	—
Sacrum	1	1	—	—	—	—
Innominate	7	1	1	—	1	—
Caudal vertebra	3	1	—	—	—	—
Sternal segment	2	—	—	—	—	—
Scapula	11	—	1	—	—	1
Humerus	21	2	1	1	1	—
Radius	8	2	2	—	1	1
Ulna	10	1	1	—	—	1
Carpals	22	3	2	2	—	—
Metacarpal	~8	1	4	—	1	—
Femur	20	2	2	1	2	—
Patella	5	2	—	—	—	—
Tibia	12	4	4	—	—	1
Astragalus	28	—	4	—	—	2
Calcaneum	15	—	—	—	—	—
Naviculo-cuboid	5	—	—	—	—	—
Other tarsals	4	1	—	—	—	—
Metatarsal	~8	1	~5	—	—	1
1st phalanx	68	3	16	2	—	5
2nd phalanx	62	2	15	2	—	3
3rd phalanx	64	—	6	2	—	—
Sesamoids	18	11	—	—	3	—
Indet. major metapods.	26	—	—	—	—	—
I^{upper}	1	—	—	1	—	—
P^2	5	—	3	—	—	—
P^3	2	—	2	—	—	—
P^4	6	—	1	1	—	—
M^1	4	—	1	—	—	—
M^2	11	—	2	—	—	—
M^3	9	—	2	—	—	1
I_{lower}–C	12	2	10	2	—	1
P_2	7	—	3	—	—	—
P_3	4	—	2	—	—	1
P_4	6	—	3	—	—	—
M_1	4	—	4	1	—	1
M_2	10	—	5	—	—	2
M_3	~10	—	~4	—	—	—

Continued on the next page

APPENDIX 18 (CONTINUED)

Element	SU and Gazelle	MU	Fallow Deer	Pig	LU	Aurochs
I	—	—	—	1	—	—
C	—	—	—	1	—	—
UP/M	5	—	2	—	—	1
LP/M	—	—	1	2	—	?
dP^3	?	—	—	—	—	—
dP^4	1	—	—	—	—	—
dI$_{lower}$–C	—	—	1	1	—	—
dP$_3$	4	—	?	—	—	1

NOTE: MNE is calculated from the most common portion of each element.

KEY: (SU) Small, gazelle-sized ungulate; (MU) medium, deer-sized ungulate; (LU) large, aurochs- or horse-sized ungulate. (~) Approximation, because morphologically unique, countable portions were absent.

APPENDIX 19: MNE COUNTS FOR TORTOISES AND LEGLESS LIZARDS IN THE KEBARAN AND PRE-KEBARAN LAYERS OF HAYONIM CAVE AND MEGED ROCKSHELTER

Element	Hayonim Kebaran	Meged Early Kebaran	Meged Pre-Kebaran UP
Tortoise (Testudo graeca)			
Carapace (nuccal or anal plates)	7	6	2
Plastron (bridges)	8	6	4
Innominate	—	1	—
Scapula	12	8	—
Humerus	~15	12	4
Radius	3	1	—
Ulna	—	—	—
Femur	~8	10	2
Tibia	1	2	1
Edge fragments[a]	50	41	9
Legless Lizard (Ophisaurus apodus)			
Cranium (maxilla)	—	—	—
Mandible	2	—	—
Vertebra	28	4	—
Ossified scale plates	1	—	—

NOTE: MNE is calculated from the most common portion of each element. Portions or elements in parentheses are morphological features used to estimate MNE.

KEY: (~) Approximation, because morphologically unique, countable portions were absent.

[a] "Edge fragments" refer to rims of the carapace or plastron, which are useful for examining the distribution of burning damage on tortoise remains.

APPENDIX 20: MNE COUNTS FOR UNGULATE TAXA IN THE EARLY KEBARAN AND PRE-KEBARAN LAYERS OF MEGED ROCKSHELTER

Element	Early Kebaran (0–200 cm bd)				Pre-Kebaran (UP) (>200 cm bd)			
	SU and Gazelle	Fallow Deer	MU	Goat	SU and Gazelle	Fallow Deer	MU	Goat
Horn core/antler	2	1	—	1	2	1	—	—
Half cranium	5	3	—	—	1	—	1	—
Hemimandible	7	2	—	1	4	1	1	1
Atlas	1	—	1	—	—	—	—	—
Axis	1	—	3	—	2	—	—	—
Other cervical vertebra	4	—	3	—	4	—	2	—
Thoracic vertebra	5	—	1	—	3	—	1	—
Rib	10	—	4	—	2	—	5	—
Lumbar vertebra	3	—	3	—	2	—	2	—
Innominate	6	—	2	—	4	—	2	—
Caudal vertebra	1	—	—	—	1	—	—	—
Scapula	3	—	1	—	5	—	1	—
Humerus	11	1	1	—	5	1	2	—
Radius	9	—	1	—	4	—	1	—
Ulna	11	—	1	—	3	—	2	—
Carpals	22	1	2	—	5	1	1	—
Metacarpal	11	5	1	—	9	—	1	—
Femur	5	1	3`	—	5	—	2	—
Tibia	9	—	3	—	3	1	2	—
Astragalus	22	1	1	1	7	2	—	—
Calcaneum	9	—	1	—	5	1	3	—
Naviculo-cuboid	4	2	1	—	1	—	—	—
Other tarsals	4	—	—	—	1	—	—	—
Metatarsal	21	5	—	1	10	1	—	—
1st phalanx	27	3	11	1	14	1	4	—
2nd phalanx	29	2	5	1	15	1	2	—
3rd phalanx	26	—	—	1	11	—	—	—
Sesamoids	—	—	2	—	2	—	1	—
P^2	1	—	—	—	2	—	—	—
P^3	—	—	—	1	1	—	—	—
P^4	1	—	—	1	—	—	—	—
M^1	4	1	—	—	—	—	—	—
M^2	5	1	—	—	2	—	—	—
M^3	7	4	—	—	—	—	—	—
I^{lower}–C	6	1	—	—	—	—	—	—
P_2	1	—	—	—	—	—	—	—
P_3	2	1	—	—	1	—	—	—
P_4	5	—	—	—	—	—	—	—
M_1	3	1	—	1	1	—	—	1
M_2	4	1	—	1	1	—	—	1
M_3	2	1	—	—	2	1	—	1

Continued on the next page

APPENDIX 20 (CONTINUED)

Element	Early Kebaran (0–200 cm bd)				Pre-Kebaran (UP) (>200 cm bd)			
	SU and Gazelle	Fallow Deer	MU	Goat	SU and Gazelle	Fallow Deer	MU	Goat
UP/M	6	1	—	—	2	—	—	—
LP/M	1	1	—	1	1	—	—	—
dP2	1	—	—	—	—	—	—	—
dP3	1	—	—	—	—	—	—	—
dP4	—	—	—	—	1	—	—	—

NOTE: MNE is calculated from the most common portion of each element. At Meged Rockshelter, extensive calcite concretions limited the visibility of long bone shaft features such as foraminae on gazelle long bones, as well as very small elements such as sesamoids, although some were recognized. Maximum counts for metapodials were obtained from distal ends; to the extent that front and rear members could be distinguished, there may be a slight counting bias toward metatarsals. Small ungulates are gazelle *(Gazella gazella)* almost exclusively; medium ungulates are a combination of goat *(Capra aegagrus)* and fallow deer *(Dama mesopotamica)*. Because no substantial differences in element representation are apparent for goat and deer on the basis of species-specific identifications, they are combined in the analysis of body part representation.

KEY: (SU) Small, gazelle-sized ungulate; (MU) medium, deer-sized ungulate; (LU) large, aurochs- or horse-sized ungulate. (~) Approximation, because morphologically unique, countable portions are absent.

APPENDIX 21: MNE COUNTS FOR UNGULATE TAXA IN MOUSTERIAN (MP) UNIT 1 OF THE CENTRAL TRENCH IN HAYONIM CAVE

Element	SU	Gazelle	MU	Fallow Deer	Red Deer	Pig	LU	Aurochs
Half cranium	—	1	—	1	—	—	—	—
Hemimandible	—	2	—	1	—	—	—	1
Axis	—	1	—	—	—	—	—	—
Other cervical vertebra	1	—	—	—	—	—	—	—
Thoracic vertebra	2	—	—	—	—	—	—	—
Rib	2	—	—	—	—	—	—	—
Lumbar vertebra	3	—	1	—	—	—	—	—
Innominate	—	2	—	—	—	—	—	—
Caudal vertebra	2	—	—	—	—	—	—	—
Scapula	1	—	—	—	—	—	1	—
Humerus	—	1	1	—	—	—	—	—
Radius	1	1	1	1	—	—	—	—
Ulna	—	—	—	~1	—	—	—	—
Carpals	—	—	—	1	—	—	—	—
Metacarpal	—	—	—	1	—	—	—	—
Femur	—	—	1	—	—	—	—	—
Tibia	1	3	~1	—	—	—	—	—
Calcaneum	—	—	1	—	—	—	—	—
Naviculo-cuboid	—	1	—	—	—	—	—	—
1st phalanx	1	4	—	1	—	—	—	—
2nd phalanx	—	2	—	—	2	—	—	—
3rd phalanx	—	1	—	—	—	—	—	—
Indet. major metapods.	1	1	—	—	—	—	—	—
M^3	—	—	—	—	—	—	—	1
I_{lower}–C	—	1	—	1	—	—	—	—
P_2	—	1	—	1	—	—	—	—
P_3	—	1	—	1	—	—	—	—
P_4	—	1	—	1	1	—	—	—
M_1	—	1	—	1	—	—	—	—
M_2	—	1	—	1	—	—	—	—
M_3	—	1	—	1	—	—	—	1
UP/M	—	—	—	?	—	?	—	?
LP/M	—	—	—	—	—	?	—	?

NOTE: MNE is calculated from the most common portion of each element.

KEY: (SU) Small, gazelle-sized ungulate; (MU) medium, deer-sized ungulate; (LU) large, aurochs- or horse-sized ungulate. (~) Approximation, because morphologically unique, countable portions are absent.

APPENDIX 22: MNE COUNTS FOR UNGULATE TAXA IN MOUSTERIAN (MP) UNIT 2 OF THE CENTRAL TRENCH IN HAYONIM CAVE

Element	SU	Gazelle	MU	Fallow Deer	Red Deer	Pig	Goat	LU	Aurochs
Horn core/antler	—	1	—	~1	—	—	—	—	1
Half cranium	—	2	—	—	—	—	—	—	1
Hemimandible	—	1	—	1	—	—	—	~1	—
Other cervical vertebra	3	—	—	—	—	—	—	1	—
Thoracic vertebra	1	—	1	—	—	—	—	—	—
Rib	4	—	1	—	—	—	—	—	—
Lumbar vertebra	2	—	—	—	—	—	—	—	—
Sacrum	—	—	1	—	—	—	—	—	—
Innominate	1	1	—	—	—	—	—	—	—
Sternal segment	1	—	1	—	—	—	—	—	—
Scapula	—	2	—	—	—	—	—	—	—
Humerus	—	2	~1	—	—	—	—	—	—
Radius	—	2	~1	—	—	—	—	1	—
Ulna	—	1	—	—	—	—	—	—	—
Carpals	—	2	—	1	—	—	—	—	—
Metacarpal	—	1	—	—	1	—	—	—	—
Femur	—	2	—	1	—	—	—	—	—
Patella	—	2	—	—	—	—	—	—	—
Tibia	1	1	—	—	—	—	—	—	—
Astragalus	—	1	—	—	—	—	—	—	—
Calcaneum	—	3	1	—	1	—	—	—	—
Naviculo-cuboid	—	1	—	—	—	—	—	—	—
Other tarsals	—	2	—	—	—	—	—	—	—
Metatarsal	—	—	—	2	?	—	—	1	—
1st phalanx	—	2	—	2	1	1	—	—	—
2nd phalanx	—	5	—	—	—	—	1	—	—
3rd phalanx	—	1	—	1	—	—	—	—	1
Indet. major metapods.	—	1	—	—	—	—	—	—	—
P^3	—	—	—	—	—	—	—	—	2
P^4	—	—	—	—	—	—	—	—	1
M^1	—	1	—	—	—	—	—	—	—
M^2	—	3	—	—	—	1	—	—	—
M^3	—	3	—	—	—	1	—	—	1
I$_{lower}$–C	—	—	—	2	—	—	—	—	—
P$_2$	—	—	—	2	—	—	—	—	—
P$_3$	—	—	—	1	—	—	—	—	—
P$_4$	—	—	—	1	—	1	—	—	—
M$_1$	—	—	—	1	—	—	—	—	—
M$_2$	—	1	—	—	—	—	—	—	1
M$_3$	—	2	—	—	—	—	—	—	—
C	—	—	—	—	—	1	—	—	—
UP/M	—	~2	—	—	—	—	—	—	—
LP/M	—	~1	—	—	—	—	—	—	1
dP^4	—	2	—	—	—	—	—	—	—

NOTE: MNE is calculated from the most common portion of each element.

KEY: (SU) Small, gazelle-sized ungulate; (MU) medium, deer-sized ungulate; (LU) large, aurochs- or horse-sized ungulate. (~) Approximation, because morphologically unique, countable portions are absent.

APPENDIX 23: MNE COUNTS FOR UNGULATE TAXA IN MOUSTERIAN (MP) UNIT 3 OF THE CENTRAL TRENCH IN HAYONIM CAVE

Element	SU	Gazelle	MU	Fallow Deer	Red Deer	Pig	Wild Ass	LU	Aurochs	Goat and Horse
Horn core/antler	—	2	—	?	—	—	—	—	1	—
Half cranium	—	2	—	2	—	—	—	—	1	—
Hemimandible	—	5	—	6	~1	1	—	—	—	—
Atlas	—	1	1	1	—	—	—	—	—	—
Axis	—	3	—	2	—	—	—	—	—	—
Other cervical vertebra	8	—	3	—	—	—	—	—	—	—
Thoracic vertebra	9	—	2	—	1	—	—	—	—	—
Rib	11	—	6	—	—	—	—	2	—	—
Lumbar vertebra	1	—	2	—	—	—	—	1	—	—
Sacrum	1	—	1	—	—	—	—	—	—	—
Innominate	1	3	1	1	—	—	—	—	—	—
Caudal vertebra	—	—	1	—	—	—	—	—	—	—
Sternal segment	1	—	1	—	—	—	—	—	—	—
Scapula	2	3	—	1	—	—	—	—	1	—
Humerus	1	3	—	2	—	—	—	1	1	—
Radius	—	3	—	3	—	—	—	—	—	—
Ulna	2	2	1	2	—	—	—	—	1	—
Carpals	—	1	1	4	1	—	—	—	1	—
Metacarpal	—	1	1	4	—	—	—	—	—	—
Femur	—	5	2	2	—	—	—	—	—	—
Patella	2	2	—	1	—	—	—	—	—	—
Tibia	—	6	1	4	—	—	—	~1	—	—
Astragalus	—	2	—	2	2	—	—	—	—	—
Calcaneum	—	3	1	3	1	1	—	—	—	—
Naviculo-cuboid	—	—	—	1	—	—	—	—	—	—
Other tarsals	1	1	2	—	—	—	—	—	—	—
Metatarsal	—	1	—	3	1	1	—	—	—	horse=1
1st phalanx	—	8	1	7	4	1	1	—	1	—
2nd phalanx	—	11	—	5	2	1	—	—	1	—
3rd phalanx	—	12	—	7	1	—	—	—	—	—
Sesamoids	4	—	6	—	—	—	—	2	1	—
Indet. major metapods.	—	10	—	1	—	1	—	—	1	—
I^{upper}	—	—	—	—	—	?	—	—	—	—
P^2	—	4	—	—	—	—	—	—	—	—
P^3	—	1	—	—	—	—	—	—	—	—
P^4	—	—	—	1	—	—	—	—	1	—
M^1	—	1	—	2	—	—	—	—	—	—
M^2	—	5	—	1	—	—	—	—	—	—
M^3	—	~2	—	1	1	—	—	—	—	—
I_{lower}–C	—	3	—	4	—	1	—	—	3	—
C_1	—	—	—	—	—	1	—	—	—	—
P_2	—	3	—	2	—	1	—	—	—	—
P_3	—	3	—	2	—	—	—	—	—	—
P_4	—	3	—	1	—	1	—	—	1	—
M_1	—	3	—	~2	—	—	—	—	—	—
M_2	—	7	—	2	—	—	—	—	—	—

Continued on the next page

APPENDIX 23 (CONTINUED)

Element	SU	Gazelle	MU	Fallow Deer	Red Deer	Pig	Wild Ass	LU	Aurochs	Goat and Horse
M_3	1	11	—	—	—	—	—	—	—	—
I	—	—	—	—	—	1	—	—	—	—
UP/M	—	4	—	1	?	?	—	—	—	goat=1
LP/M	—	5	—	2	?	1	—	—	~2	—
dI^{upper}	—	1	—	—	—	—	—	—	—	—
dP^2	—	1	—	1	—	—	—	—	—	—
dP^3	—	1	—	2	—	—	—	—	—	—
dP^4	—	1	—	2	—	—	—	—	—	—
dP$_3$	—	—	—	—	—	1	—	—	—	—
dP$_4$	—	1	—	3	—	1	—	—	—	—

NOTE: MNE is calculated from the most common portion of each element.

KEY: (SU) Small, gazelle-sized ungulate; (MU) medium, deer-sized ungulate; (LU) large, aurochs- or horse-sized ungulate. (~) Approximation, because morphologically unique, countable portions are absent.

APPENDIX 24: MNE COUNTS FOR UNGULATE TAXA IN MOUSTERIAN (MP) UNIT 4 OF THE CENTRAL TRENCH IN HAYONIM CAVE

Element	SU	Gazelle	MU	Fallow Deer	Red Deer	Pig	Wild Ass	LU	Aurochs	Horse
Horn core/antler	—	3	—	1	~1	—	—	—	1	—
Half cranium	—	16	—	12	1	2	—	—	4	—
Hemimandible	—	18	—	14	1	2	—	—	1	—
Atlas	2	3	2	4	1	—	—	1	1	—
Axis	—	11	2	4	1	~1	—	—	1	—
Other cervical vertebra	22	—	18	—	—	~1	—	2	—	—
Thoracic vertebra	23	—	29	—	—	—	—	3	—	—
Rib	38	—	34	—	—	—	—	3	—	—
Lumbar vertebra	16	—	16	—	—	—	—	1	—	—
Sacrum	3	—	2	—	—	—	—	—	—	—
Innominate	5	5	4	4	1	—	—	1	—	—
Caudal vertebra	2	—	3	—	—	—	—	—	—	—
Sternal segment	6	—	4	—	—	—	—	—	—	—
Scapula	4	15	—	15	1	—	—	2	—	—
Humerus	5	11	3	5	3	—	—	1	1	?
Radius	1	13	1	9	1	2	—	—	2	—
Ulna	—	8	3	4	1	—	—	—	2	—
Carpals	1	21	2	52	2	2	—	—	2	—
Metacarpal	1	12	—	7	1	—	—	—	1	—
Femur	—	13	2	3	1	—	—	1	—	—
Patella	3	5	3	2	—	—	—	—	—	—
Tibia	2	13	2	12	—	3	—	3	2	—
Astragalus	—	14	?	15	7	—	—	—	2	—
Calcaneum	—	12	1	8	2	1	—	1	1	—
Naviculo-cuboid	—	8	1	4	—	2	—	—	—	—
Other tarsals	1	2	1	5	—	—	—	—	1	—
Metatarsal	—	7	—	8	1	—	—	—	1	—
1st phalanx	—	25	2	31	8	5	—	—	7	1
2nd phalanx	—	32	1	29	5	2	1	—	2	—
3rd phalanx	—	30	1	36	—	1	1	—	2	—
Sesamoids	18	—	51	—	—	1	—	7	1	—
Indet. major metapods.	—	—	—	2	—	6	1	—	—	1
P2	—	7	—	9	1	—	—	—	2	—
P3	—	7	—	8	1	—	—	—	—	—
P4	—	8	—	8	2	—	—	—	—	—
M1	—	14	—	10	3	—	—	—	2	—
M2	—	15	—	7	1	—	—	—	3	—
M3	—	11	—	9	4	—	—	—	2	—
I_{lower}–C	—	8	—	9	—	2	—	—	2	—
P_2	—	10	—	6	1	1	—	—	—	—
P_3	—	10	—	9	—	—	—	—	1	—
P_4	—	10	—	7	—	1	—	—	—	—
M_1	—	15	—	7	—	—	—	—	1	—
M_2	—	17	—	7	—	1	—	—	1	—
M_3	—	12	—	7	—	—	—	—	—	—
I	—	—	—	—	—	3	—	—	—	—

Continued on the next page

APPENDIX 24 (CONTINUED)

Element	SU	Gazelle	MU	Fallow Deer	Red Deer	Pig	Wild Ass	LU	Aurochs	Horse
UP/M	—	2	—	2	—	1	—	—	3	—
LP/M	—	4	—	~5	?	2	—	—	1	—
dP^2	—	1	—	—	—	—	—	—	2	—
dP^3	—	1	—	1	—	1	—	—	—	—
dP^4	—	1	—	2	—	1	—	—	—	—
dI$_{lower}$–C	—	2	—	—	—	3	—	—	—	—
dP$_2$	—	1	—	2	—	—	—	—	—	—
dP$_3$	—	3	—	6	—	—	—	—	—	—
dP$_4$	—	4	—	5	—	—	—	—	—	—

NOTE: MNE is calculated from the most common portion of each element.

KEY: (SU) Small, gazelle-sized ungulate; (MU) medium, deer-sized ungulate; (LU) large, aurochs- or horse-sized ungulate. (~) Approximation, because morphologically unique, countable portions are absent.

APPENDIX 25: MNE COUNTS FOR UNGULATE TAXA IN MOUSTERIAN (MP) UNIT 5 OF THE CENTRAL TRENCH IN HAYONIM CAVE

Element	SU	Gazelle	MU	Fallow Deer	Red Deer	Pig	Goat	Wild Ass	LU	Aurochs
Horn core/antler	—	5	—	~1	—	—	—	—	—	~1
Half cranium	—	8	—	5	—	—	—	—	—	1
Hemimandible	—	7	—	4	—	1	1	—	—	1
Atlas	—	1	—	1	—	—	—	—	—	1
Axis	—	1	1	—	—	—	—	—	—	—
Other cervical vertebra	6	—	6	—	—	—	—	—	2	—
Thoracic vertebra	8	—	4	—	—	—	—	—	—	—
Rib	9	—	4	—	—	—	—	—	3	—
Lumbar vertebra	9	—	2	—	—	—	—	—	1	—
Sacrum	2	—	1	—	—	—	—	—	1	—
Innominate	1	5	1	—	—	—	—	—	—	1
Caudal vertebra	—	—	2	—	—	—	—	—	—	—
Sternal segment	1	—	—	—	—	—	—	—	—	—
Scapula	2	5	3	2	—	—	—	—	1	1
Humerus	1	4	—	2	—	—	—	—	—	3
Radius	2	2	1	1	—	1	—	—	2	—
Ulna	—	1	—	1	—	—	—	—	1	1
Carpals	—	14	—	15	—	3	—	—	—	—
Metacarpal	—	2	—	1	1	—	—	—	—	—
Femur	—	3	1	—	—	~1	—	—	1	—
Patella	—	6	—	—	—	—	—	—	—	—
Tibia	—	4	—	4	1	1	—	—	—	2
Astragalus	—	18	—	1	2	—	—	—	—	—
Calcaneum	—	13	—	4	—	1	—	—	—	—
Naviculo-cuboid	—	4	—	~1	—	—	—	—	—	—
Other tarsals	—	—	—	6	—	—	—	—	—	1
Metatarsal	1	8	—	5	1	—	—	—	—	—
1st phalanx	—	23	—	8	2	1	—	—	—	2
2nd phalanx	—	29	—	4	1	3	—	—	—	1
3rd phalanx	1	21	—	2	—	3	—	1	~1	—
Sesamoids	6	—	15	—	—	—	—	—	1	2
Indet. major metapods.	—	6	—	—	—	5	—	—	—	1
I^{upper}	—	—	—	—	—	1	—	—	—	—
C^1	—	—	—	—	—	1	—	—	—	—
P^2	—	2	—	1	—	—	—	—	—	—
P^3	—	2	—	—	—	—	—	—	—	—
P^4	—	4	—	—	—	—	—	—	—	—
M^1	—	6	—	1	—	—	—	—	—	—
M^2	—	10	—	1	—	—	—	—	—	—
M^3	—	5	—	2	—	1	—	—	—	2
I$_{lower}$ –C	—	6	—	—	—	—	—	—	—	3
C$_1$	—	—	—	—	—	?	—	—	—	—
P$_2$	—	1	—	3	—	1	—	—	—	1
P$_3$	—	2	—	2	—	—	—	—	—	—
P$_4$	—	2	—	2	—	—	—	—	—	—
M$_1$	—	1	—	3	—	—	—	—	—	—

Continued on the next page

APPENDIX 25 (CONTINUED)

Element	SU	Gazelle	MU	Fallow Deer	Red Deer	Pig	Goat	Wild Ass	LU	Aurochs
M_2	—	1	—	2	—	1	—	—	—	1
M_3	—	7	—	~1	—	1	—	—	—	1
C	—	—	—	—	—	?	—	—	—	—
UP/M	—	~4	—	1	—	?	—	—	—	2
LP/M	—	~12	—	2	—	1	—	—	—	3
dP^2	—	1	—	—	—	—	—	—	—	—
dP^3	—	1	—	—	—	—	—	—	—	—
dP^4	—	1	—	—	—	—	—	—	—	—
dP_3	—	—	—	?	—	—	—	—	—	—
dP_4	—	1	—	—	—	—	—	—	—	1

NOTE: MNE is calculated from the most common portion of each element.

KEY: (SU) Small, gazelle-sized ungulate; (MU) medium, deer-sized ungulate; (LU) large, aurochs- or horse-sized ungulate. (~) Approximation, because morphologically unique, countable portions are absent.

APPENDIX 26: MNE COUNTS FOR UNGULATE TAXA IN MOUSTERIAN (MP) UNIT 6 OF THE CENTRAL TRENCH IN HAYONIM CAVE

Element	SU	Gazelle	MU	Fallow Deer	Red Deer	Pig	Wild Ass	LU	Aurochs
Horn core/antler	—	3	—	?	—	—	—	—	?
Half cranium	—	3	—	3	—	—	—	—	3
Hemimandible	—	5	—	2	—	~1	—	—	1
Atlas	—	1	—	—	—	—	—	—	—
Axis	—	1	~1	—	—	—	—	1	—
Other cervical vertebra	7	—	5	—	—	—	—	—	—
Thoracic vertebra	9	—	4	—	—	—	—	—	—
Rib	4	—	7	—	—	—	—	1	—
Lumbar vertebra	7	—	3	—	—	—	—	1	—
Sacrum	2	—	—	—	—	—	—	—	—
Innominate	—	4	—	1	—	—	—	—	1
Caudal vertebra	1	—	—	—	—	—	—	—	—
Sternal segment	—	—	1	—	—	—	—	—	—
Scapula	1	5	1	2	—	—	—	1	—
Humerus	—	5	1	1	—	—	—	1	—
Radius	—	4	1	1	—	1	—	—	1
Ulna	1	4	1	1	—	—	—	—	—
Carpals	—	6	1	—	—	—	—	—	1
Metacarpal	—	5	—	1	—	—	—	—	1
Femur	—	3	—	1	—	1	—	—	—
Patella	—	4	—	—	—	—	—	—	—
Tibia	1	4	—	2	—	2	—	—	1
Astragalus	—	10	—	1	—	—	—	—	—
Calcaneum	—	3	1	1	—	—	—	—	—
Naviculo-cuboid	—	3	—	1	—	—	—	—	—
Other tarsals	—	—	—	2	—	—	—	—	—
Metatarsal	—	7	—	2	—	—	—	—	1
1st phalanx	—	11	2	3	—	1	1	—	2
2nd phalanx	—	10	—	3	—	1	—	—	1
3rd phalanx	—	16	1	8	—	1	—	—	—
Sesamoids	—	—	3	—	—	1	—	—	2
Indet. major metapods.	—	—	—	—	—	1	—	—	—
P^3	—	1	—	—	—	—	—	—	—
P^4	—	1	—	—	—	—	—	—	—
M^1	—	4	—	1	—	—	—	—	—
M^2	—	2	—	—	—	—	—	—	1
M^3	—	2	—	—	—	—	—	—	—
I_{lower}–C	—	3	—	4	—	—	—	—	1
P_2	—	4	—	1	—	—	—	—	—
P_3	—	3	—	—	—	—	—	—	—
P_4	—	1	—	—	2	—	—	—	—
M_1	—	2	—	1	1	—	—	—	—
M_2	—	3	—	2	1	—	—	—	—
M_3	—	1	—	—	1	—	—	—	—

Continued on the next page

APPENDIX 26 (CONTINUED)

Element	SU	Gazelle	MU	Fallow Deer	Red Deer	Pig	Wild Ass	LU	Aurochs
UP/M	—	1	—	~2	—	?	—	—	—
LP/M	—	1	—	~1	—	?	—	—	2
dP$_3$	—	—	—	—	—	—	—	—	1
dP$_4$	—	—	—	1	—	—	—	—	1

NOTE: MNE is calculated from the most common portion of each element.

KEY: (SU) Small, gazelle-sized ungulate; (MU) medium, deer-sized ungulate; (LU) large, aurochs- or horse-sized ungulate. (~) Approximation, because morphologically unique, countable portions are absent.

APPENDIX 27: MNE COUNTS FOR UNGULATE TAXA IN MOUSTERIAN (MP) UNIT 7 OF THE CENTRAL TRENCH IN HAYONIM CAVE

Element	SU	Gazelle	MU	Fallow Deer	Red Deer	Goat	LU	Aurochs
Horn core/antler	—	~1	—	—	—	—	—	—
Half cranium	—	1	—	1	—	—	—	—
Hemimandible	—	1	—	1	—	—	—	—
Axis	—	1	—	—	—	—	—	—
Other cervical vertebra	1	—	2	—	—	—	—	—
Thoracic vertebra	1	—	1	—	—	—	—	—
Rib	2	—	3	—	—	—	—	—
Lumbar vertebra	1	—	2	—	—	—	—	—
Innominate	—	1	—	—	—	—	—	—
Scapula	—	—	1	—	—	—	—	—
Humerus	—	~1	~1	—	—	—	—	—
Radius	—	1	~1	—	—	—	—	—
Ulna	—	—	—	—	—	—	1	—
Carpals	—	1	—	—	—	—	—	—
Metacarpal	~1	—	~1	—	~1	—	—	—
Femur	—	—	2	—	—	—	—	—
Tibia	1	1	~1	—	—	—	—	—
Astragalus	—	1	1	—	—	—	—	—
Calcaneum	—	2	1	—	—	—	—	—
Metatarsal	—	1	—	1	—	—	—	—
1st phalanx	—	2	—	~1	—	—	—	—
2nd phalanx	—	—	—	2	—	—	—	—
3rd phalanx	—	1	—	3	—	—	—	—
P^2	—	—	—	—	1	—	—	—
I_{lower}–C	–-	1	—	1	—	—	—	—
P_2	—	—	—	1	—	—	—	—
M_1	—	—	—	1	—	—	—	—
M_2	—	—	—	—	—	1	—	—
M_3	—	—	—	1	—	—	—	—
UP/M	—	—	—	—	—	—	—	1
LP/M	—	1	—	—	—	—	—	?

NOTE: MNE is calculated from the most common portion of each element.

KEY: (SU) Small, gazelle-sized ungulate; (MU) medium, deer-sized ungulate; (LU) large, aurochs- or horse-sized ungulate. (~) Approximation, because morphologically unique, countable portions are absent.

APPENDIX 28: MNE COUNTS FOR TORTOISES AND LEGLESS LIZARDS IN THE MOUSTERIAN (MP) UNITS OF THE CENTRAL TRENCH IN HAYONIM CAVE

Element	MP Unit							
	1	2	3	4	5	6	7	All
Tortoise (Testudo graeca)								
Carapace (nuccal or anal plates)	—	2	7	67	16	9	2	103
Plastron (bridges)	1	3	13	73	38	35	3	166
Innominate	—	1	1	8	1	1	1	13
Scapula	—	1	6	16	7	7	1	38
Humerus	—	—	12	23	10	12	2	59
Radius	—	1	3	8	2	—	2	16
Ulna	—	—	—	—	1	1	1	3
Femur	—	1	2	28	11	6	1	49
Tibia	—	1	2	8	1	1	1	14
Edge fragments[a]	—	9	39	276	246	219	9	798
Legless lizard (Ophisaurus apodus)								
Cranium (maxilla)	—	—	1	—	—	—	—	1
Mandible	—	1	2	4	—	1	—	8
Vertebra	—	1	18	140	26	11	3	199
Ossified scale plates	—	—	—	14	—	—	—	14

NOTE: MNE is calculated from the most common portion of each element. Portions or elements in parentheses are morphological features used to estimate MNE.

[a] "Edge fragments" refer to rims of the carapace or plastron, which are useful for examining the distribution of burning damage on tortoise remains.

APPENDIX 29: STANDARDIZED MNE (MAU) VALUES FOR UNGULATE ASSEMBLAGES FROM THE NATUFIAN, KEBARAN, AND MOUSTERIAN LAYERS OF HAYONIM CAVE AND MEGED ROCKSHELTER

Anatomical Region and Code	Hay Nat.	Hay Keb.	Meg Keb.	Meg UP	Hay MP1	Hay MP2	Hay MP3	Hay MP4	Hay MP5	Hay MP6	Hay MP7	Hay MP, All
Gazelle (small ungulate)												
Horn (1)	13.00	2.00	1.00	1.00	0.00	0.50	1.00	1.50	2.50	1.50	0.50	7.50
Head (2)	12.00	4.25	3.00	1.00	0.75	0.75	1.75	8.50	3.75	2.00	0.50	18.00
Neck (3)	4.29	2.00	0.86	0.86	0.29	0.43	1.71	5.43	1.14	1.29	0.29	10.57
Axial (4)	3.41	1.29	0.49	0.22	0.18	0.18	0.53	1.84	0.69	0.53	0.10	4.06
Upper front (5)	26.00	8.00	3.50	2.50	0.50	1.00	2.25	8.75	3.00	2.75	0.25	18.50
Lower front (6)	17.33	6.50	5.17	2.67	0.33	0.67	1.33	5.83	1.17	2.33	0.33	12.00
Upper hind (7)	10.00	10.00	2.50	2.50	0.00	1.00	2.50	6.50	1.50	1.50	0.00	13.00
Lower hind (8)	26.12	9.50	5.12	3.12	0.62	0.75	1.50	6.00	5.50	3.12	0.75	18.25
Feet (9)	26.62	8.08	3.42	1.67	0.33	0.33	1.29	3.62	3.08	1.54	0.12	10.33
Fallow Deer (medium ungulate)												
Horn (1)	0.00	0.50	1.00	0.50	0.00	0.50	0.00	0.50	0.50	0.00	0.00	1.50
Head (2)	0.00	1.75	0.75	0.75	0.50	0.25	2.00	3.50	2.25	1.25	0.50	10.25
Neck (3)	0.00	1.00	1.00	0.29	0.00	0.00	1.00	4.29	1.14	0.86	0.29	7.57
Axial (4)	0.00	0.43	0.20	0.20	0.02	0.06	0.26	1.82	0.24	0.31	0.12	2.84
Upper front (5)	0.00	1.00	0.75	1.00	0.25	0.25	0.75	5.75	1.75	1.25	0.50	10.50
Lower front (6)	0.00	1.83	1.33	0.67	0.67	0.17	1.83	4.00	0.67	0.83	0.33	8.50
Upper hind (7)	0.00	2.00	2.00	1.00	0.50	0.50	2.00	2.50	0.50	0.50	1.00	7.50
Lower hind (8)	0.00	2.62	1.62	1.25	0.25	0.37	1.75	5.75	1.75	0.87	0.50	11.25
Feet (9)	0.00	1.83	1.00	0.37	0.04	0.12	0.83	4.17	0.58	0.71	0.25	6.71
Aurochs (large ungulate)												
Horn (1)	0.00	0.00	0.00	0.00	0.00	0.50	0.50	0.50	0.50	0.00	0.00	2.00
Head (2)	0.00	0.50	0.00	0.00	0.25	0.50	0.25	0.25	0.50	1.00	0.00	2.75
Neck (3)	0.00	0.14	0.00	0.00	0.00	0.14	0.00	0.71	0.43	0.14	0.00	1.43
Axial (4)	0.00	0.04	0.00	0.00	0.00	0.00	0.06	0.16	0.12	0.06	0.00	0.41
Upper front (5)	0.00	0.50	0.00	0.00	0.25	0.00	0.75	1.00	1.25	0.50	0.00	3.75
Lower front (6)	0.00	0.67	0.00	0.00	0.00	0.17	0.17	0.83	0.67	0.33	0.17	2.33
Upper hind (7)	0.00	1.00	0.00	0.00	0.00	0.00	0.00	0.50	0.50	0.00	0.00	1.00
Lower hind (8)	0.00	0.50	0.00	0.00	0.00	0.12	0.12	1.25	0.25	0.25	0.00	2.00
Feet (9)	0.00	0.33	0.00	0.00	0.00	0.04	0.08	0.46	0.17	0.12	0.00	0.87

References

Abildgard, F., J. Anderson, and O. Barndorf-Nielsen
  1972 The hare population (*Lepus europaeus* Pallas) of Illumo, Denmark. *Danish Review of Game Biology* 6:1–32.

Adovasio, J. M., O. Soffer, and B. Klima
  1996 Paleolithic fiber technology: Data from Pavlov I, ca. 26,000 B.P. *Antiquity* 70:526–534.

Akazawa, T.
  1979 Middle Paleolithic assemblages from Douara Cave. In *Paleolithic Site of Douara Cave and Paleography of Palmyra Basin in Syria*, K. Hanihara and T. Akazawa, eds., pp. 1–30. University Museum, University of Tokyo, Tokyo.
  1987 The ecology of the Middle Paleolithic occupation at Douara Cave, Syria. *Bulletin of the University Museum* (University of Tokyo) 29: 155–166.

Albert, R. M., O. Bar-Yosef, L. Meignen, and S. Weiner
  2003 Quantitative phytolith study of hearths from the Natufian and Middle Palaeolithic levels of Hayonim Cave (Galilee, Israel). *Journal of Archaeological Science* 30:461–480.

Albert, R. M., O. Lavi, L. Estroff, S. Weiner, A. Tsatskin, A. Ronen, and S. Lev-Yadun
  1999 Occupation of Tabun Cave, Mount Carmel, Israel, during the Mousterian period: A study of the sediments and phytoliths. *Journal of Archaeological Science* 26:1249–1260.

Albert, R. M., S. Weiner, O. Bar-Yosef, and L. Meignen
  2000 Phytoliths in the Middle Palaeolithic deposits of Kebara Cave, Mount Carmel, Israel: Study of the plant materials used for fuel and other purposes. *Journal of Archaeological Science* 27: 931–947.

Albrecht, G., and H. Berke
  1982– Site catalogue: Petersfels, Engen, Baden-
  1983 Wurttemberg BRD—Magdalenian. *Early Man News: Newsletter for Human Paleoecology, Tubingen* 7–8:7–14.

Ali, S., and S. Ripley
  1969 *Handbook of the Birds of India and Pakistan*, vol. 2. Oxford University Press, Oxford.

Alkon, P.
1983   Nesting and brood production in an Israeli population of chukars, *Alectoris chukar* (Aves: Phasianidae). *Israel Journal of Zoology* 32: 185–193.

Aloni, R.
1984   The vegetation of the Lower Galilee. In *Vegetation of Israel: Plants and Animals of the Land of Israel* 8 (Hebrew), A. Alon, ed., pp. 143–151. Ministry of Defense/Society for the Protection of Nature, Tel Aviv.

Ambrose, S.
1998   Chronology of the Later Stone Age and food production in East Africa. *Journal of Archaeological Science* 25:377–392.

Ammerman, A. J., and L. S. Cavalli-Sforza
1984   *The Neolithic Transition and the Genetics of Populations in Europe.* Princeton University Press, Princeton, New Jersey.

Andrews, P.
1990   *Owls, Caves, and Fossils.* University of Chicago Press, Chicago.

Arensburg, G., O. Bar-Yosef, A. Belfer-Cohen, and Y. Rak
1990   Mousterian and Aurignacian human remains from Hayonim Cave, Israel. *Paléorient* 16: 107–109.

Ash, J. S.
1970   Bag records as indicators of population trends in partridges. Seventh International Congress of Game Biologists. *Finnish Game Research* 30:357–360.

Audouze, F.
1987   The Paris Basin in Magdalenian times. In *The Pleistocene Old World: Regional Perspectives,* O. Soffer, ed., pp. 183–200. Plenum Press, New York.

Avellino, E., A. Bietti, L. Giacopini, and A. Lo Pinto
1989   A new Dryas II site in southern Lazio, Riparo Salvini: Thoughts on the Late Epigravettian in middle and southern Tyrrhenian Italy. In *The Mesolithic in Europe,* C. Bonsall, ed., pp. 516–532. John Donald, Edinburgh.

Ayal, Y., and D. Baharav
1983   Conservation and management plan for the mountain gazelle in Israel. In *Game Harvest Management,* S. L. Beason and S. F. Roberton, eds., pp. 269–387. Caesar Kleburg Wildlife Research Institute, Kingsville, Texas.

Baharav, D.
1974   Notes on the population structure and biomass of the mountain gazelle, *Gazella gazella gazella*. *Israel Journal of Zoology* 23:39–44.
1983a   Reproductive strategies in female mountain and dorcas gazelles *(Gazella gazella gazella* and *Gazella dorcas*). *Journal of Zoology,* London 200:445–453.
1983b   Observation on the ecology of the mountain gazelle in the upper Galilee, Israel. *Mammalia* 47(1):60–69.

Bailey, G. N.
1983   Economic change in late Pleistocene Cantabria. In *Hunter-Gatherer Economy in Prehistory,* G. N. Bailey, ed., pp. 149–165. Cambridge University Press, Cambridge.

Barbujani, G., and G. Bertorelle
2001   Genetics and the population history of Europe. *Proceedings of the National Academy of Sciences* 98:22–25.

Bar-El, T., and E. Tchernov
2001   Lagomorph remains at prehistoric sites in Israel and southern Sinai. *Paléorient* 26: 93–109.

Bar-Matthews, M., A. Ayalon, M. Gilmour, A. Mathews, and C. Hawkesworth
2003   Sea-land oxygen isotopic relationships from planktonic foraminifera and speleothems in the eastern Mediterranean region and their implications for paleorainfall during interglacial intervals. *Geochimica et Cosmochimica Acta* 66(17):3181–3200.

Bar-Matthews, M., A. Ayalon, A. Kaufman, and G. J. Wasserburg
1999   The eastern Mediterranean paleoclimate as a reflection of regional events: Soreq Cave, Israel. *Earth and Planetary Science Letters* 166:85–95.

Bar-Oz, G., and T. Dayan
2002　"After 20 years": A taphonomic reevaluation of Nahal Hadera V, an Epipalaeolithic site on the Israeli coastal plain. *Journal of Archaeological Science* 29:145–156.

Bar-Oz, G., T. Dayan, and D. Kaufman
1999　The Epipaleolithic faunal sequence of Israel: A view from Neve David. *Journal of Archaeological Science* 26:67–82.

Barton, R. N. E.
1999　Colonization and resettlement of Europe in the Late Glacial: A view from the western periphery. In *Post-pleniglacial Re-colonization of the Great European Lowland,* M. Kobusiewicz and J. K. Kozlowski, eds., pp. 71–86. Folia Quaternaria 70, Polska Akademia Umiejêtnoœci, Komisja Paleogeografii Czwartorzêdu, Kraków, Poland.

Barton, R. N. E., A. P. Currant, Y. Fernandez-Jalvo, J. C. Finlayson, P. Goldberg, R. Macphail, P. B. Pettitt, and C. B. Stringer
1999　Gilbraltar Neanderthals and results of recent excavations in Gorham's, Vanguard and Ibex Caves. *Antiquity* 73:13–23.

Bartram, L. E., and C. W. Marean
1999　Explaining the "Klasies pattern": Kua ethnoarchaeology, the Die Kelders Middle Stone Age archaeofauna, long bone fragmentation and carnivore ravaging. *Journal of Archaeological Science* 26:9–20.

Baruch, U., and S. Bottema
1991　Palynological evidence for climatic changes in the Levant circa 17,000–9,000 B.P. In *The Natufian Culture in the Levant,* O. Bar-Yosef and F. R. Valla, eds., pp. 11–20. International Monographs in Prehistory, Ann Arbor, Michigan.

Bar-Yosef, D. E.
1989　Late Paleolithic and Neolithic marine shells in the southern Levant as cultural markers. In *Proceedings of the 1986 Shell Bead Conference: Selected Papers,* O. Bar-Yosef and F. R. Valla, eds., pp. 169–174. Research Records no. 20, Rochester Museum and Science Center, Rochester, New York.

Bar-Yosef, O.
1970　The Epi-Palaeolithic cultures of Palestine. Ph.D. dissertation, Institute of Archaeology, Hebrew University of Jerusalem.
1981　The Epi-Palaeolithic complexes in the southern Levant. In *Préhistoire du Levant,* J. Cauvin and P. Sanlaville, eds., pp. 389–408. Éditions du CNRS, Paris.
1983　The Natufian of the southern Levant. In *The Hilly Flanks and Beyond,* C. T. Young, Jr., P. E. L. Smith, and P. Mortensen, eds., pp. 11–42. Studies in Ancient Oriental Civilizations 36, Oriental Institute, Chicago.
1989a　Geochronology of the Levantine Middle Palaeolithic. In *The Human Revolution: Behavioural and Biological Perspectives on the Origins of Modern Humans,* P. Mellars and C. Stringer, eds., pp. 589–610. Princeton University Press, Princeton, New Jersey.
1989b　Upper Pleistocene cultural stratigraphy in southwest Asia. In *Patterns and Processes in Later Pleistocene Human Emergence,* E. Trinkaus, ed., pp. 154–179. Cambridge University Press, Cambridge.
1991　The archaeology of the Natufian layer at Hayonim Cave. In *The Natufian Culture in the Levant,* O. Bar-Yosef and F. R. Valla, eds., pp. 81–93.
1992　Middle Palaeolithic chronology and the transition to the Upper Palaeolithic in Southwest Asia. In *Continuity or Replacement: Controversies in* Homo sapiens *Evolution,* G. Brauer and F. H. Smith, eds., pp. 261–272. A. A. Balkema, Rotterdam, Netherlands.
1994　The Lower Paleolithic of the Near East. *Journal of World Prehistory* 8:211–266.
1995　The role of climate in the interpretation of human movements and cultural transformations in western Asia. In *Paleoclimate and Evolution with Emphasis on Human Origins,* E. S. Vrba, G. H. Denton, T. C. Partridge, and L. H. Buckle, eds., pp. 507–523. Yale University Press, New Haven, Connecticut.
1996　The impact of late Pleistocene–early Holocene climatic changes on humans in southwest Asia. In *Humans at the End of the Ice Age: The Archaeology of the Pleistocene-Holocene Transition,* L. G. Straus, B. V. Eriksen, J. M. Erlandson, and D. R. Yesner, eds., pp. 61–76. Plenum Press, New York.

1998 The chronology of the Middle Paleolithic of the Levant. In *Neanderthals and Modern Humans in Western Asia,* T. Akazawa, K. Aoki, and O. Bar-Yosef, eds., pp. 39–56. Plenum Press, New York.

2000 The middle and early Upper Paleolithic in Southwest Asia and neighboring regions. In *The Geography of Neandertals and Modern Humans in Europe and the Greater Mediterranean,* O. Bar Yosef and D. Pilbeam, eds., pp. 107–156. Peabody Museum, Harvard University, Cambridge, Massachusetts.

2002 Natufian: A complex society of foragers. In *Beyond Foraging and Collecting: Evolutionary Change in Hunter-Gatherer Settlement Systems,* B. Fitzhugh and J. Habu, eds., pp. 91–149. Kluwer Academic, Plenum Press, New York.

Bar-Yosef, O., M. Arnold, A. Belfer-Cohen, P. Goldberg, R. Houseley, H. Laville, L. Meignen, N. Mercier, J. C. Vogel, and B. Vandermeersch
1996 The dating of the Upper Paleolithic layers in Kebara Cave, Mount Carmel. *Journal of Archaeological Science* 23:297–307.

Bar-Yosef, O., and A. Belfer-Cohen
1988 The early Upper Palaeolithic in Levantine caves. In *The Early Upper Paleolithic in Europe and the Near East,* J. Hoffecker and C. Wolf, eds., pp. 23–41. British Archaeological Reports, International Series 437, Oxford.

1989 The origins of sedentism and farming communities in the Levant. *Journal of World Prehistory* 3(4):447–498.

1996 Another look at the Levantine Aurignacian. In *Colloquium XI: The Late Aurignacian,* A. Montet-White and A. Palma di Cesnola, eds., pp. 139–150. Proceedings of the Thirteenth Conference of the International Congress of Prehistoric and Protohistoric Sciences, UISPP. ABACO Edizioni, Forlì, Italy.

1998 Natufian imagery in perspective. *Rivista di Scienze Prehistoriche* 49:247–263.

1999 Encoding information: Unique Natufian objects from Hayonim Cave, western Galilee, Israel. *Antiquity* 73:402–410.

2002 Facing environmental crisis: Societal and cultural changes at the transition from the Younger Dryas to the Holocene in the Levant. In *The Dawn of Farming in the Near East,* R. T. J. Cappers and S. Bottema, eds. *Ex oriente* (Berlin) 6:55–66.

Bar-Yosef, O., and N. Goren
1973 Natufian remains in Hayonim Cave. *Paléorient* 1(1):49–68.

Bar-Yosef, O., H. Laville, L. Meignen, A. M. Tillier, B. Vandermeersch, B. Arensburg, A. Belfer-Cohen, P. Goldberg, Y. Rak, and E. Tchernov
1988 La sepulture neanderthalienne de Kébara (unite XII). In *L'homme de Neanderthal,* vol. 5, M. Otte, ed., pp. 17–24. ERAUL, Liège, Belgium.

Bar-Yosef, O., and R. H. Meadow
1995 The origins of agriculture in the Near East. In *Last Hunters—First Farmers: New Perspectives on the Prehistoric Transition to Agriculture,* T. D. Price and A. G. Gebauer, eds., pp. 39–94. School of American Research Press, Santa Fe, New Mexico.

Bar-Yosef, O., and L. Meignen
1992 Insights into Levantine Middle Paleolithic cultural variability. In *The Middle Paleolithic: Adaptation, Behavior, and Variability,* H. L. Dibble and P. Mellars, eds., pp. 163–182. Monograph no. 78, University Museum, University of Pennsylvania, Philadelphia.

Bar-Yosef, O., and L. Meignen, eds.
n.d. *The Middle and Upper Paleolithic in Kebara Cave (Mt. Carmel), Part 1: The Site Stratigraphy, Hearths, and the Faunal Assemblages.* American School of Prehistoric Research, Peabody Museum Press, Harvard University, Cambridge, Mass. In press.

Bar-Yosef, O., and E. Tchernov
1967 Archaeological finds and fossil faunas of the Natufian and microlithic industries at Hayonim Cave (western Galilee, Israel). *Israel Journal of Zoology* 1(5):104–140.

Bar-Yosef, O., and B. Vandermeersch
1981 Notes concerning the possible age of the Mousterian layers at Qafzeh Cave. In *Préhistoire du Levant,* J. Cauvin and P. Sanlaville, eds., pp. 555–569. Éditions du CNRS, Paris.

Bar-Yosef, O., B. Vandermeersch, B. Arensburg, A. Belfer-Cohen, P. Goldberg, H. Laville, L. Meignen, Y. Rak, J. D. Speth, E. Tchernov, A.-M. Tillier, and S. Weiner
1992 The excavations in Kebara Cave, Mount Carmel. *Current Anthropology* 33:497–550.

Bar-Yosef, O., B. Vandermeersch, B. Arensburg, P. Goldberg, H. Laville, L. Meignen, Y. Rak, E. Tchernov, and A. M. Tillier
1986 New data concerning the origins of modern man in the Levant. *Current Anthropology* 27: 63–64.

Bar-Yosef Mayer, D. E.
2002 The shells of the Nawamis in southern Sinai. In *Archaeozoology of the Near East,* vol. 5, H. Buitenhuis, A. M. Choyke, M. Mashkour, and A. H. Al-Shiyab, eds., pp. 166–180. ARC Publications, Groningen, Netherlands.

Baryshnikov, G.
1996 The dhole, *Cuon alpinus* (Carnivora, Canidae), from the Upper Pleistocene of the Caucasus. *Acta Zoologica Cracova* 39(1):67–73.

Bate, D. M. A.
1937a New Pleistocene mammals from Palestine. *Annals and Magazine of Natural History* 10: 397–400.
1937b Palaeontology: The fossil fauna of the Wady el-Mughara caves. In *The Stone Age of Mount Carmel,* Part 2, D. E. E. Garrod and D. M. A. Bate, eds., pp. 137–240. Clarendon Press, Oxford.
1940 The fossil antelopes of Palestine in Natufian (Mesolithic) times, with descriptions of new species. *Geological Magazine* 77(6):418–443.
1942 Pleistocene Murinae from Palestine. *Annals and Magazine of Natural History* 9:465–486.
1943 Pleistocene Cricetinae from Palestine. *Annals and Magazine of Natural History* 11:813–836.

Bayham, F. E.
1979 Factors influencing the archaic pattern of animal utilization. *Kiva* 44:219–235.
1982 A diachronic analysis of prehistoric animal exploitation at Ventana Cave. Ph.D. dissertation, Department of Anthropology, Arizona State University, Tempe, Arizona.

Beaver, J. E.
2004 Identifying necessity and sufficiency relationships in skeletal-part representation using fuzzy-set theory. *American Antiquity* 69(1): 131–140.

Behrensmeyer, A. K.
1975 Taphonomy and paleoecology in the hominid fossil record. *Yearbook of Physical Anthropology* 19:36–50.
1978 Taphonomic and ecologic information from bone weathering. *Paleobiology* 4:150–162.
1991 Terrestrial vertebrate accumulations. In *Taphonomy: Releasing the Data Locked in the Fossil Record,* P. A. Allison and D. E. G. Briggs, eds., pp. 291–335. Plenum Press, New York.

Belfer-Cohen, A.
1988a The Natufian graveyard in Hayonim Cave. *Paléorient* 14:297–308.
1988b The Natufian settlement at Hayonim Cave: A hunter-gatherer band on the threshold of agriculture. Ph.D. dissertation, Institute of Archaeology, Hebrew University of Jerusalem.
1991 The Natufian of the Levant. *Annual Review of Anthropology* 20:167–186.
1995 Rethinking social stratification in the Natufian culture: The evidence from the burials. In *The Archaeology of Death in the Near East,* S. Campbell and A. Green, eds., pp. 9–16. Oxbow Monograph 51, Oxford.

Belfer-Cohen, A., and O. Bar-Yosef
1981 The Aurignacian at Hayonim Cave. *Paléorient* 7:19–42.
2000 Early sedentism in the Near East: A bumpy road to village life. In *Life in Neolithic Farming Communities: Social Organization, Identity, and Differentiation,* I. Kuijt, ed., pp. 19–37. Plenum Press, New York.

Belfer-Cohen, A., and N. Goring-Morris
1986 Har Horesha I: An Upper Paleolithic occupation in the western Negev. *Mitekufat Ha'even, Journal of the Israel Prehistory Society* 19:43–57.

Bellomo, R. V., and J. W. K. Harris
1990 Preliminary report of actualistic studies of fire within Virunga National Park, Zaire: Towards an understanding of archaeological occurrences. In *Evolution of Environments and Hominidae in the African Western Rift Valley,* N. T. Boaz, ed., pp. 317–338. Memoir no. 1, Virginia Museum of Natural History, Martinsville, Virginia.

Belovsky, G. E.
1988 An optimal-foraging-based model of hunter-gatherer population dynamics. *Journal of Anthropological Archaeology* 7(4):329–372.

Bennett, J. L.
1999 Thermal alteration of buried bone. *Journal of Archaeological Science* 26(1):1–8.

Berke, H.
1984 The distributions of bones from large mammals at Petersfels. In *Upper Palaeolithic Settlement Patterns in Europe,* H. Berke, J. Hahn, and C.-J. Kind, eds., pp. 103–108. Verlag Archaeologica Venatoria, Institut fur Urgeschichte der Universitat Tubingen, Tubingen, Germany.

Berna, F., A. Matthews, and S. Weiner
2004 Solubilities of bone mineral from archaeological sites: The recrystallization window. *Journal of Archaeological Science* 31(7):867–882.

Berner, R. A.
1971 *Principles of Chemical Sedimentology.* McGraw Hill, New York.

Bietti, A.
1976– The excavation 1955–1959 in the Upper Pale-
1977 olithic deposit of the rockshelter at Palidoro (Rome, Italy). *Quaternaria* 19:149–155.

Bietti, A., S. L. Kuhn, A. G. Segre, and M. C. Stiner
1990– Grotta Breuil: A general introduction and
1991 stratigraphy. *Quaternaria Nova* (new series) 1:305–323.

Bietti, A., G. Manzi, P. Passarello, A. G. Segre, and M. C. Stiner
1988 The 1986 excavation campaign at Grotta Breuil (Monte Circeo LT). *Quaderno Archeologico Laziale* 16:372–388.

Bietti, A., and Stiner, M. C.
1992 Les modèles de la subsistance et de l'habitat de l'Epigravettien Italien: L'exemple de Riparo Salvini (Terracina, Latium). In *Colloque International: Le Peuplement Magdalénien,* H. Laville, J.-F. Rigaud, and B. Vandermeersch, eds., pp. 137–152. Édition CTHS, Paris.

Binford, L. R.
1968 Post-Pleistocene adaptations. In *New Perspectives in Archaeology,* S. R. Binford and L. R. Binford, eds., pp. 313–341. Aldine, Chicago.
1978 *Nunamiut Ethnoarchaeology.* Academic Press, New York.
1981 *Bones: Ancient Men and Modern Myths.* Academic Press, New York.

1991 When the going gets tough, the tough get going: Nunamiut local groups, camping patterns, and economic organization. In *Ethnoarchaeological Approaches to Mobile Campsites: Hunter-Gatherer and Pastoralist Case Studies,* C. S. Gamble and W. A. Boismer, eds., pp. 25–137. International Monographs in Prehistory, Ann Arbor, Michigan.
1998 Hearth and home: The spatial analysis of ethnographically documented rock shelter occupations as a template for distinguishing between human and hominid use of sheltered space. In *Thirteenth UISPP Congress Proceedings: Forli,* 8–14 September 1996, N. J. Conard and F. Wendorf, eds., pp. 229–239. ABACO Edizioni, Forlì, Italy.
1999 Time as a clue to cause? *Proceedings of the British Academy* 101:1–35.
2001 *Constructing Frames of Reference: An Analytical Method for Archaeological Theory Building Using Hunter-Gatherer and Environmental Data Sets.* University of California Press, Berkeley.

Binford, L. R., and J. Bertram
1977 Bone frequencies and attritional processes. In *For Theory Building in Archaeology,* L. R. Binford, ed., pp. 77–156. Academic Press, New York.

Binford, S. R.
1968 Early Upper Pleistocene adaptations in the Levant. *American Anthropologist* 70(4):707–717.

Blanc, A. C.
1953 *Il Riparo Mochi ai Balzi Rossi di Grimaldi: Le industrie.* Paleontographia Italica, Paleontologia et Ecologia del Quaternario, vol. 3. Pisa, Italy.
1958– Industria musteriana su calcare e su valve di
1961 Meretrix chione associata con fossili di elefante e rinocerante. In *Nuovi giacimenti costieri del Capo de Leuca. Quaternaria* 5:308–313.

Blank, T. H., and J. S. Ash
1962 Fluctuations in a partridge population. In *The Exploitation of Natural Animal Populations,* E. D. Le Cren and M. W. Holdgate, eds., pp. 118–133. Blackwell Scientific, Oxford.

Blasco, M., E. Crespillo, and J. M. Sanchez
1986– The growth dynamics of *Testudo graeca* L. (Rep-
1987 tilia: Testudinidae) and other data on its pop-
      ulations in the Iberian Peninsula. *Israel Jour-
      nal of Zoology* 34:139–147.

Blondel, J.
1991 Invasions and range modification of birds in
      the Mediterranean Basin. In *Biogeography of
      Mediterranean Invasions,* R. H. Groves and F. di
      Castri, eds., pp. 311–326. Cambridge Univer-
      sity Press, Cambridge.

Blondel, J., and J. Aronson
1999 *Biology and Wildlife of the Mediterranean Region.*
      Oxford University Press, Oxford.

Blumenschine, R. J., and M. M. Selvaggio
1991 On the marks of marrow processing by ham-
      merstones and hyenas: Their anatomical pat-
      terning and archaeological implications. In
      *Cultural Beginnings: Approaches to Understanding
      Early Hominid Lifeways in the African Savanna,*
      L. R. Binford, ed., pp. 17–32. Rudolf Habelt,
      Bonn, Germany.

Boëda, E.
1995 Levallois: A volumetric construction, meth-
      ods, a technique. In *The Definition and Inter-
      pretation of Levallois Technology,* H. L. Dibble and
      O. Bar-Yosef, eds., pp. 41–68. Prehistory Press,
      Madison, Wisconsin.

Boëda, E., L. Bourguignon, and C. Griggo
1998 Activités de subsistence au Paléolithic moyen:
      Couche V13 b' du gisement d'Umm el Tlel
      (Syrie). In *Économie préhistorique: Les comporte-
      ments de subsistence au Paléolithique,* J.-P. Brugal,
      L. Meignen, and M. Patou-Mathis, eds., pp.
      243–258. XVIIIᵉ Rencontres Internationales
      d'Archéologie et d'Histoire d'Antibes. Édi-
      tions du APDCA, Sophia Antipolis, France.

Boëda, E., J.-M. Geneste, and L. Meignen
1990 Identification de chaînes opératoires lithiques
      du Paléolithique ancien et moyen. *Paleo* 2:
      43–80.

Boëda, E., and S. Muhesen
1993 Umm El Tlel (El Kown, Syrie): Étude prélim-
      inaire des industries lithiques du Paléo-
      lithique moyen et supérieur 1991–1992. In
      Cahiers de l'Euphrate, J. Cauvin, ed., pp. 47–92.
      Éditions Recherche sur les Civilisations, Paris.

Bohl, W.
1957 *Chukars in New Mexico, 1931–1957.* Bulletin no.
      6, New Mexico Department of Game and
      Fish, Santa Fe, New Mexico.

Bordes, F.
1961 *Typologie du paléolithique ancien et moyen.*
      Cahiers du Quaternarie no. 1, Institut du
      Quaternarie, Université de Bordeaux, Bor-
      deaux, France.

Borziyak, I. A.
1993 Subsistence practices of late Paleolithic
      groups along the Dneister River and its trib-
      utaries. In *From Kostenki to Clovis: Paleoindian
      and Upper Paleolithic Adaptations,* O. Soffer and
      N. Praslov, eds., pp. 67–84. Plenum Press, New
      York.

Botkin, S.
1980 Effects of human exploitation on shellfish
      populations at Malibu Creek, California. In
      *Modeling Change in Prehistoric Subsistence
      Economies,* T. K. Earle and A. L. Christenson,
      eds., pp. 121–139. Academic Press, New York.

Boutin, S.
1992 Predation and moose population dynamics: A
      critique. *Journal of Wildlife Management* 56:
      116–127.

Brain, C. K.
1967 Hottentot food remains and their bearing on
      the interpretation of fossil bone assemblages.
      *Scientific Papers of the Namib Desert Research Sta-
      tion* 32:1–7.
1969 The contribution of Namib Desert Hottentots
      to an understanding of Australopithecine
      bone accumulations. *Scientific Papers of the
      Namib Desert Research Station* 39:13–22.
1981 *The Hunters or the Hunted?* University of
      Chicago Press, Chicago.
1993 The occurrence of burnt bones at Swartkrans
      and their implications for the control of fire
      by early hominids. In *Swartkrans: A Cave's
      Chronicle of Early Man,* C. K. Brain, ed., pp.
      229–242. Transvaal Museum Monograph no.
      8, Pretoria, South Africa.

Brain, C. K., and A. Sillen
1988 Evidence from the Swartkrans cave for the
      earliest use of fire. *Nature* 336:464–466.

Bratlund, B.
1996 Hunting strategies in the Late Glacial of Northern Europe: A survey of the faunal evidence. *Journal of World Prehistory* 10(1):1–48.
1999 Review of the faunal evidence from the Late Glacial of northern Europe. In *Post-pleniglacial Re-colonization of the Great European Lowland*, M. Kobusiewicz and J. K. Kozlowski, eds., pp. 31–37. Folia Quaternaria 70, Polska Akademia Umiejêtnoœci, Komisja Paleogeografii Czwartorzêdu, Kraków.

Brink, J. W.
1997 Fat content in leg bones of *Bison bison* and applications to archaeology. *Journal of Archaeological Science* 24:259–274.

Broekhuizen, S.
1979 Survival in adult European hares. *Acta Theriologica* 34:465–473.

Bronson, F. H., and O. W. Tiemeier
1958 Reproduction and age distribution of black-tailed jackrabbits in Kansas. *Journal of Wildlife Management* 22(4):409–414.

Brooks, A.
1984 San land-use patterns, past and present: Implications for southern African prehistory. In *Frontiers: Southern African Archaeology Today*, M. Hall, G. Avery, D. M. Avery, M. L. Wilson, and A. J. B. Humphreys, eds., pp. 40–52. British Archaeological Reports, International Series 207, Oxford.

Brooks, A., C. M. Helgren, J. S. Cramer, A. Franklin, W. Hornyak, J. M. Keating, R. G. Klein, W. J. Rink, H. P. Schwarcz, J. N. L. Smith, K. Stewart, N. Todd, and J. Verniers
1995 Dating and context of three Middle Stone Age sites with bone points in the upper Semliki Valley, Zaire. *Science* 268:548–553.

Broughton, J. M.
1994 Declines in mammalian foraging efficiency during the Late Holocene, San Francisco Bay, California. *Journal of Anthropological Archaeology* 13:371–401.
1997 Widening diet breadth, declining foraging efficiency, and prehistoric harvest pressure: Ichthyofaunal evidence from the Emeryville Shellmound, California. *Antiquity* 71:845–862.

Brugal, J.-P., and J. Jaubert
1991 Les gisements paléontologiques Pléistocènes à indices de fréquentation humaine: Un nouveau type de comportement de prédation? *Paléo* 3:15–41.

Bunn, H. T., L. E. Bartram, and E. M. Kroll
1988 Variability in bone assemblage formation from Hadza hunting, scavenging, and carcass processing. *Journal of Anthropological Archaeology* 7:412–457.

Bunn, H. T., and E. M. Kroll
1986 Systematic butchery by Plio/Pleistocene hominids at Olduvai Gorge, Tanzania. *Current Anthropology* 27:431–452.
1988. Reply to Binford. *Current Anthropology* 29:135–149.

Burke, A.
2000 The view from Starosele: Faunal exploitation at the Middle Palaeolithic site in western Crimea. *International Journal of Osteoarchaeology* 10:325–335.

Byrd, B. F.
1994 Later Quaternary hunter-gatherer complexes in the Levant between 20,000 and 10,000 B.P. In *Late Quaternary Chronology and Paleoclimates of the Eastern Mediterranean*, O. Bar-Yosef and R. S. Kra, eds., pp. 205–226. RADIOCARBON, University of Arizona, Tucson.

Byrd, B., and A. Garrard
1990 The Last Glacial Maximum in the Jordanian desert. In *The World at 18,000 B.P.*, vol. 2, *Low Latitudes*, C. Gamble and O. Soffer, eds., pp. 78–96. Plenum Press, New York.

Campana, D. V.
1989 *Natufian and Protoneolithic Bone Tools: The Manufacture and Use of Bone Implements in the Zagros and the Levant*. British Archaeological Reports, International Series 494, Oxford.

Capaldo, S. D.
1997 Experimental determinations of carcass processing by Plio-Pleistocene hominids and carnivores at FLK 22 *(Zinjanthropus)*, Olduvai Gorge, Tanzania. *Journal of Human Evolution* 33:555–597.

Cardini, L., and I. Biddittu
1967 L'attivita scientifica dell'Istituto Italiano di Paleontologia Umana dalla sua fondazione. *Quaternaria* 9:353–383.

Casevitz-Weulersse, J.
1992 Analyse biogéographique de le myrméco-fauna Corse et comparaison avec celle des régions voisines. *Compte Rendus des Séances del la Société de Biogéographie* 68:105–129.

Cashden, E. A.
1985 Coping with risk: Reciprocity among the Basarwa of northern Botswana. *Man* 20(3): 454–474.

Cassoli, P.
1976– Upper Paleolithic fauna at Palidoro (Rome):
1977 1955 excavations. *Quaternaria* 19:187–195.

Castanet, J., and M. Cheylan
1979 Les marques de croissance des os et des ecailles comme indicateur de l'age chez *Testudo hermanni* et *Testudo graeca* (Reptilia, Chelonia, Testudinidae). *Canadian Journal of Zoology* 57:1649–1665.

Caughley, G.
1977 *Analysis of Vertebrate Populations*. John Wiley and Sons, London.

Chapman, D., and N. Chapman
1975 *Fallow Deer: Their History, Distribution and Biology*. Terrence Dalton, Levanham, Suffolk.

Chase, P. G.
1986 *The Hunters of Combe Grenal: Approaches to Middle Paleolithic Subsistence in Europe*. British Archaeological Reports, International Series 286, Oxford.
1989 How different was Middle Palaeolithic subsistence? A zooarchaeological perspective on the Middle to Upper Palaeolithic transition. In *The Human Revolution: Behavioural and Biological Perspectives on the Origins of Modern Humans*, P. Mellars and C. Stringer, eds., pp. 321–337. Princeton University Press, Princeton, New Jersey.

Chase, P. G., and H. Dibble
1987 Middle Paleolithic symbolism: A review of current evidence and interpretations. *Journal of Anthropological Archaeology* 6:263–293.

Cheylan, G.
1991 Patterns of Pleistocene turnover, current distribution and speciation among Mediterranean mammals. In *Biogeography of Mediterranean Invasions*, R. H. Groves and F. di Castri, eds., pp. 227–262. Cambridge University Press, Cambridge.

Cheylan, G., and F. Poitevin
1998 Conservazione di retilli e anfibi. In *La gestione degli ambienti costieri e insulari del Mediterraneo*, X. Monbailliu and A. Torre, eds., pp. 275–336. Edizione del Sole, Alghero.

Christensen, G. C.
1970 *The Chukar Partridge*. Biological Bulletin no. 4, Nevada Department of Fish and Game, Carson City, Nevada.

Christenson, A. L.
1980 Change in the human niche in response to population growth. In *Modeling Change in Prehistoric Subsistence Economies*, T. K. Earle and A. L. Christenson, eds., pp. 31–72. Academic Press, New York.

Clark, G. A.
1987 From the Mousterian to the Metal Ages: Long-term change in the human diet of northern Spain. In *The Pleistocene Old World: Regional Perspectives*, O. Soffer, ed., pp. 293–316. Plenum Press, New York.

Clark, G. A., and J. M. Lindly
1989a Modern human origins in the Levant and Western Asia. *American Anthropologist* 91:962–985.
1989b The case for continuity: Observations on the biocultural transition in Europe and Western Asia. In *The Human Revolution: Behavioural and Biological Perspectives on the Origins of Modern Humans*, P. Mellars and C. Stringer, eds., pp. 626–676. Princeton University Press, Princeton, New Jersey.

Clark, G., J. Lindly, M. Donaldson, A. Garrard, N. Coinman, J. Schuldenrein, S. Fish, and D. Olszewski
1988 Excavations at Middle, Upper and Epipalaeolithic Sites in the Wadi Hasa, west-central Jordan. In *The Prehistory of Jordan*, A. N. Garrard and H. G. Gebel, eds., pp. 209–285. British Archaeological Reports, International Series 396, Oxford, England.

Clark, G. A., and L. G. Straus
1983 Late Pleistocene hunter-gatherer adaptations in Cantabrian Spain. In *Hunter-Gatherer Economy in Prehistory,* G. Bailey, ed., pp. 131–148. Cambridge University Press, Cambridge.

Cogger, H. G., and R. G. Zweifel
1998 *Encyclopedia of Reptiles and Amphibians.* 2nd ed. Academic Press, San Diego, California.

Cohen, M. N.
1977 *The Food Crisis in Prehistory: Overpopulation and the Origins of Agriculture.* Yale University Press, New Haven, Connecticut.
1985 Prehistoric hunter-gatherers: The meaning of social complexity. In *Prehistoric Hunter-Gatherers: The Emergence of Cultural Complexity,* T. D. Price and J. A. Brown, eds., pp. 99–119. Academic Press, San Diego, California.

Coles, B., ed.
1992 *The Wetland Revolution in Prehistory.* The Prehistoric Society, Exeter, UK.

Conard, N. J., ed.
2001 *Settlement Dynamics of the Middle Paleolithic and Middle Stone Age.* Kerns Verlage, Tübingen, Germany.

Conkey, M.
1985 Ritual communication, social elaboration, and the variable trajectories of Paleolithic material culture. In *Prehistoric Hunter-Gatherers: The Emergence of Cultural Complexity,* T. D. Price and J. A. Brown, eds., pp. 299–323. Academic Press, San Diego, California.

Cope, C.
1991 The evolution of Natufian megafaunal economics. Ph.D. dissertation, Department of Evolution, Systematics, and Ecology, Hebrew University of Jerusalem.
1992 Gazelle hunting strategies in the southern Levant. In *The Natufian Culture in the Levant,* O. Bar-Yosef and F. Valla, eds., pp. 341–358. International Monographs in Prehistory, Ann Arbor, Michigan.

Copeland, L.
1975 The Middle and Upper Palaeolithic of Lebanon and Syria in the light of recent research. In *Problems in Prehistory: North Africa and the Levant,* F. Wendorf and A. E. Marks, eds., pp. 317–350. Southern Methodist University Press, Dallas, Texas.

Copeland, L., and F. Hours
1983 Le Yabroudien d'El Kowm (Syrie) et sa place dans le Paléolithique du Levant. *Paléorient* 9(1):21–37.

Covas, R., and J. Blondel
1998 Biogeography and history of the Mediterranean bird fauna. *Ibis* 140:395–407.

Crégut-Bonnoure, E.
1996 Ordre des Carnivores. In *Les Grands Mammifères Plio-Pléistocènes d'Europe,* C. Guérin and M. Patou-Mathis, eds., pp. 156–167. Collection Préhistoire, Masson, Paris.

Crew, H. L.
1976 The Mousterian site of Rosh Ein Mor. In *Prehistory and Paleoenvironments in the Central Negev,* A. E. Marks, ed., pp. 75–112. Southern Methodist University Press, Dallas, Texas.

Currey, J.
1984 *The Mechanical Adaptations of Bones.* Princeton University Press, Princeton, New Jersey.

Dan, J., and D. H. Yaalon.
1971 On the origin and nature of the paleopedological formations in the coastal desert fringe areas of Israel. In *Paleopedology: Origin, Nature and Dating of Paleosols,* D. H. Yaalon, ed., pp. 245–260. International Society of Soil Science, Jerusalem.

David, F., and T. Poulain
1990 La faune de grands mammifères des niveaux XI et XC de la Grotte du Renne a Arcy-sur-Cure (Yonne): Etude preliminaire. *In Paleolithique Moyen Recent et Paleolithique Superieur Ancien en Europe,* C. Farizy, ed., pp. 319–323. Mémoires du Musée de Préhistoire d'Ile de France 3. Éditions de l'Association pour la Promotion de la Recherche Archéologique, Ile de France.

Davidson, I.
1983 Site variability and prehistoric economy in Levant. In *Hunter-Gatherer Economy in Prehistory: A European Perspective,* G. Bailey, ed., pp. 79–95. Cambridge University Press, Cambridge.

Davidson, I., and W. Noble
1989 The archaeology of perception: Traces of depiction and language. *Current Anthropology* 30:125–155.

Davis, S. J. M.
1978 The large mammals of the Upper Pleistocene–Holocene in Israel. Ph.D. dissertation, Hebrew University of Jerusalem.
1980a Late Pleistocene–Holocene gazelles of northern Israel. *Israel Journal of Zoology* 29:135–140.
1980b A note on the dental and skeletal ontogeny of *Gazella. Israel Journal of Zoology* 29:129–134.
1981 The effects of temperature change and domestication on the body size of Late Pleistocene to Holocene mammals. *Paleobiology* 7:101–114.
1982 Climatic change and the advent of domestication: The succession of ruminant artiodactyls in the late Pleistocene–Holocene in the Israel region. *Paléorient* 8(2):5–14.
1983 The age profiles of gazelles predated by ancient man in Israel: Possible evidence for a shift from seasonality to sedentism in the Natufian. *Paléorient* 9:55–62.

Davis, S. J. M., O. Lernau, and J. Pichon
1994 The animal remains: New light on the origin of animal husbandry. In *Le site de Hatoula en Judée occidentale, Israel,* M. Lechevallier and A. Ronen, eds., pp. 83–100. Mémoires et Travaux du Centre de Recherche Francais de Jérusalem 8. Association Paléorient, Paris.

Davis, S. J. M., R. Rabinovich, and N. Goren-Inbar
1988 Quaternary extinctions and population increase in western Asia: The animal remains from Biq'at Quneitra. *Paléorient* 14(1):95–105.

Dayan, T., and D. Simberloff
1995 Natufian gazelles: Proto-domestication reconsidered. *Journal of Archaeological Science* 22:671–675.

Dean, R. M.
1997 *Ungulate Ethnoarchaeology: Interpreting Late Pleistocene and Early Holocene Archaeological Ungulate Assemblages from Southwest Asia.* Master's thesis, Department of Anthropology, University of Arizona, Tucson.

Delage, C.
2001 Les resources lithiques dans le nord d'Israël: La question des territories d'approvisionnement natoufiens confrontée à l'hypothèse de leur sédentarité. 3 vols. Ph.D. dissertation, University of Paris I–Sorbonne.

Delage, C., L. Meignen, and O. Bar-Yosef
2000 Chert procurement and the organization of lithic production in the Mousterian of Hayonim Cave (Israel). (Abstract). *Journal of Human Evolution* 38:A10–11.

Delaugerre, M., and M. Cheylan
1992 *Atlas de répartition des Batraciens et reptiles de Corse.* Parc Naturel Regional de la Corese. E. Ph. E., Ajaccio.

Delpech, F.
1998 Comment on Marean and Kim, Mousterian large-mammal remains from Kobeh Cave: Behavioral implications. *Current Anthropology* 39:594–595.

de Lumley-Woodyear, H.
1969 *Le Paléolithique Inférieur et Moyen di midi Mediterranéen dans son cadre géologique,* vol. 1, Ligurie-Provence. Éditions du CNRS, Paris.

Dement'ev, G. P., and N. A. Gladkov
1967 *The Birds of the Soviet Union,* vol. 4. Israel Program of Scientific Translation, Jerusalem.

DeNiro, M. J.
1985 Postmortem preservation and alteration of in vivo bone collagen isotope ratios in relation to palaeodietary reconstruction. *Nature* 317:806–809.

DeNiro, M. J., and S. Weiner
1988a Chemical, enzymatic and spectroscopic characterization of "collagen" and other organic fractions from prehistoric bones. *Geochimica et Cosmochimica Acta* 52:2197–2206.

1988b Use of collagenase to purify collagen from prehistoric bones for stable isotopic analysis. *Geochimica et Cosmochimica Acta* 52:2425–2431.

1988c Organic matter within crystalline aggregates of hydroxyapatite: A new substrate for stable isotopic and possibly other biogeochemical analyses of bone. *Geochimica et Cosmochimica Acta* 52:2415–2423.

Deniz, E., and S. Payne
1982 Eruption and wear in the mandibular dentition as a guide to ageing Turkish angora goats. In *Aging and Sexing Animal Bones from Archaeological Sites,* B. Wilson, C. Grigson, and S. Payne, eds., pp. 155–205. British Archaeological Reports, British Series 109, Oxford.

d'Errico, F., C. Henshilwood, and P. Nilssen
2001 Engraved bone fragment from c. 70,000-year-old Middle Stone Age levels at Blombos Cave, South Africa: Implications for the origin of symbolism and language. *Antiquity* 75:309–318.

d'Errico, F., J. Zilháo, M. Julien, D. Baffier, and J. Pelegrin
1998 Neanderthal acculturation in western Europe? A critical review of the evidence and its interpretation. *Current Anthropology* 39 (supplement):S1–44.

Diamond, J.
1994 Zebras and the Anna Karenina principle. *Natural History* 103(9):4–10.

Dibble, H.
1987 Reduction sequences in the manufacture of Mousterian implements in France. In *The Pleistocene Old World: Regional Perspectives,* O. Soffer, ed., pp. 33–46. Plenum Press, New York.

di Castri, F.
1981 Mediterranean-type shrublands of the world. In *Mediterranean-Type Shrublands,* F. di Castri, D. W. Goodall, and R. L. Specht, eds., pp. 1–52. Elsevier, Amsterdam.

1991 The biogeography of Mediterranean animal invasions. In *Biogeography of Mediterranean Invasions,* R. H. Groves and F. di Castri, eds., pp. 439–452. Cambridge University Press, Cambridge.

Doak, D., P. Kareiva, and B. Klepetka
1994 Modeling population viability for the desert tortoise in the western Mojave desert. *Ecological Applications* 4:446–460.

Ducos, P.
1968 *L'origine des animaux domestiques en Palestine.* Publication de l'Institut de Prehistoire de l'Universite de Bordeaux 6, Delmas, Bordeaux, France.

Dunning, J. B., Jr.
1993 *CRC Handbook of Avian Body Masses.* CRC Press, Boca Raton, Florida.

Dye, A. H., G. M. Branch, J. C. Castilla, and B. A. Bennett
1994 Biological options for the management of the exploitation of intertidal and subtidal resources. In *Rocky Shores: Exploitation in Chile and South Africa,* W. R. Siegfried, ed., pp. 131–154. Springer-Verlag, Berlin.

Earle, T. K.
1980 A model of subsistence change. In *Modeling Change in Prehistoric Subsistence Economies,* T. K. Earle and A. L. Christenson, eds., pp. 1–29. Academic Press, New York.

Edwards, P. C.
1989 Revising the Broad Spectrum Revolution: Its role in the origins of Southwest Asian food production. *Antiquity* 63:225–246.

1991 Wadi Hammeh 27: An Early Natufian site at Pella, Jordan. In *The Natufian Culture in the Levant,* O. Bar-Yosef and F. R. Valla, eds., pp. 123–148. International Monographs in Prehistory, Ann Arbor, Michigan.

Elkin, D. C., and J. R. Zanchetta
1991 Densitometria osea de camélidos: Aplicaciones arqueológicas. *Actas del X Congreso Nacional de Arqueológia Argentina (Catamarca)* 3:195–204.

Emerson, A. M.
1990 Archaeological implications of variability in the economic anatomy of *Bison bison.* Ph.D. dissertation, Washington State University. University Microfilms, Ann Arbor, Michigan.

Emlen, J.
1966   The role of time and energy in food preference. *American Naturalist* 100:611–617.

Enghoff, I. B.
1991   Mesolithic eel-fishing at Bjørnsholm, Denmark, spiced with exotic species. *Journal of
Danish Archaeology* 10:105–118.

Enloe, J. G.
1997   Seasonality and age structure in remains of
Rangifer tarandus: Magdalenian hunting strategy at Verberie. *Anthropozoologica* 25–26:95–
102.

Epifanio, C. E., and C. A. Mootz
1976   Growth of oysters in recirculating maricul-
tural system. *Proceedings of the National Shell-
fisheries Association* 65:32–37.

Ernst, C. H., and R. W. Barbour
1989   *Turtles of the World.* Smithsonian Institution
Press, Washington, DC.

Eswaran, V.
2002   A diffusion wave out of Africa: The mechanisms of the modern human revolution? *Current Anthropology* 43(5):749–774.

Ewer, R. F.
1973   *The Carnivores.* Cornell University Press,
Ithaca, New York.

Fanshawe, J. H.
1989   Serengeti's painted wolves. *Natural History*
89:56–67.

Farizy, C.
1990   The transition from Middle to Upper Palaeolithic at Arcy-cur-Cure (Yonne, France):
Technological, economic, and social aspects.
In *The Emergence of Modern Humans,* P. Mellars,
ed., pp. 303–326. Cornell University Press,
Ithaca, New York.

Featherstone, J. D. B., S. Pearson, and R. Z. LeGeros
1984   An infrared method for quantification of carbonate in carbonated apatites. *Caries Research*
18:63–66.

Fisher, R. A., A. S. Corbet, and C. B. Williams
1943   The relation between the number of species
and the number of individuals in a random
sample of an animal population. *Journal of
Animal Ecology* 12:42–58.

Flannery, K. V.
1969   Origins and ecological effects of early domestication in Iran and the Near East. In *The
Domestication and Exploitation of Plants and Animals,* P. J. Ucko and G. W. Dimbleby, eds., pp.
73–100. Aldine, Chicago.

Fleisch, S. J.
1970   Les habitats du Paléolithique Moyen à Naamé
(Liban). *Bulletin du Museé Beyrouth* 23:25–98.

Flessa, K. W., A. H. Cutler, and K. H. Meldahl
1993   Time and taphonomy: Quantitative estimates
of time-averaging and stratigraphic disorder
in a shallow marine habitat. *Paleobiology*
19:266–286.

Flux, J. E.
1981   Reproductive strategies in the genus Lepus.
In *Proceedings of the World Lagomorph Conference,*
K. Myers and C. D. MacInnes, eds., pp. 155–
174. University of Guelph, Guelph, Ontario,
Canada.

Frankenburg, E.
1992   Management of mountain gazelles in Israel.
Ungulates 91:353–355.

Freund, R.
1978   Geology. In *The Lower Galilee and the Kinneret,*
A. Itzhaki, ed., pp. 9–14. Keter/Ministry of
Defense, Jerusalem.

Frylestam, B.
1979   Structure, size, and dynamics of three European hare populations in southern Sweden.
Acta Theriologica 33:449–464.

Fuller, T. K., and P. W. Kat
1990   Movements, activity, and prey relationships
of African wild dogs *(Lycaon pictus)* near
Aitong, southwestern Kenya. *African Journal
of Ecology* 28:330–350.

Galanidou, N.
2000   Patterns in caves: Foragers, horticulturalists, and the use of space. *Journal of Anthropological Archaeology* 19:243–275.

Galbreath, D. S., and R. Moreland
1953   The chukar partridge in Washington. *Washington State Game Department Biological Bulletin* 11:1–54.

Gamble, C.
1983   Caves and faunas from Last Glacial Europe. In *Animals and Archaeology,* vol. 1, *Hunters and Their Prey,* J. Clutton-Brock and C. Grigson, eds., pp. 163–172. British Archaeological Reports, International Series 163, Oxford.
1986   *The Palaeolithic Settlement of Europe.* Cambridge University Press, Cambridge.
1999   The Hohlenstein-Stadel revisited. In *The role of early humans in the accumulation of European Lower and Middle Palaeolithic bone assemblages.* E. Turner and S. Gaudzinski, eds., pp. 305–324. Monographien des Römisch-Germanischen Zentralmuseums (Mainz) 42.

Garrard, A. N.
1980   Man-animal-plant relationships during the Upper Pleistocene and Early Holocene of the Levant. Ph.D. dissertation, Darwin College, Cambridge University, Cambridge.

Garrod, D. A. E.
1954   Excavations at the Mugharet Kebara, Mount Carmel, 1931: The Aurignacian industries. *Proceedings of the Prehistoric Society* 20:155–192.
1957   Notes sur le Paléolithique supérior du Moyen-Orient. *Bulletin de la Société Préhistorique Francaise* 54:439–446.

Garrod, D. A. E., and D. M. A. Bate
1937   *The Stone Age of Mount Carmel,* vol. 1. Clarendon Press, Oxford.

Gaudzinski, S.
1995   Wallertheim revisited: A reanalysis of the fauna from the Middle Palaeolithic site of Wallertheim (Rheinhessen/Germany). *Journal of Archaeological Science* 22:51–66.

Gavin, A.
1991   Why ask "why": The importance of evolutionary biology in wildlife science. *Journal of Wildlife Management* 55:760–766.

Geneste, J.-M.
1985   Analyse lithique d'industries mousteriennes du Perigord: Une approche technologique du comportement des groupes humains au Paleolithique Moyen. Ph.D. dissertation, University of Bordeaux, Bordeaux, France.

Gifford, D. P.
1977   Observations of modern human settlements as an aid to archaeological interpretation. Ph.D. dissertation, Department of Anthropology, University of California, Berkeley.

Gifford, D. P., G. L. Isaac, and C. M. Nelson
1980   Evidence for predation and pastoralism at Prolonged Drift, a pastoral Neolithic site in Kenya. *Azania* 15:57–108.

Gifford-González, D. P.
1991   Examining and refining the quadratic crown height method of age estimation. In *Human Predators and Prey Mortality,* M. C. Stiner, ed., pp. 41–78. Westview Press, Boulder, Colorado.

Girman, D. J., P. W. Kat, M. G. L. Mills, J. R. Ginsberg, M. Borner, V. Wilson, J. H. Fanshawe, C. Fitzgibbon, L. M. Lau, and R. K. Wayne
1993   Molecular genetic and morphological analyses of the African wild dog *(Lycaon pictus).* *Journal of Heredity* 84:450–459.

Goldberg, P.
1979   Micromorphology of sediments from Hayonim Cave, Israel. *Catena* 6:167–181.

Goldberg, P., and O. Bar-Yosef
1998   Site formation processes in Kebara and Hayonim Caves and their significance in Levantine prehistoric caves. In *Neanderthals and Modern Humans in Western Asia,* T. Akazawa, K. Aoki, and O. Bar-Yosef, eds., pp. 107–125. Plenum Press, New York.

Goldberg, P., and H. Laville
1988   Le contexte stratigraphique des occupations paleolithiques de Kebara (Israel). *Paléorient* 14:117–123.

Gomez-Campo, C.
1985   *Plant conservation in the Mediterranean area.* W. Junk, Dordrecht, Netherlands.

Goodale, J. C.
1957   Alonga bush: A Tiwi hunt. *University Museum Bulletin* (University of Pennsylvania) 21:3–38.

Gordon, D.
1993   Mousterian tool selection, reduction, and discard at Ghar, Israel. *Journal of Field Archaeology* 20:205–218.

Gorecki, P. P.
1991   Horticulturalists as hunter-gatherers: Rock shelter usage in Papua New Guinea. In *Ethnoarchaeological Approaches to Mobile Campsites: Hunter-Gatherer and Pastoralist Case Studies,* C. S. Gamble and W. A. Boismer, eds., pp. 237–262. International Monographs in Prehistory, Ann Arbor, Michigan.

Goring-Morris, N.
1987   *At the Edge: Terminal Hunter-Gatherers in the Negev and Sinai.* British Archaeological Reports, International Series 361, Oxford.
1995   Complex hunter-gatherers at the end of the Paleolithic, 20,000–10,000 B.P. In *The Archaeology of Society in the Holy Land,* T. E. Levy, ed., pp. 141–168. Leicester University Press, London.

Goring-Morris, A. N., O. Marder, A. Davidzon, and F. Ibrahim
1998   Putting Humpty Dumpty together again: Preliminary observations on refitting studies in the eastern Mediterranean. In *From Raw Material Procurement to Tool Production: The Organisation of Lithic Technology in Late Glacial and Early Postglacial Europe,* S. Milliken, ed., pp. 149–182. British Archaeological Reports, International Series 700, Oxford.

Gould, R. A.
1980   *Living Archaeology.* Cambridge University Press, Cambridge.

Gramsch, B., and K. Kloss
1989   Excavations near Friesack: An early Mesolithic marshland site on the northern plain of central Europe. In *The Mesolithic in Europe,* C. Bonsall, ed., pp. 313–324. John Donald, Edinburgh.

Grant, A.
1982   The use of tooth wear as a guide to the age of domestic animals. In *Aging and Sexing Animal Bones from Archaeological Sites,* B. Wilson, C. Grigson, and S. Payne, eds., pp. 91–108. British Archaeological Reports, British Series 109, Oxford.

Grave, P., and L. Kealhofer
1999   Assessing bioturbation in archaeological sediments using soil morphology and phytolith analysis. *Journal of Archaeological Science* 26: 1239–1248.

Grayson, D. K.
1984   *Quantitative Zooarchaeology.* Academic Press, Orlando, Florida.
1989   Bone transport, bone destruction, and reverse utility curves. *Journal of Archaeological Science* 16:643–652.
1996   Review of *Honor among Thieves: A Zooarchaeological Study of Neandertal Ecology,* by M. C. Stiner. American Antiquity 61:815–816.

Grayson, D. K., and F. Delpech.
1998   Changing diet breadth in the early Upper Palaeolithic of southwestern France. *Journal of Archaeological Science* 25:1119–1129.

Greuter, W.
1991   Botanical diversity, endemism, rarity, and extinction in the Mediterranean area: An analysis based on the published volumes of Med-Checklist. *Botanica Chronica* 10:63–79.
1994   Extinction in the Mediterranean areas. *Philosophical Transactions of the Royal Society,* London, Series B, 344:41–46.

Gross, J. E., L. C. Stoddart, and F. H. Wagner
1974   *Demographic Analysis of a Northern Utah Jackrabbit Population.* Wildlife Monographs, no. 40.

Groves, R. H.
1991   The biogeography of Mediterranean plant invasions. In *Biogeography of Mediterranean Invasions,* R. H. Groves and F. di Castri, eds., pp. 427–438. Cambridge University Press, Cambridge.

Grun, R., and C. B. Stringer
1991   Electron-spin-resonance dating and the evo-
       lution of modern humans. *Archaeometry* 33:
       153–199.

Grun, R., C. B. Stringer, and H. P. Schwarcz
1991   ESR dating of teeth from Garrod's Tabun
       Cave collection. *Journal of Human Evolution*
       20:231–248.

Güleç, E., S. L. Kuhn, and M. C. Stiner
2002   The 2000 excavation at Üçağızlı Cave.
       Araşterma Sonuçları Toplantısı (Ankara) 23:
       255–264.

Hahn, J.
1972   Aurignacian signs, pendants, and art objects
       in central and eastern Europe. *World Archaeol-
       ogy* 3:252–266.

Hailey, A.
1988   Population ecology and conservation of tor-
       toises: The estimation of density, and dynam-
       ics of a small population. *Herpetological Journal*
       1:263–271.
1990   Adult survival and recruitment and the expla-
       nation of an uneven sex ratio in a tortoise
       population. *Canadian Journal of Zoology* 68:
       547–555.

Hailey, A., and N. S. Loumbourdis
1988   Egg size and shape, clutch dynamics, and
       reproductive effort in European tortoises.
       Canadian Journal of Zoology 66:1527–1536.

Hailey, A., J. Wright, and E. Steer
1988   Population ecology and conservation of tor-
       toises: The effects of disturbance. *Herpetologi-
       cal Journal* 1:294–301.

Hames, R.
1992   Time allocation. In *Evolutionary Ecology and
       Human Behavior,* E. A. Smith and B. Winter-
       halder, eds., pp. 203–235. Aldine de Gruyter,
       New York.

Hansen, K.
1992   Reproduction in European hare in a Danish
       farmland. *Acta Theriologica* 37(1–2):27–40.

Harpending, H., and J. Bertram
1975   Human population dynamics in archaeologi-
       cal time: Some simple models. In *Population
       Studies in Archaeology and Biological Anthropol-
       ogy,* A. C. Swedlund, ed., pp. 82–91. Society of
       American Archaeology Memoir no. 30, Wash-
       ington, DC.

Harper, H., B. Harry, and W. Bailey
1958   The chukar partridge in California. *California
       Fish and Game* 44:5–50.

Hawkes, K.
1987   Limited needs and hunter-gatherer time allo-
       cation. *Ethology and Sociobiology* 8:87–91.

Hawkes, K., J. F. O'Connell, and N. Blurton Jones
1997   Hadza women's time allocation, offspring
       provisioning, and the evolution of long post-
       menopausal life spans. *Current Anthropology*
       38(4):551–577.

Hayden, B.
1981   Research and development in the Stone Age:
       Technological transitions among hunter-gath-
       erers. *Current Anthropology* 22:519–548.

Hedges, R. E. M., and A. Millard
1995   Bones and groundwater: Towards the model-
       ling of diagenetic processes. *Journal of Archae-
       ological Science* 22:155–164.

Hedges, R. E. M., A. Millard, and A. W. G. Pike
1995   Measurements and relationships of diage-
       netic alteration of bone from three archaeo-
       logical sites. *Journal of Archaeological Science*
       22:201–209.

Henry, D. O.
1985   Preagricultural sedentism: The Natufian
       example. In *Prehistoric Hunter-Gatherers: The
       Emergence of Cultural Complexity,* T. D. Price and
       J. A. Brown, eds., pp. 365–384. Academic
       Press, San Diego, California.
1989   *From Foraging to Agriculture: The Levant at the
       End of the Ice Age.* University of Pennsylvania
       Press, Philadelphia.
1992   Transhumance during the late Levantine
       Mousterian. In *The Middle Paleolithic: Adapta-
       tion, Behavior, and Variability,* H. Dibble and P.
       Mellars, eds., pp. 143–162. Monograph no. 78,
       University Museum, University of Pennsylva-
       nia, Philadelphia.

1998a Intrasite spatial patterns and behavioral modernity: Indications from the late Levantine Mousterian rockshelter of Tor Faraj, southern Jordan. In *Neandertals and Modern Humans in Western Asia,* T. Akazawa, K. Aoki, and O. Bar-Yosef, eds., pp. 127–142. Plenum Press, New York.

1998b The Middle Paleolithic of Jordan. In *The Prehistoric Archaeology of Jordan,* D. O. Henry, ed., pp. 23–38. British Archaeological Reports, International Series 705, Oxford.

Henry, D. O., and A. N. Garrard
1988 Tor Hamar: An Epipaleolithic rockshelter in southern Jordan. *Palestine Exploration Quarterly* 120:1–25.

Henry, D. O., A. Leroi-Gourhan, and S. Davis
1981 The excavation of Hayonim Terrace: An examination of terminal Pleistocene climatic and adaptive changes. *Journal of Archaeological Science* 8:33–58.

Heppel, S. S., L. B. Crowder, and D. T. Crouse
1996 Models to evaluate headstarting as a management tool for long-lived turtles. *Ecological Applications* 6:556–565.

Heppel, S. S., C. J. Limpus, D. T. Crouse, N. B. Frazer, and L. B. Crowder
1996 Population model analysis for the loggerhead sea turtle, *Caretta caretta,* in Queensland. *Wildlife Research* 23:143–159.

Hewitt, G.
2000 The genetic legacy of the Quaternary ice ages. *Nature* 405:907–913.

Hillman, G. C., S. Colledge, and D. R. Harris
1989 Plant food economy during the Epi-Palaeolithic period at Tell Abu Hureyra, Syria: Dietary diversity, seasonality and modes of exploitation. In *Foraging and Farming: The Evolution of Plant Exploitation,* G. C. Hillman and D. R. Harris, eds., pp. 240–266. Unwin-Hyman, London.

Hillson, S.
1986 *Teeth.* Cambridge University Press, Cambridge.

Hockett, B. S.
1991 Toward distinguishing human and raptor patterning on leporid bones. *American Antiquity* 56(4):667–679.

1994 A descriptive reanalysis of the leporid bones from Hogup Cave, Utah. *Journal of California and Great Basin Anthropology* 16(1):106–117.

Hockett, B., and N. Bicho
2000 The rabbits of Picareiro Cave: Small mammal hunting during the late Upper Palaeolithic in the Portuguese Estremadura. *Journal of Archaeological Science* 27:715–723.

Hockey, P. A. R.
1994 Man as a component of the littoral predator spectrum: A conceptual overview. In *Rocky Shores: Exploitation in Chile and South Africa,* W. R. Siegfried, ed., pp. 17–31. Springer-Verlag, Berlin.

Hoffecker, J. F., G. Baryshnikov, and O. Potapova
1991 Vertebrate remains from the Mousterian site of Il'skaya I (northern Caucasus, USSR): New analysis and interpretation. *Journal of Archaeological Science* 18:113–147.

Hooijer, D.
1961 The fossil vertebrates of Ksar 'Akil, a Paleolithic rockshelter in Lebanon. *Zoologische Verhandelingen* 49:4–65.

Hopf, M., and O. Bar-Yosef
1987 Plant remains from Hayonim Cave, western Galilee. *Paléorient* 13(1):117–120.

Horwitz, L. K.
1996 The impact of animal domestication on species richness: A pilot study from the Neolithic of the southern Levant. *Archaeozoologia* 8:53–70.

1998 The influence of prey body size on patterns of bone distribution and representation in a striped hyena den. In *Économie préhistorique: Les comportements de subsistance au Paléolithique,* J.-P. Brugal, L. Meignen, and M. Patou-Mathis, eds., pp. 29–40. XVIIIᵉ Rencontres Internationales d'Archéologie et d'Histoire d'Antibes. Éditions du APDCA, Sophia Antipolis, France.

Horwitz, L. K., C. Cope, and E. Tchernov
1990   Sexing the bones of mountain-gazelle *(Gazella gazella)* from prehistoric sites in the southern Levant. *Paléorient* 16(2):1–12.

Horwitz, L. K., and N. Goring-Morris
2001   Fauna from the early Natufian site of Upper Besor 6 in the central Negev, Israel. *Paléorient* 26(1):111–128.

Horwitz, L. K., and P. Smith
1988   The effects of striped hyaena activity on human remains. *Journal of Archaeological Science* 15:471–481.

Hovers, E.
1997   Variability of Levantine Mousterian assemblages and settlement patterns: Implications for the development of human behavior. Ph.D. dissertation, Hebrew University of Jerusalem.
1998   The lithic assemblages of Amud Cave: Implications for understanding the end of the Mousterian in the Levant. In *Neandertals and Modern Humans in Western Asia*, T. Akazawa, K. Aoki, and O. Bar-Yosef, eds., pp. 143–163. Plenum Press, New York.

Hovers, E., L. K. Horowitz, et al.
1988   The site of Urkan-E-Rub IIa: A case study of subsistence and mobility patterns in the Kebaran period in the lower Jordan Valley. *Mitekufat Haeven, Journal of the Israel Prehistoric Society* 21:20–48.

Howell, F. C.
1952   Pleistocene glacial ecology and the evolution of "classic Neandertal" man. *Southwest Journal of Anthropology* 8:377–410.
1959   Upper Pleistocene stratigraphy and early man in the Levant. *Proceedings of the American Philosophical Society* 103:1–65.

Howells, W. W.
1976   Explaining modern man: Evolutionists versus migrationists. *Journal of Human Evolution* 5:477–495.

Hudson, J.
1991   Nonselective small game hunting strategies: An ethnoarchaeological study of Aka Pygmy sites. In *Human Predators and Prey Mortality*, M. C. Stiner, ed., pp. 105–120. Westview Press, Boulder, Colorado.

Hutchinson, G. E.
1957   Concluding remarks. *Cold Spring Harbor Symposium in Quantative Biology* 22:415–427.

Issar, A., and U. Kafri
1972   *Neogene and Pleistocene geology of the western Galilee coastal plain.* Bulletin 53, Ministry of Development Geological Survey, State of Israel.

Jacob-Friesen, K. H.
1956   Eiszeitliche Elephantenjäger in der Lüneburger Heide. *Jarbüch des Römisch-Germanischen Zentralmuseums Mainz* 3:1–22.

James, T. R., and R. W. Seabloom
1969   Reproductive biology of the white-tailed jackrabbit in North Dakota. *Journal of Wildlife Management* 33(3):558–568.

Jaubert, J., M. Lorblanchet, H. Laville, R. Slott-Moller, A. Turq, and J.-P. Brugal
1990   *Les chasseurs d'aurochs de la Borde.* Éditions de la Maison des Sciences de l'Homme, Documents d'Archéologie Francaise, no. 27. Paris.

Jelinek, A. J.
1981   The Middle Palaeolithic in the Southern Levant from the Perspective of the Tabun Cave. In *Préhistoire du Levant*, J. Cauvin and P. Sanlaville, eds., pp. 265–280. Éditions du CNRS, Paris.
1982   The Middle Palaeolithic in the southern Levant with comments on the appearance of modern *Homo sapiens*. In *The Transition from Lower to Middle Palaeolithic and the Origin of Modern Man*, A. Ronen, ed., pp. 57–104. British Archaeological Reports, International Series 151, Oxford.
1991   Observations on reduction patterns and raw material in some Middle Paleolithic industries in the Perigord. In *Raw Material Economies among Prehistoric Hunter-Gatherers*, A. Monte-White and S. Holen, eds., pp. 7–31. University of Kansas Publications in Anthropology no. 19, Lawrence, Kansas.

Jelinek, A. J., W. R. Farrand, G. Haas, A. Horowitz, and P. Goldberg
1973   New excavations at the Tabun Cave, Mount Carmel, Israel, 1967–1972: A preliminary report. *Paléorient* 1:151–183.

Jenkins, D.
1961 Population control in protected partridges (*Perdix perdix*). *Journal of Animal Ecology* 30: 235–258.

Jerardino, A.
1997 Changes in shellfish species composition and mean shell size from a late-Holocene record of the west coast of southern Africa. *Journal of Archaeological Science* 24:1031–1044.

Jochim, M.
1998 *A Hunter-Gatherer Landscape: Southwest Germany in the Late Paleolithic and Mesolithic.* Plenum Press, New York.

Jones, K. T.
1984 Hunting and scavenging by early hominids: A study in archaeological method and theory. Ph.D. dissertation, Department of Anthropology, University of Utah, Salt Lake City.

Josephson, S. C., K. E. Juell, and A. R. Rogers
1996 Estimating sexual dimorphism by method-of-moments. *American Journal of Physical Anthropology* 100:191–206.

Karkanas, P., O. Bar-Yosef, P. Goldberg, and S. Weiner
2000 Diagenesis in prehistoric caves: The use of minerals that form in situ to assess the completeness of the archaeological record. *Journal of Archaeological Science* 27:915–929.

Karkanas, P., N. Kyparissi-Apostolika, O. Bar-Yosef, and S. Weiner
1999 Mineral assemblages in Theopetra, Greece: A framework for understanding diagenesis in a prehistoric cave. *Journal of Archaeological Science* 26:1171–1180.

Karkanas, P., J.-P. Rigaud, J. F. Simek, R. M. Albert, and S. Weiner
2002 Ash, bones and guano: A study of the minerals and phytoliths in the sediments of Grotte XVI, Dordogne, France. *Journal of Archaeological Science* 29(7):721–732.

Keck, R., D. Maurer, W. Daisey, and L. Sterling
1973 *Marine Invertebrate Resources: Annual Report 1972–73.* Field Station, College of Marine Studies, University of Delaware–Lewes, submitted to the Delaware Department of Natural Resources and Environmental Control.

Keeley, L. H.
1988 Hunter-gatherer economic complexity and population pressure. *Journal of Anthropological Archaeology* 7:373–411.
1995 Protoagricultural practices among hunter-gatherers: A cross-cultural survey. In *Last Hunters—First Farmers: New Perspectives on the Prehistoric Transition to Agriculture,* T. D. Price and A. B. Gebauer, eds., pp. 243–272. School of American Research Press, Santa Fe, New Mexico.

Keith, L. B.
1981 Population dynamics of hares. In *Proceedings of the World Lagomorph Conference,* K. Myers and C. D. MacInnes, eds., pp. 395–440. University of Guelph, Guelph, Ontario, Canada.

Kelly, R.,
1995 *The Foraging Spectrum: Diversity in Hunter-Gatherer Lifeways.* Smithsonian Institution Press, Washington, DC.

Kersten, A. M. P.
1987 Age and sex composition of Epipalaeolithic fallow deer and wild goat from Ksar 'Akil. *Palaeohistoria* 29:119–131.

Khalaily, H., Y. Goren, and F. R. Valla
1993 A late Pottery Neolithic assemblage from Hayonim Terrace, Western Galilee. *Journal of the Israel Prehistoric Society* 25:132–144.

Kingery, D. W., P. Vandiver, and M. Pickett
1988 The beginnings of pyrotechnology, Part 2: Production and use of lime plaster and gypsum plaster in the Pre-Pottery Neolithic of the Near East. *Journal of Field Archaeology* 15(2): 219–244.

Kintigh, K. W.
1984 Measuring archaeological diversity by comparison with simulated assemblages. *American Antiquity* 49:44–54.

Klein, R. G.
1978 Stone Age predation of large African bovids. *Journal of Archaeological Science* 5:195–217.
1979 Stone Age exploitation of animals in southern Africa. *American Scientist* 67:151–160.

1989   *The Human Career: Human Biological and Cultural Origins.* University of Chicago Press, Chicago.
1995   The Tor Hamar fauna. In *Prehistoric Cultural Ecology and Evolution: Insights from Southern Jordan,* D. O. Henry, ed., pp. 405–416. Plenum Press, New York.
1999   *The Human Career: Human Biological and Cultural Origins.* 2nd ed. University of Chicago Press, Chicago.

Klein, R. G., and K. Cruz-Uribe
1983   Stone Age population numbers and average tortoise size at Bynesdranskop Cave 1 and Die Kelders Cave 1, southern Cape Province, South Africa. *South African Archaeological Bulletin* 38:26–30.
1984   *The Analysis of Animal Bones from Archaeological Sites.* University of Chicago Press, Chicago.

Klein, R. G., and K. Scott
1986   Reanalysis of faunal assemblages from the Haua Fteah and other Late Quaternary archaeological sites in Cyrenaican Libya. *Journal of Archaeological Science* 13:515–542.

Knecht, H.
1997   *Projectile Technology.* Plenum Press, New York.

Knight, J. A.
1985   Differential preservation of calcined bone at the Hirundo site, Alton, Maine. Master's thesis, Quaternary Studies, University of Maine, Orono.

Koslow, J. A.
1997   Seamounts and the ecology of deep-sea fisheries. *American Scientist* 85(2):168–176.

Kovacs, G.
1983   Survival pattern in adult European hares. *Acta Zoologica Fennica* 174:69–70.

Kowalewski, M., G. A. Goodfriend, and K. W. Flessa
1998   High-resolution estimates of temporal mixing within shell beds: The evils and virtues of time-averaging. *Paleobiology* 24:287–304.

Kozlowski, J. K., ed.
1982   *Excavation in the Bacho Kiro Cave, Bulgaria.* Paristwowe Wydarunictwo, Nawkowe, Warsaw, Poland.

1999   The evolution of the Balkan Aurignacian. In *Dorothy Garrod and the Progress of the Palaeolithic: Studies in the Prehistoric Archaeology of the Near East and Europe,* W. Davies and R. Charles, eds., pp. 97–117. Oxbow Books, Oxford.
2000   The problem of cultural continuity between the Middle and the Upper Paleolithic in central and eastern Europe. In *The Geography of Neandertals and Modern Humans in Europe and the Greater Mediterranean,* O. Bar-Yosef and D. Pilbeam, eds., pp. 77–105. Peabody Museum Bulletin 8, Harvard University, Cambridge, Massachusetts.

Kreutzer, L. A.
1992   Bison and deer bone mineral densities: Comparisons and implications for the interpretation of archaeological faunas. *Journal of Archaeological Science* 19:271–294.

Kruuk, H.
1972   *The Spotted Hyaena.* University of Chicago Press, Chicago.

Kuhn, S. L.
1992a   Blank form and reduction as determinants of Mousterian scraper morphology. *American Antiquity* 57(1):115–128.
1992b   On planning and curated technologies in the Middle Paleolithic. *Journal of Anthropological Research* 48:185–214.
1993   Mousterian technology as adaptive response: A case study. In *Hunting and Animal Exploitation in the Later Palaeolithic and Mesolithic of Eurasia,* G. L. Peterkin, H. Bricker, and P. Mellars, eds., pp. 25–31. Archaeological Papers of the American Anthropological Association 4, Washington, DC.
1995   *Mousterian Lithic Technology: An Ecological Perspective.* Princeton University Press, Princeton, New Jersey.

Kuhn, S. L., A. Belfer-Cohen, O. Bar-Yosef, B. Vandermeersch, B. Arensburg, and M. C. Stiner
2004   The Last Glacial Maximum at Meged Rockshelter, upper Galilee, Israel. *Mitekufat Ha'even, Journal of the Israel Prehistory Society* 34:5–47.

Kuhn, S. L., and M. C. Stiner
1992   New research on Riparo Mochi, Balzi Rossi (Liguria): Preliminary results. *Quaternaria Nova* 2:77–90.

1998a The earliest Aurignacian of Riparo Mochi (Liguria). *Current Anthropology* 39(supplement):S175–189.

1998b Middle Paleolithic "creativity": Reflections on an oxymoron? In *Creativity and Human Evolution and Prehistory*, S. Mithen, ed., pp. 143–164. Routledge, London.

2001 The antiquity of hunter-gatherers. In *Hunter-Gatherers: Interdisciplinary Perspectives*, C. Panter-Brick, R. H. Layton, and P. A. Rowley-Conwy, eds., pp. 99–142. Cambridge University Press, Cambridge.

Kuhn, S. L., M. C. Stiner, and E. Güleç

1999 Initial Upper Paleolithic in south-central Turkey and its regional context: A preliminary report. *Antiquity* 73(281):505–517.

2004 New perspectives on the Initial Upper Paleolithic: The view from Üçağızlı Cave, Turkey. In *The Early Upper Paleolithic beyond Western Europe*, P. J. Brantingham, S. L. Kuhn, and K. W. Kerry, eds., pp. 113–128. University of California Press, Berkeley.

Kuhn, S. L., M. C. Stiner, K. W. Kerry, and E. Güleç

2003 The early Upper Paleolithic at Üçağızlı Cave (Hatay, Turkey): Preliminary results. In *More Than Meets the Eye: Studies on Upper Palaeolithic Diversity in the Near East*, N. Goring-Morris and A. Belfer-Cohen, eds., pp. 106–117. Oxbow Books, Oxford.

Kuhn, S. L., M. C. Stiner, D. S. Reese, and E. Güleç

2001 Ornaments in the earliest Upper Paleolithic: New results from the Levant. *Proceedings of the National Academy of Sciences* 98(13):7641–7646.

Kurtén, B.

1968 *Pleistocene Mammals of Europe*. Aldine, London.

Lam, Y. M., X. Chen, C. W. Marean, and C. J. Frey

1998 Bone density and long bone representation in archaeological faunas: Comparing results from CT and photon densitometry. *Journal of Archaeological Science* 25:559–570.

Lam, Y. M., X. Chen, and O. M. Pearson

1999 Intertaxonomic variability in patterns of bone density and the differential representation of bovid, cervid, and equid elements in the archaeological record. *American Antiquity* 64:343–362.

Lambert, M. R. K.

1982 Studies on the growth, structure and abundance of the Mediterranean spur-thighed tortoise, *Testudo graeca*, in field populations. *Journal of Zoology*, London 196:165–189.

1984 Threats to Mediterranean (West Palaearctic) tortoises and their effects on wild populations: An overview. *Amphibia-Reptilia* 5:5–15.

Laplace, G.

1977 Il Riparo Mochi ai Balzi Rossi de Grimaldi (Fouilles 1938–1949): Les industries leptolithiques. *Rivista di Scienze Prehistoriche* 32(1–2):3–131.

Latham, R. E., and R. E. Ricklefs

1993 Continental comparisons of temperate-zone tree species diversity. In *Species Diversity in Ecological Communities: Historical and Geographical Perspectives*, R. E. Ricklefs and D. Schluter, eds., pp. 294–314. University of Chicago Press, Chicago.

Lechleitner, R. R.

1959 Sex ratio, age classes and reproduction of the black-tailed jackrabbit. *Journal of Mammalogy* 40(1):63–81.

Lee, R. B.

1979 *The !Kung San: Men, Women, and Work in a Foraging Society*. Cambridge University Press, Cambridge.

Lemonnier, P.

1992 *Elements for an Anthropology of Technology*. Anthropological Papers 88, Museum of Anthropology, University of Michigan, Ann Arbor.

Leroi-Gourhan, A.

1973 Fouilles de Pincevent: Essai d'analyse ethnographique d'un habitat magdalénien. *Supplément à Gallia Préhistoire*, no. 7. Éditions du CNRS, Paris.

Levine, M. A.

1983 Mortality models and the interpretation of horse population structure. In *Hunter-Gatherer Economy in Prehistory*, G. Bailey, ed., pp. 23–46. Cambridge University Press, Cambridge.

Levins, R.
1968   *Evolution in Changing Environments: Some The-oretical Explorations.* Princeton University Press, Princeton, New Jersey.

Levinton, J. S.
1995   *Marine Biology: Function, Biodiversity, Ecology.* Oxford University Press, New York.

Lieberman, D. E.
1991   Seasonality and gazelle hunting at Hayonim Cave: New evidence for "sedentism" during the Natufian. *Paléorient* 17(1):47–57.
1993   The rise and fall of seasonal mobility among hunter-gatherers: The case of the southern Levant. *Current Anthropology* 34(5):599–631.

Lindly, J. M.
1988   Hominid and carnivore activity at Middle and Upper Paleolithic cave sites in eastern Spain. *Munibe* 40:45–70.

Lindly, J., and G. Clark.
1987   A preliminary lithic analysis of the Mouster-ian site of 'Ain Difla (WHS Site 634) in the Wadi Ali, west-central Jordan. *Proceedings of the Prehistoric Society* 53:279–292.

Little, C., and J. A. Kitching
1998   *The Biology of Rocky Shores.* 2nd ed. Oxford University Press, Oxford.

Lowe, V. P.
1967   Teeth as indicators of age, with special refer-ence to red deer *(Cervus elaphus)* of known age from Rhum. *Journal of Zoology, London* 152:137–153.

Lowenstam, H. A., and S. Weiner
1989   *On Biomineralization.* Oxford University Press, New York.

Lupo, K. D.
1995   Hadza bone assemblages and hyena attrition: An ethnographic example of the influence of cooking and mode of discard on the intenstiy of scavenger ravaging. *Journal of Anthropologi-cal Archaeology* 14:288–314.

Lupo, K. D., and D. N. Schmitt
1997   Experiments in bone boiling: Nutritional returns and archaeological reflections. *Anthro-pozoologica* 25–26:137–144.

Lyman, R. L.
1984   Bone density and differential survivorship of fossil classes. *Journal of Anthropological Archae-ology* 3:259–299.
1991   Taphonomic problems with archaeological analyses of animal carcass utilization and transport. In *Beamers, Bobwhites, and Blue-Points: Tributes to the Career of Paul W. Parmalee,* J. R. Purdue, W. E. Klippel, and B. W. Styles, eds., pp. 125–138. Illinois State Museum Sci-entific Papers no. 23, Springfield, Illinois.
1994   *Vertebrate Taphonomy.* Cambridge University Press, Cambridge.

Lyman, R. L., and G. L. Fox
1989   A critical evaluation of bone weathering as an indication of bone assemblage formation. *Journal of Archaeological Science* 16:293–317.

Lyman, R. L., L. E. Houghton, and A. L. Chambers
1992   The effect of structural density on marmot skeletal part representation in archaeological sites. *Journal of Archaeological Science* 19:557–573.

MacArthur, R. H., and R. Levins
1967   The limiting similarity, convergence, and divergence of coexisting species. *American Naturalist* 101:377–385.

MacArthur, R. H., and E. Pianka
1966   On optimal use of a patchy environment. *American Naturalist* 100:603–609.

MacArthur, R. H., and E. O. Wilson
1967   *The Theory of Island Biogeography.* Princeton University Press, Princeton, New Jersey.

Mackie, R., and H. Buechner
1963   The reproductive cycle of the chukar. *Journal of Wildlife Management* 27(2):246–260.

Maddock, A. H., and M. G. L. Mills
1994   Population characteristics of African wild dogs, *Lycaon pictus,* in the Eastern Transvaal lowveld, South Africa, as revealed through photograph records. *Biological Conservation* 67:57–62.

Madeyska, T.
1999   Palaeogeography of European lowland during the late Vistulian. In *Post-pleniglacial Re-colonization of the Great European Lowland,* M. Kobusiewicz and J. K. Kozlowski, eds., pp. 7–30. Folia Quaternaria 70, Polska Akademia Umiejêtnoœci, Komisja Paleogeografii Czwartorzêdu, Kraków, Poland.

Madsen, D. B., and D. N. Schmitt
1998   Mass collecting and the diet breadth model: A Great Basin example. *Journal of Archaeological Science* 25:445–455.

Marboutin, E., and R. Peroux
1995   Survival pattern of the European hare in a decreasing population. *Journal of Applied Ecology* 32:809–816.

Marean, C. W.
1995   Of taphonomy and zooarchaeology. *Evolutionary Anthropology* 4:64–72.

Marean, C. W., and S. Y. Kim
1998   Mousterian large-mammal remains from Kobeh Cave: Behavioral implications. *Current Anthropology* 39:S79–113.

Marean, C. W., and L. M. Spencer
1991   Impact of carnivore ravaging on zooarchaeological measures of element abundance. *American Antiquity* 56:645–658.

Marean, C. W., L. M. Spencer, R. J. Blumenschine, and S. D. Capaldo
1992   Captive hyaena bone choice and destruction, the Schlepp Effect and Olduvai archaeofaunas. *Journal of Archaeological Science* 19:101–121.

Margaris, A. V.
2000   A mineralogical analysis of sediments from the Tabun Cave (Israel) using Fourier-transform infrared spectroscopy. Master's thesis, Department of Anthropology, University of Arizona, Tucson.

Marks, A., and K. Monigal
1995   Modeling the production of elongated blanks from the early Levantine Mousterian at Rosh Ein Mor. In *The Definition and Interpretation of Levallois Technology,* H. Dibble and O. Bar-Yosef, eds., pp. 267–278. Prehistory Press, Madison, Wisconsin.

Marks, A. E., and P. Volkman
1986   The Mousterian of Ksar 'Akil. *Paléorient* 12:5–20.

Marshack, A.
1997   Paleolithic image making and symboling in Europe and the Middle East: A comparative review. In *Beyond Art: Pleistocene Image and Symbol,* M. Conkey, O. Soffer, D. Stratmann, and N. G. Jablonski, eds., pp. 53–91. Memoirs of California Academy of Sciences 23. San Francisco.

Martin, H.
1907–   *Recherches sur l'evolution du Moustérien dans le*
1910   *gisement de la Quina (Charente),* Vol. *Industrie Osseuse.* Schleicher Freres, Paris.

Martínez-Navarro, B., and P. Palmqvist
1995   Presence of the African machairodont *Megantereon whitei* (Broom, 1937) (Felidae, Carnivora, Mammalia) in the Lower Pleistocene site of Venta Micena (Orce, Granada, Spain), with some considerations on the origin, evolution and dispersal of the genus. *Journal of Archaeological Science* 22(4):569–582.
1996   Presence of the African saber-toothed felid *Megantereon whitei* (Broom, 1937) (Mammalia, Carnivora, Machairodontinae) in Apollonia-1 (Mygdonia Basin, Macedonia, Greece). *Journal of Archaeological Science* 23:869–872.

Martinson, D. G., N. G. Pisias, J. D. Hays, J. Imbrie, T. C. Moore, and N. J. Shackleton
1987   Age dating and the orbital theory of the ice ages: Development of a high-resolution 0 to 300,000–year chronostratigraphy. *Quaternary Research* 27:1–29.

Masters, P. M.
1987   Preferential preservation of noncollagenous protein during bone diagenesis: Implications for chronometric and stable isotopic measurements. *Geochimica et Cosmochimica Acta* 51:3209–3214.

McBrearty, S., and A. Brooks
2000   The revolution that wasn't: A new interpretation of the origin of modern human behavior. *Journal of Human Evolution* 39:456–463.

McConnell, D.
1952 The crystal chemistry of carbonate apatites and their relationship to the composition of calcified tissues. *Journal of Dental Research* 31: 53–63.

McCullough, D. R., D. S. Pine, D. L. Whitmore, T. M. Mansfield, and R. H. Decker
1990 *Linked Sex Harvest Strategy for Big Game Management with a Test Case on Black-Tailed Deer.* Wildlife Monographs, no. 112.

McNutt, J. W.
1996 Sex-biased dispersal in African wild dogs, *Lycaon pictus. Animal Behaviour* 52:1067–1077.

Meek, R.
1989 The comparative population ecology of Hermann's tortoise, *Testudo hermanni,* in Croatia and Montenegro, Yugoslavia. *Herpetological Journal* 1:404–414.

Meignen, L.
1988 Un example de comportement technologique differentiel selon des matieres premieres: Marillac, couches 9 et 10. In *L'homme de Neandertal: La technique,* M. Otte, ed., pp. 71–88. Universite de Liège, Liège, Belgium.
1994 Le Paléolithique Moyen au Proche-Orient: Le phénomène laminaire. In *Les industries laminaires au Paléolithique Moyen,* S. Révillion and A. Tuffreau, eds., pp. 125–159. Dossier de Documentation Archéologique 19, Éditions du CNRS, Paris.
1995 Levallois lithic production systems in the Middle Paleolithic of the Near East: The case of the unidirectional method. In *The Definition and Interpretation of Levallois Technology,* H. Dibble and O. Bar-Yosef, eds., pp. 361–380. Prehistory Press, Madison, Wisconsin.
1998 Hayonim Cave lithic assemblages in the context of the Near Eastern Middle Paleolithic: A preliminary report. In *The Origins of Modern Humans in Western Asia,* T. Akazawa, K. Aoki, and O. Bar-Yosef, eds., pp. 165–180. Plenum Press, New York.
2000 Early Middle Palaeolithic blade technology in southwestern Asia. *Acta Anthropologica Sinica* 19(Supplement):158–168.

Meignen, L., and O. Bar-Yosef
1991 Les outillage lithiques moustériens de Kébara. In *Le Squelette moustérienne de Kébara 2, Mount Carmel, Israël,* O. Bar-Yosef and B. Vandermeersch, eds., pp. 49–76. Éditions du CNRS, Paris.
1992 Middle Palaeolithic variability in Kebara Cave (Mount Carmel, Israel). In *The Evolution and Dispersal of Modern Humans in Asia,* T. Akazawa, K. Aoki, and T. Kimura, eds., pp. 129–148. Hokusen-Sha, Tokyo.

Meignen, L., O. Bar-Yosef, and P. Goldberg
1989 Les structures de combustion moustériennes de la grotte de Kébara (Mont Carmel, Israël). In *Nature et fonctions des foyers préhistoriques,* M. Olive and Y. Taborin, eds., pp. 141–146. Memoires du Musée de Préhistoire d'Ile de France, vol. 2. APRAIF, Nemours, France.

Meignen, L., O. Bar-Yosef, N. Mercier, H. Valladas, P. Goldberg, and B. Vandermeersch
2001 Apport des datations su probléme de l'origine des hommes modernes au Proche-Orient. In *Datation,* J.-N. Barrandon, P. Guibert, and V. Michel, eds., pp. 295–313. XXI^c Rencontres Internationals d'Archéologie et d'Histoire d'Antibes. Éditions du APDCA, Sophia Antipolis, France.

Meignen, L., O. Bar-Yosef, J. D. Speth, and M. C. Stiner
n.d. Middle Paleolithic settlement patterns in the Levant. In *Transitions before the Transition: Evolution and Stability in the Middle Paleolithic and Middle Stone Age,* E. Hovers and S. L. Kuhn, eds. Kluwer, New York. In press.

Meliadou, A., and A. Troumbis
1997 Aspects of heterogeneity in the distribution of diversity of the European herpetofauna. *Acta Oecologica* 18:393–412.

Mellars, P.
1989 Major issues in the emergence of modern humans. *Current Anthropology* 30:349–385.

Mellars, P., H. M. Bricker, J. A. Gowlett, and E. E. M. Hedges
1987 Radiocarbon-accelerator dating of French Upper Paleolithic sites. *Current Anthropology* 28:128–133.

Mendelssohn, H.
1974 The development of the populations of gazelles in Israel and their behavoural adaptations. In *The Behaviour of Ungulates and Its Relation to Management,* pp. 722–743. Papers of the International Union for Conservation of Nature and Natural Resources, no. 40. Morges, Switzerland.

Mercier, N., and H. Valladas
2003a Chronologie par la thermoluminescence de gisements du Paléolithique Moyen du Proche-Orient. In *Échanges et diffusion dans la préhistoire méditerranéenne,* B. Vandermeersch, ed., pp. 29–39. Comité des Travaux Historiques et Scientifiques, Paris.
2003b Reassessment of TL age estimates of burnt flints from the Paleolithic site of Tabun Cave, Israel. *Journal of Human Evolution* 45(5): 401–409.

Mercier, N., H. Valladas, L. Froget, J.-L. Joron, J.-L. Reyss, S. Weiner, P. Goldberg, L. Meignen, O. Bar-Yosef, S. Kuhn, M. Stiner, A.-M. Tillier, B. Arensburg, and B. Vandermeersch
n.d. Hayonim Cave: A TL-based chronology for the Levantine Mousterian sequence. Manuscript in preparation.

Mercier, N., H. Valladas, J.-L. Joron, S. Schiegl, O. Bar-Yosef, and S. Weiner
1995a Thermoluminescence dating and the problem of geochemical evolution of sediments: A case study, the Mousterian levels at Hayonim. *Israel Journal of Chemistry* 35:137–141.

Mercier, N., H. Valladas, G. Valladas, J.-L. Reyss, A. Jelinek, L. Meignen, and J.-L. Joron
1995b TL dates of burnt flints from Jelinek's excavations at Tabun and their implications. *Journal of Archaeological Science* 22:495–510.

Middleton, A. D.
1935 The population of partridges *(Perdix perdix)* in 1933 and 1934 in Great Britain. *Journal of Animal Ecology* 4:137–145.

Miller, N. F.
1992 The origins of plant cultivation in the Near East. In *The Origins of Agriculture: An International Perspective,* C. W. Cowan and P. J. Watson, eds., pp. 39–58. Smithsonian Institution Press, Washington, DC.

Mirazón Lahr, M., and R. Foley
2003 Demography, dispersal and human evolution in the Last Glacial period. In *Neandertals and Modern Humans in the European Landscape during the Last Glaciation,* T. H. van Andel and W. Davies, eds., pp. 241–256. McDonald Institute for Archaeological Research, University of Cambridge, Cambridge.

Mithen, S.
1993 Simulating mammoth hunting and extinction: Implications for the Late Pleistocene of the central Russian Plain. In *Hunting and Animal Exploitation in the Later Palaeolithic and Mesolithic of Eurasia,* G. L. Peterkin, H. Bricker, and P. Mellars, eds., pp. 163–178. Archaeological Papers of the American Anthropological Association 4, Washington, DC.

Moncel, M.-H., M. Patou-Mathis, and M. Otte
1998 Halte de chasse au chamois au Paléolithique Moyen: La couche 5 de al grotte Scladina (Sclayn, Namur, Belgique). In *Économie préhistorique: Les comportements de subsistance au Paléolithique,* J.-P. Brugal, L. Meignen, and M. Patou-Mathis, eds., pp. 291–308. XVIIIᵉ Rencontres Internationales d'Archéologie et d'Histoire d'Antibes. Éditions du APDCA, Sophia Antipolis, France.

Mordant, C., and D. Mordant
1992 Noyen-sur-Seine: A Mesolithic waterside settlement. In *The Wetland Revolution in Prehistory,* B. Coles, ed., pp. 55–64. Prehistoric Society, Exeter, UK.

Morlan, R. E.
1994a Bison bone fragmentation and survivorship: A comparative method. *Journal of Archaeological Science* 21:797–807.
1994b Oxbow bison procurement as seen from the Harder Site, Saskatchewan. *Journal of Archaeological Science* 21:757–777.

Müller-Beck, H.
1988 The ecosystem of the "Middle Paleolithic" (late Lower Paleolithic) in the Upper Danube region: A stepping stone to the Upper Paleolithic. In *Upper Pleistocene Prehistory in Western Eurasia,* H. Dibble and A. Montet-White, eds., pp. 233–254. Monograph no. 54, University Museum, University of Pennsylvania, Philadelphia.

Munro, N. D.

1999 Small game as indicators of sedentization during the Natufian period at Hayonim Cave in Israel. In *Zooarchaeology of the Pleistocene/Holocene Boundary,* J. Driver, ed., pp. 37–45. British Archaeological Reports, International Series 800, Oxford.

2001 A prelude to agriculture: Game use and occupation intensity during the Natufian period in the southern Levant. Ph.D. dissertation. Department of Anthropology, University of Arizona, Tucson.

2004 Zooarchaeological measures of human hunting pressure and site occupation intensity in the Natufian of the southern Levant and the implications for agricultural origins. *Current Anthropology.* 45 Supplement: S5–33.

Munson, P. J.

1991 Mortality profiles of white-tailed deer from archaeological sites in eastern North America: Selective hunting or taphonomy? In *Beamers, Bobwhites, and Blue-Points: Tributes to the Career of Paul W. Parmalee,* J. R. Purdue, W. E. Klippel, and B. W. Styles, eds., pp. 139–151. Scientific Papers no. 23, Illinois State Museum, Springfield.

2000 Age-correlated differential destruction of bones and its effect on archaeological mortality profiles of domestic sheep and goats. *Journal of Archaeological Science* 27:391–407.

Myers, N.

1990 The biodiversity challenge: Expanded hotspots analysis. *Environmentalist* 10:243–256.

Nadel, D.

1997 The spatial organization of prehistoric sites in the Jordan Valley: Kebaran, Natufian and Neolithic case studies. Ph.D. dissertation, Hebrew University of Jerusalem.

Nadel, D., A. Danin, E. Werker, T. Schick, M. E. Kislev, and K. Stewart

1994 Nineteen-thousand–year-old twisted fibers from Ohalo II. *Current Anthropology* 35(4): 451–458.

Naveh, Z., and R. H. Whittaker

1979 Structural and floristic diversity of shrublands and woodlands in northern Israel and other Mediterranean areas. *Vegetatio* 41:171–190.

Neeley, M. P., and G. A. Clark

1993 The human food niche in the Levant over the past 150,000 years. In *Hunting and Animal Exploitation in the Later Palaeolithic and Mesolithic of Eurasia,* G. L. Peterkin, H. Bricker, and P. Mellars, eds., pp. 221–240. Archaeological Papers of the American Anthropological Association 4, Washington, DC.

Neuville, R.

1951 *Le Paléolithique et le Mésolithique de Désert de Judée.* Archives de l'Institut de Paléontologie Humaine, Mémoire 24. Masson et Cie, Paris.

Newell, R. R., D. Kielman, T. S. Constandse-Westermann, W. A. B. van der Sanden, and A. van Gijn

1990 *An Inquiry into the Ethnic Resolution of Mesolithic Regional Groups: The Study of Their Decorative Ornaments in Time and Space.* E. J. Brill, Leiden, Netherlands.

Nicholson, R. A.

1993 A morphological investigation of burnt animal bone and an evaluation of its utility in archaeology. *Journal of Archaeological Science* 20:411–428.

Nielsen-Marsh, C. M., and R. E. M. Hedges

2000a Patterns of diagenesis in bone, 1: The effects of site environments. *Journal of Archaeological Science* 27:1139–1150.

2000b Patterns of diagenesis in bone, 2: Effects of acetic acid treatment and the removal of diagenetic CO_3^{2-}. *Journal of Archaeological Science* 27:1151–1159.

Nishiaki, Y.

1989 Early blade industries in the Levant: The placement of Douara IV industry in the context of the Levantine early Middle Paleolithic. *Paléorient* 15:215–229.

Nowak, R. L.

1991 *Walker's Mammals of the World.* 5th ed. Johns Hopkins University Press, Baltimore, Maryland.

O'Brien, S. J., and E. Mayr

1991 Bureaucratic mischief: Recognizing endangered species and subspecies. *Science* 251: 1187–1188.

O'Connell, J. F., K. Hawkes, and N. Blurton Jones
1988a Hadza hunting, butchering, and bone transport and their archaeological implications. *Journal of Anthropological Research* 44:113–161.
1988b Hadza scavenging: Implications for Plio/Pleistocene hominid subsistence. *Current Anthropology* 29:356–363.

Odum, E. P.
1971 *Fundamentals of Ecology.* 3rd ed. Saunders, Philadelphia.

Odum, E. P., and H. T. Odum
1959 *Fundamentals of Ecology.* 2nd ed. Saunders, Philadelphia.

Ohnuma, K.
1992 The significance of layer B (square 8-19) of the Amud Cave (Israel) in the Levantine Levalloiso-Mousterian: A technological study. In *The Evolution and Dispersal of Modern Humans in Asia,* T. Akazawa, K. Aoki, and T. Kimura, eds., pp. 83–106. Hokusen-sha, Tokyo.

Olszewski, T.
1999 Taking advantage of time averaging. *Paleobiology* 25:226–238.

Oswalt, W. H.
1976 *An Anthropological Analysis of Food-Getting Technology.* John Wiley, New York.

Ozenda, P.
1975 Sur les étages de végétation dans les montagnes du Bassin Mediterranéen. *Documents de Cartographie Ecologique* 16:1–32.

Palma di Cesnola, A.
1965 Notizie preliminari sulla terza campagna di scavi nella Grotta del Cavallo (Lecce). *Rivista di Scienze Prehistoriche* 25:3–87.
1969 Il Musteriano della Grotta del Poggio a Marina di Camerota (Salerno). In *Estratto dagli Scritti sul Quaternario in onore di Angelo Pasa,* pp. 95–135. Museo Civico di Storia Naturale, Verona, Italy.

Parker, S. P., ed.
1994 *McGraw-Hill Dictionary of Scientific and Technical Terms.* 5th ed. McGraw-Hill, New York.

Pavao, B., and P. W. Stahl
1999 Structural density assays of leporid skeletal elements with implications for taphonomy, actualistic and archaeological research. *Journal of Archaeological Science* 26(1):53–67.

Payne, S.
1973 Kill-off patterns in sheep and goats: The mandibles from Asvan Kale. *Anatolian Studies* 23:281–303.

Paz, U.
1987 *The Birds of Israel.* Stephen Greene Press, Lexington, Massachusetts.

Pepin, D.
1987 Dynamics of a heavily exploited population of brown hare in a large-scale farming area. *Journal of Applied Ecology* 24:725–734.

Person, A., H. Bocherens, A. Mariotti, and M. Renard
1996 Diagenetic evolution and experimental heating of bone phosphate. *Palaeogeography, Palaeoclimatology and Palaeoecology* 126:135–149.

Peters, J.
1991 Mesolithic fishing along the central Sudanese Nile and lower Atbara. *Sahara* 4:33–40.

Petruesewicz, K.
1970 Dynamics and production of the hare population in Poland. *Acta Theriologica* 15:413–445.

Pfaffenberger, B.
1992 Social anthropology of technology. *Annual Review of Anthropology* 21:491–516.

Pianka, E. R.
1978 *Evolutionary Ecology.* 2nd ed. Harper and Row, New York.

Pichon, J.
1983 Parures natoufiennes en os de perdrix. *Paléorient* 9(1):91–98.
1984 L'avifaune natoufienne du Levant. These de 3e cycle, Université Pierre et Marie-Curie, Paris IV.
1987 L'avifaune de Mallaha. In *La faune du GISEMENT Natufien de Mallaha (Eynan) Israel,* J. Bouchud, ed., pp. 115–150. Memoires et Travaux du Centre de Recherche Français de Jerusalem, Association Paléorient, Paris.

Pidoplichko, I. H.
1998   *Upper Palaeolithic Dwellings of Mammoth Bones in the Ukraine: Kiev-Kirillovskii, Gontsy, Dobranichevka, Mezin, and Mezhirich.* British Archaeological Reports, International Series 712, Oxford.

Pielowski, Z.
1971   Length of life of the hare. *Acta Theriologica* 16(6):89–94.
1976   Number of young born and dynamics of the European hare population. In *Ecology and Management of European Hare Populations,* Z. Pielowski and Z. Pucek, eds., pp. 75–78. Państwowa Wydawnictwo Rolnicze i Leśne, Warsaw, Poland.

Pike-Tay, A., V. Cabrera Valdés, and F. Bernaldo de Quirós
1999   Seasonal variations of the Middle–Upper Paleolithic transition at El Castillo, Cueva Morin and El Pendo (Cantabria, Spain). *Journal of Human Evolution* 36:283–317.

Plisson, H., and S. Beyries
1998   Pointes ou outils triangulaires? Données fonctionnelles dans le Moustérien Levantin. *Paléorient* 24(1):5–24.

Pope, C. H.
1956   *The Reptile World: A Natural History of the Snakes, Lizards, Turtles, and Crocodilians.* Alfred A. Knopf, New York.

Potts, G. R.
1986   *The Partridge: Pesticides, Predation, and Conservation.* Collins Professional and Technical, London.

Potts, R.
1982   Lower Pleistocene site formation and hominid activities at Olduvai Gorge, Tanzania. Ph.D. dissertation, Harvard University, Cambridge, Massachusetts.

Price, T. D.
1991   The Mesolithic of northern Europe. *Annual Review of Anthropology* 20:211–233.

Price, T. D., and A. B. Gebauer
1995   New perspectives on the transition to agriculture. In *Last Hunters—First Farmers: New Perspectives on the Prehistoric Transition to Agriculture,* T. D. Price and A. B. Gebauer, eds., pp. 3–19. School of American Research Press, Santa Fe, New Mexico.

Quézel, P.
1976   Les forêts du pourtour méditerranéen. *Note Technique MAB* 2:9–34.
1985   Definition of the Mediterranean region and origin of its flora. In *Plant Conservation in the Mediterranean Area,* C. Gomez-Campo, ed., pp. 9–24. W. Junk, Dordrecht, Netherlands.

Rabinovich, R.
1998   Patterns of animal exploitation and subsistence in Israel during the Upper Paleolithic and Epipaleolithic (40,000–12,500 BP), as studied from selected case studies. Ph.D. dissertation, Department of Evolution, Systematics, and Ecology, Hebrew University of Jerusalem.

Rabinovich, R., and E. Tchernov
1995   Chronological, paleoecological and taphonomical aspects of the Middle Paleolithic site of Qafzeh, Israel. In *Archaeozoology of the Near East,* vol. 2, H. Buitenhuis and H.-P. Uerpmann, eds., pp. 5–44. Backhuys Publishers, Leiden, Netherlands.

Raczynski, J.
1964   Studies on the European hare, Part 5: Reproduction. *Acta Theriologica* 9(19):305–352.

Radmilli, A. M.
1974   *Gli scavi nella Grotta Polesini a Ponte Lucano di Tivoli e la piu antica arte nel Lazio.* Sansoni Editore, Florence, Italy.

Redding, R.
1988   A general explanation of subsistence change: From hunting and gathering to food production. *Journal of Anthropological Archaeology* 7: 56–97.

Reich, D. E., and D. B. Goldstein
1998   Genetic evidence for a Paleolithic human population expansion in Africa. *Proceedings of the National Academy of Sciences* 95:8119–8123.

Relethford, J. H.
1998   Genetics of modern human origins and diversity. *Annual Review of Anthropology* 27:1–23.

Richter, D., H. B. Schroeder, et al.
2001   The Middle to Upper Palaeolithic transition in the Levant and new thermoluminescence dates for a Late Mousterian assemblage from Jerf al-Ajla Cave (Syria). *Paléorient* 27(2): 29–46.

Rick, J., and K. M. Moore
2001   Specialized meat-eating in the Holocene: An archaeological case from the frigid tropics of high-altitude Peru. In *Meat-Eating and Human Evolution*, C. Stanford and H. Bunn, eds., pp. 237–260. Oxford University Press, Oxford.

Rink, W. J.
2001   Beyond C14 dating: A user's guide to long-range dating methods in archaeology. In *Earth Sciences and Archaeology*, P. Goldberg, V. T. Holliday, and C. R. Ferring, eds., pp. 385–417. Kluwer Academic, Plenum Press, New York.

Rink, W. J., J. Bartoll, P. Goldberg, and A. Ronen
2003   ESR dating of archaeologically relevant authigenic terrestrial apatite veins from Tabun Cave, Israel. *Journal of Archaeological Science* 30:1127–1138.

Rink, W. J., H. P. Schwarcz, S. Weiner, P. Goldberg, L. Meignen, and O. Bar-Yosef
2004   Age of the Mousterian industry at Hayonim Cave, northern Israel, using electron spin resonance and 230Th/234U methods. *Journal of Archaeological Science* 31:953–964.

Robbins, G. E.
1984   *Partridge: Their Breeding and Management*. Boydell Press, Suffolk, UK.

Robinson, R. A.
1952   An electron microscope study of the crystalline inorganic component of bone and its relationship to the organic matrix. *Journal of Bone Joint Surgery* 34A:389–434.

Roettcher, D., and R. R. Hofmann
1970   The ageing of impala from a population in the Kenya Rift Valley. *East African Wildlife Journal* 8:37–42.

Rogers, A.
2000   On the value of soft bones in faunal analysis. *Journal of Archaeological Science* 27:635–639.

Ronen, A.
1979   Paleolithic Industries. In *The Quaternary of Israel*, A. Horowitz, ed., pp. 296–307. Academic Press, New York.
1984   *Sefunim Prehistoric Sites, Mount Carmel, Israel*. British Archaeological Reports, International Series 230, Oxford.

Russell, E. S.
1942   *The Overfishing Problem*. Cambridge University Press, Cambridge.

Sabelli, B.
1980   *Simon and Schuster's Guide to Shells*. Simon and Schuster, New York.

Saxon, E. C.
1974   The mobile herding economy of Kebarah Cave, Mount Carmel: An economic analysis of the faunal remains. *Journal of Archaeological Science* 1:27–45.

Schick, T., and M. Stekelis
1977   Mousterian assemblages in Kebara Cave, Mount Carmel. *Eretz Israel* 13:97–150.

Schiegl, S., P. Goldberg, O. Bar-Yosef, and S. Weiner
1996   Ash deposits in Hayonim and Kebara Caves, Israel: Macroscopic, microscopic and mineralogical observations and their archaeological implications. *Journal of Archaeological Science* 23:763–781.

Schiegl, S., S. Lev-Yadun, O. Bar-Yosef, A. El Goresy, and S. Weiner
1994   Siliceous aggregates from prehistoric wood ash: A major component of sediments in Kebara and Hayonim Caves (Israel). *Israel Journal of Earth Sciences* 43:267–278.

Schiffer, M. B.
1983   Toward the identification of formation processes. *American Antiquity* 48:675–706.

Schmitt, D. N., and K. E. Juell
1994   Toward the identification of coyote scatological faunal accumulations in archaeological contexts. *Journal of Archaeological Science* 21: 249–262.

Schroeder, B.
1969   The lithic industries from Jerf Ajla and their bearing on the problem of the Middle to Upper Paleolithic transition. Ph.D. dissertation, Columbia University, New York.

Schwarcz, H. P., W. Buhay, R. Grun, M. C. Stiner, S. Kuhn, and G. H. Miller
1991   Absolute dating of sites in coastal Lazio. *Quaternaria Nova* (new series) 1:51–67.

Severinghaus, C. W.
1949   Tooth development and wear as criteria of age in white-tailed deer. *Journal of Wildlife Management* 13:195–216.

Shackleton, N. J., and N. D. Opdyke
1973   Oxygen isotope and palaeomagnetic stratigraphy of equatorial Pacific core, V28–238. *Quaternary Research* 3:39–55.

Shahack-Gross, R., O. Bar-Yosef, and S. Weiner
1997   Black-coloured bones in Hayonim Cave, Israel: Differentiating between burning and oxide staining. *Journal of Archaeological Science* 24(5):439–446.

Shahack-Gross, R., F. Berna, P. Karkanas, and S. Weiner
2004   Bat guano and preservation of archaeological remains in cave sites. *Journal of Archaeological Science* 31:953–964.

Shea, J.
1989   A functional study of the lithic industries associated with hominid fossils in the Kebara and Qafzeh caves, Israel. In *The Human Revolution: Behavioural and Biological Perspectives on the Origins of Modern Humans,* P. Mellars and C. Stringer, eds., pp. 611–625. Princeton University Press, Princeton, New Jersey.
1991   The behavioral significance of Levantine Mousterian industrial variability. Ph.D. dissertation, Department of Anthropology, Harvard University, Cambridge, Massachusetts.
1993   Lithic use-wear evidence for hunting by Neandertals and early modern humans from the Levantine Mousterian. In *Hunting and Animal Exploitation in the Later Palaeolithic and Mesolithic of Eurasia,* G. L. Peterkin, H. Bricker, and P. Mellars, eds., pp. 189–197. Archaeological Papers of the American Anthropological Association 4, Washington, DC.

Shine, R., and J. B. Iverson
1995   Patterns of survival, growth and maturation in turtles. *Oikos* 72:343–348.

Shipman, P., G. F. Foster, and M. Schoeninger
1984   Burnt bones and teeth: An experimental study of colour, morphology, crystal structure and shrinkage. *Journal of Archaeological Science* 11:307–325.

Shy, E., E. Frankenburg, D. Kaplan, P. Giladi, A. Lachman, and M. Har-Zion
1998   The effect of management on mountain gazelle *(Gazella g. gazella)* populations in Israel. *Gibier Faune Sauvage* 15:617–634.

Sillen, A.
1981   Postdepositional changes in Natufian and Aurignacian faunal bones from Hayonim Cave. *Paléorient* 7:81–85.

Sillen, A., and T. Hoering
1993   Chemical characterization of burnt bones from Swartkrans. In *Swartkrans: A Cave's Chronicle of Early Man,* C. K. Brain, ed., pp. 243–249. Transvaal Museum Monograph no. 8, Pretoria, South Africa.

Sillen, A., and J. Parkington
1996   Diagenesis of bones from Eland's Bay Cave. *Journal of Archaeological Science* 23:535–542.

Silva, M., and J. A. Downing
1995   *CRC Handbook of Mammalian Body Masses.* CRC Press, Boca Raton, Florida.

Simek, J. F., and L. M. Snyder
1988   Changing assemblage diversity in Perigord archaeofaunas. In *Upper Pleistocene Prehistory of Western Eurasia,* H. L. Dibble and A. Montet-White, eds., pp. 321–332. Monograph no. 54, University Museum, University of Pennsylvania, Philadelphia.

Simmons, T., and D. Nadel
1998   The avifauna of the early Epipalaeolithic site of Ohalo II (19,400 years BP), Israel: Species diversity, habitat and seasonality. *International Journal of Osteoarchaeology* 8:79–96.

Simmons, T., and F. H. Smith
1991   Human population relationships in the Late Pleistocene. *Current Anthropology* 32:623–627.

Simpson, E. H.
1949 Measurement of diversity. *Nature* 163:688.

Sinclair, A. R. E.
1991 Science and the practice of wildlife management. *Journal of Wildlife Management* 55:767–773.

Sinclair, A. R. E., and M. Norton-Griffiths, eds.
1979 *Serengeti: Dynamics of an Ecosystem.* University of Chicago Press, Chicago.

Smith, A.
1998 Intensification and transformation processes towards food production in Africa. In *Before Food Production in North Africa,* S. Lernia and G. Manzi, eds., pp. 19–33. Union Internationale des Sciences Préhistorique et Protohistorique, Forlì, Italy.

Smith, B. C.
1996 *Fundamentals of Fourier Transform Infrared Spectroscopy.* CRC Press, Boca Raton, Florida.

Smith, B. D.
1974 Predator-prey relationships in the southeastern Ozarks, A.D. 1300. *Human Ecology* 2(1):31–43.

Smith, F. H.
1984 Fossil hominids from the Upper Pleistocene of central Europe and the origin of modern Europeans. In *The Origins of Modern Humans: A World Survey of the Fossil Evidence,* F. H. Smith and F. Spencer, eds., pp. 137–210. Plenum Press, New York.

Soffer, O.
1985 Patterns of intensification as seen from the Upper Paleolithic of the central Russian Plain. In *Prehistoric Hunter-Gatherers: The Emergence of Cultural Complexity,* T. D. Price and J. A. Brown, eds., pp. 235–270. Academic Press, San Diego, California.
1989a The Middle to Upper Palaeolithic transition on the Russian Plain. In *The Human Revolution: Behavioural and Biological Perspectives on the Origins of Modern Humans,* P. Mellars and C. Stringer, eds., pp. 714–742. Princeton University Press, Princeton, New Jersey.
1989b Storage, sedentism and the Eurasian Palaeolithic record. *Antiquity* 63:719–732.

1990 The central Russian Plain at the Last Glacial Maximum. In *The World at 18,000 BP,* vol. 2: *Low Latitudes,* C. Gamble and O. Soffer, eds., pp. 228–242. Plenum Press, New York.

Solecki, R. L., and R. S. Solecki
1995 The Mousterian industries of Yabrud Shelter I: A reconsideration. In *The Definition and Interpretation of Levallois Technology,* H. Dibble and O. Bar-Yosef, eds., pp. 381–398. Prehistory Press, Madison, Wisconsin.

Speth, J. D., and K. A. Spielmann
1983 Energy source, protein metabolism, and hunter-gatherer subsistence strategies. *Journal of Anthropological Archaeology* 2:1–31.

Speth, J. D., and E. Tchernov
1998 The role of hunting and scavenging in Neanderthal procurement strategies: New evidence from Kebara Cave (Israel). In *Neanderthals and Modern Humans in Western Asia,* T. Akazawa, K. Aoki, and O. Bar-Yosef, eds., pp. 223–239. Plenum Press, New York.
2001 Neandertal hunting and meat-processing in the Near East: Evidence from Kebara Cave (Israel). In *Meat-Eating and Human Evolution,* C. Stanford and H. Bunn, eds., pp. 52–72. Oxford University Press, Oxford.
2002 Middle Paleolithic tortoise use at Kebara Cave (Israel). *Journal of Archaeological Science* 29(5):471–483.
n.d. Kebara Cave as a Middle Paleolithic settlement: A faunal perspective. In *The Middle Paleolithic Archaeology of Kebara Cave, Mount Carmel, Israel,* Part 1, O. Bar-Yosef and L. Meignen, eds. Peabody Museum Press, Cambridge, Massachusetts. In press.

Stanford, C. B.
2001 Hunting primates: A comparison of the predatory behavior of chimpanzees and human foragers. In *Meat-Eating and Human Evolution,* C. Stanford and H. Bunn, eds., pp. 122–140. Oxford University Press, Oxford.

Steele, T. E., and T. D. Weaver
2002 The modified triangular graph: A refined method for comparing mortality profiles in archaeological samples. *Journal of Archaeological Science* 29:317–322.

Stephens, D. W., and J. R. Krebs
1986 *Foraging Theory.* Princeton University Press, Princeton, New Jersey.

Stewart, K. M.
1989 *Fishing Sites of North and East Africa in the Late Pleistocene and Holocene: Environmental Change and Human Adaptation.* British Archaeological Reports, International Series 521, Oxford.

Stiner, M. C.
1990 The use of mortality patterns in archaeological studies of hominid predatory adaptations. *Journal of Anthropological Archaeology* 9:305–351.
1991a Food procurement and transport by human and nonhuman predators. *Journal of Archaeological Science* 18:455–482.
1991b A taphonomic perspective on the origins of the faunal remains of Grotta Guattari (Latium, Italy). *Current Anthropology* 32:103–117.
1992 Overlapping species "choice" by Italian Upper Pleistocene predators. *Current Anthropology* 33:433–451.
1994 *Honor among Thieves: A Zooarchaeological Study of Neandertal Ecology.* Princeton University Press, Princeton, New Jersey.
1998a Comment on Marean and Kim, Mousterian large-mammal remains from Kobeh Cave: Behavioral implications. *Current Anthropology* 39:S98–103.
1998b Mortality analysis of Pleistocene bears and its paleoanthropological relevance. *Journal of Human Evolution* 34:303–326.
1999 Trends in Paleolithic mollusk exploitation at Riparo Mochi (Balzi Rossi, Italy): Food and ornaments from the Aurignacian through Epigravettian. *Antiquity* 73(282):735–754.
2001 Thirty years on the "Broad Spectrum Revolution" and Paleolithic demography. *Proceedings of the National Academy of Sciences* 98(13): 6993–6996.
2002a Carnivory, coevolution, and the geographic spread of the genus *Homo. Journal of Archaeological Research* 10(1):1–63.
2002b On in situ attrition and vertebrate body part profiles. *Journal of Archaeological Science* 29: 979–991.
2003 Zooarchaeological evidence for resource intensification in Algarve, southern Portugal. *Promontoria* 1(1):27–61.

2004 A comparison of photon densitometry and computed tomography parameters of bone density in ungulate body part profiles. *Journal of Taphonomy* 2:117–145.

Stiner, M. C., H. Achyutan, G. Arsebuk, F. C. Howell, S. C. Josephson, K. E. Juell, J. Pigati, and J. Quade
1998 Reconstructing cave bear paleoecology from skeletons: The Middle Pleistocene case from Yarimburgas Cave, Turkey. *Paleobiology* 24(1): 74–98.

Stiner, M. C., G. Arsebuk, and F. C. Howell
1996 Cave bears and Paleolithic artifacts in Yarýmburgaz Cave, Turkey: Dissecting a palimpsest. *Geoarchaeology* 11(4):279–327.

Stiner, M. C., F. C. Howell, B. Martínez-Navarro, E. Tchernov, and O. Bar-Yosef
2001b Outside Africa: Middle Pleistocene Lycaon from Hayonim Cave, Israel. *Bolletino della Società Paleontologica Italiana* 40(2):293–302.

Stiner, M. C., and S. L. Kuhn
1992 Subsistence, technology, and adaptive variation in Middle Paleolithic Italy. *American Anthropologist* 94:12–46.

Stiner, M. C., S. L. Kuhn, T. A. Surovell, P. Goldberg, L. Meignen, S. Weiner, and O. Bar-Yosef
2001a Bone preservation in Hayonim Cave (Israel): A macroscopic and mineralogical study. *Journal of Archaeological Science* 28:643–659.

Stiner, M. C., and N. D. Munro
2002 Approaches to prehistoric diet breadth, demography, and prey ranking systems in time and space. *Journal of Archaeological Method and Theory* 9(2):181–214.

Stiner, M. C., N. D. Munro, and T. A. Surovell
2000 The tortoise and the hare: Small game use, the Broad Spectrum Revolution, and Paleolithic demography. *Current Anthropology* 41(1):39–73.

Stiner, M. C., N. D. Munro, T. A. Surovell, E. Tchernov, and O. Bar-Yosef
1999 Paleolithic population growth pulses evidenced by small animal exploitation. *Science* 283:190–194.

Stiner, M. C., C. Pehlevan, M. Sagır, and I. Özer
2002 Zooarchaeological studies at Üçağızlı Cave: Preliminary results on Paleolithic subsistence and shell ornaments. *Araşterma Sonuçları Toplantısı, Ankara* 17:29–36.

Stiner, M. C., and E. Tchernov
1998 Pleistocene species trends at Hayonim Cave: Changes in climate versus human behavior. In *Neanderthals and Modern Humans in Western Asia*, T. Akazawa, K. Aoki, and O. Bar-Yosef, eds., pp. 241–262. Plenum Press, New York.

Stiner, M. C., S. Weiner, O. Bar-Yosef, and S. L. Kuhn
1995 Differential burning, fragmentation and preservation of archaeological bone. *Journal of Archaeological Science* 22:223–237.

Straus, L. G.
1982 Carnivores and cave sites in Cantabrian Spain. *Journal of Anthropological Research* 38:75–96.
1990 The Last Glacial Maximum in Cantabrian Spain: The Solutrean. In *The World at 18,000 BP*, vol. 1, *High Latitudes*, O. Soffer and C. Gamble, eds., pp. 89–108. Unwin Hyman, London.

Stringer, C. B.
1988 Palaeoanthropology: The dates of Eden. *Nature* 331:565–566.
1989 The origin of early modern humans: A comparison of the European and non-European evidence. In *The Human Revolution: Behavioural and Biological Perspectives on the Origins of Modern Humans*, P. Mellars and C. Stringer, eds., pp. 232–244. Princeton University Press, Princeton, New Jersey.

Stringer, C., and P. Andrews
1988 Genetic and fossil evidence for the origins of modern humans. *Science* 239:1263–1268.

Stringer, C. B., J. J. Hublin, and B. Vandermeersch
1984 The origin of anatomically modern humans in western Europe. In *The Origins of Modern Humans: A World Survey of the Fossil Evidence*, F. H. Smith and F. Spencer, eds., pp. 51–136. Alan R. Liss, New York.

Stuart, A. J.
1982 *Pleistocene Vertebrates in the British Isles*. Longman, London.

1991 Mammalian extinctions in the Late Pleistocene of northern Eurasia and North America. *Biological Reviews* 66:453–562.

Stubbs, D.
1989 *Testudo graeca*, spur-thighed tortoise. In *The Conservation Biology of Tortoises*, I. R. Swigland and M. W. Klemens, eds., pp. 31–33. Occasional Papers of the World Conservation Union (IUCN) Species Survival Commission (SSC), no. 5. Morges, Switzerland.

Stuiver, M., P. J. Reimer, E. Bard, J. W. Beck, G. S. Burr, K. A. Hughen, B. Kromer, G. McCormac, J. van der Plicht, and M. Spurk
1998 INTCAL98 radiocarbon age calibration, 24,000–0 cal BP. *Radiocarbon* 40(3):1041–1084.

Stutz, A. J.
2002 Polarizing microscopy identification of chemical diagenesis in archaeological cementum. *Journal of Archaeological Science* 29(11):1327–1347.

Surovell, T. A., and M. C. Stiner
2001 Standardization of infrared measures of bone mineral crystallinity: An experimental approach. *Journal of Archaeological Science* 28:633–642.

Sutcliffe, A.
1970 Spotted hyaena: Crusher, gnawer, digestor, and collector of bones. *Nature* 227:1110–1113.

Suzuki, H., and F. Takai
1970 *The Amud Man and His Cave Site*. Academic Press of Japan, Tokyo.

Svoboda, J.
1990 Moravia during the Upper Pleniglacial. In *The World at 18,000 BP*, vol. 2: *Low Latitudes*, C. Gamble and O. Soffer, eds., pp. 193–203. Plenum Press, New York.

Swihart, R. K.
1983 Body size, breeding season length, and life history tactics of lagomorphs. *Oikos* 43:282–290.

Taber, R. D., K. J. Raedeke, and D. A. McCaughran
1982 Population characteristics. In *Elk of North America: Ecology and Management*, J. W. Thomas and D. E. Toweill, eds., pp. 279–298. Stackpole Books, Harrisburg, Pennsylvania.

Taberlet, P., L. Fumagali, A. G. Wust-Saucy, and J. F. Cosson
1998 Comparative phylogeography and postglacial colonization routes in Europe. *Molecular Ecology* 6:289–301.

Taborin, Y.
1993 Shells of the French Aurignacian and Perigordian. In *Before Lascaux: The complex record of the early Upper Paleolithic*, H. Knecht, A. Pike-Tay, and R. White, eds., pp. 211–229. CRC Press, Boca Raton, Florida.

Tchernov, E.
1968 *Succession of Rodent Faunas during the Upper Pleistocene of Israel*. Mammalia Depicta, Paul Parey.
1975 Rodent faunas and environmental changes in the Pleistocene of Israel. In *Rodents in Desert Environments*, I. Prakash and P. K. Ghosh, eds., pp. 331–362. W. Junk, The Hague, Netherlands.
1981 The biostratigraphy of the Middle East. In *Prehistoire du Levant*, J. Cauvin and P. Sanlaville, eds., pp. 67–97. Éditions du CNRS, Paris.
1984a Commensal animals and human sedentism in the Middle East. In *Animals and Archaeology*, J. Clutton-Brock and C. Grigson, eds., pp. 91–115. British Archaeological Reports, International Series 202, Oxford.
1984b Faunal turnover and extinction rate in the Levant. In *Quaternary Extinctions*, P. S. Martin and R. G. Klein, eds., pp. 528–552. University of Arizona Press, Tucson.
1988 Biochronology of the Middle Paleolithic and dispersal events of hominids in the Levant. In *L'homme de Neandertal*, M. Otte, ed., pp. 153–168. Etudes et Recherches Archeologiques de l'Universite de Liège, no. 34. Liège, Belgium.
1989 The Middle Paleolithic mammalian sequence and its bearing on the origin of *Homo sapiens* in the southern Levant. In *Investigations in South Levantine Prehistory (Préhistoire du Sud-Levant)*, O. Bar Yosef and B. Vandermeersch, eds., pp. 25–38. British Archaeological Reports, International Series 497, Oxford.

1992a The Afro-Arabian component in the Levantine mammalian fauna: A short biogeographical review. *Israel Journal of Zoology* 38:155–192.
1992b Biochronology, paleoecology, and dispersal events of hominids in the southern Levant. In *The Evolution and Dispersal of Modern Humans in Asia*, T. Akazawa, K. Aoki, and T. Kimura, eds., pp. 149–188. Hokusen-Sha, Tokyo.
1992c Eurasian-African biotic exchanges through the Levantine corridor during the Neogene and Quaternary. *Courier Forschungsinstitut Senckenberg* 153:103–123.
1992d Evolution of complexities, exploitatation of the biosphere and zooarchaeology. *Archaeozoologica* 5(1):9–42.
1993a Exploitation of birds during the Natufian and early Neolithic of the southern Levant. *Archaeofauna* 2:121–143.
1993b The impact of sedentism on animal exploitation in the southern Levant. In *Archaeozoology of the Near East*, H. Buitenhuis and A. T. Clason, eds., pp. 10–26. Universal Book Services, Leiden, Netherlands.
1994 New comments on the biostratigraphy of the Middle and Upper Pleistocene of the southern Levant. In *Late Quaternary Chronology and Paleoclimates of the Eastern Mediterranean*, O. Bar-Yosef and R. S. Kra, eds., pp. 333–350. RADIOCARBON, University of Arizona, Tucson.
1996 Rodent faunas, chronostratigraphy and paleobiogeography of the southern Levant during the Quaternary. *Acta Zoologia Cracova* 39(1):513–530.
1998a An attempt to synchronize the faunal changes with the radiometric dates and the cultural chronology in southwest Asia. In *Archaeozoology of the Near East III: Proceedings of the Third International Symposium on the Archaeozoology of Southwestern Asia and Adjacent Areas*, H. Buitenhuis, L. Bartosiewicz, and A. M. Choyke, eds., pp. 7–44. ARC Publications 18, Groningen, Netherlands.
1998b Are Late Pleistocene environmental factors, faunal changes and cultural transformations causally connected? The case of the southern Levant. *Paléorient* 23(2):209–228.
1998c The faunal sequence of the Southeast Asian Middle Paleolithic in relation to hominid dispersal events. In *Neanderthals and Modern Humans in Western Asia*, T. Akazawa, K. Aoki, and O. Bar-Yosef, eds., pp. 77–90. Plenum Press, New York.

Tchernov, E., and F. R. Valla
1997 Two new dogs, and other Natufian dogs, from the southern Levant. *Journal of Archaeological Science* 24:65–95.

Termine, J. D., and A. S. Posner
1966 Infrared analysis of rat bone: Age dependency of amorphous and crystalline mineral fractions. *Science* 153:1523–1525.

Testart, A.
1982 The significance of food storage among hunter-gatherers: Residence patterns, population densities, and social inequalities. *Current Anthropology* 23(5):523–537.

Thieme, H.
1997 Lower Palaeolithic hunting spears from Germany. *Nature* 385:807–810.

Todd, L. C., and D. Rapson
1988 Long bone fragmentation and interpretation of faunal assemblages: Approaches to comparative analysis. *Journal of Archaeological Science* 15:307–325.

Torres, T. J.
1988 *Osos (Mammalia, Carnivora, Ursidae) del Pleistoceno de la Península Ibérica.* Special publications of the *Boletín Geológico y Minero 1–4*, vol. 99.

Tortonese, E.
1985 Distribution and ecology of endemic elements in the Mediterranean faunas (fishes and echinoderms). In *Mediterranean Marine Ecosystems*, M. Moraitou-Apostolopoulo and V. Kiortsis, eds., pp. 57–83. Plenum Press, New York.

Tozzi, C.
1970 La Grotta di S. Agostino (Gaeta). *Rivista di Scienze Prehistoriche* 25:3–87.
1974 L'industria musteriana della Grotta di Gosto sulla Montagna di Cetona (Siena). *Rivista di Scienze Prehistoriche* 29:271–304.

Trabaud, L.
1991 Is fire an agent favouring plant invasions? In *Biogeography of Mediterranean Invasions*, R. H. Groves and F. di Castri, eds., pp. 179–190. Cambridge University Press, Cambridge.

Trinkaus, E.
1984 Western Asia. In *The Origins of Modern Humans*, F. H. Smith and F. Spencer, eds., pp. 251–293. Alan R. Liss, New York.
1986 The Neandertals and modern human origins. *Annual Review of Anthropology* 15:193–218.

Trueman, C. N. G., A. K. Behrensmeyer, N. Tuross, and S. Weiner
2004 Mineralogical and compositional changes in bones exposed on soil surfaces in Amboseli National Park, Kenya: Diagenetic mechanisms and the role of sediment pore fluids. *Journal of Archaeological Science* 31:21–39.

Turner, A.
1984 Hominids and fellow-travellers: Human migration into high latitudes as part of a large mammal community. In *Hominid Evolution and Community Ecology*, R. Foley, ed., pp. 193–218. Academic Press, New York.
1986 Correlation and causation in some carnivore and hominid evolutionary events. *South African Journal of Science* 82:75–76.

Turville-Petre, F.
1932 Excavations in the Mugharet el-Kebarah. *Journal of the Royal Anthropological Institute* 62:271–276.

Uerpmann, H.-P.
1981 The major faunal areas of the Middle East during the late Pleistocene and early Holocene. In *Préhistoire du Levant*, J. Cauvin and P. Sanlaville, eds., pp. 99–106. Éditions du CNRS, Paris.

Valla, F. R.
1991 Les Natoufiens de Mallaha et l'espace. In *The Natufian Culture in the Levant*, O. Bar-Yosef and F. R. Valla, eds., pp. 111–122. International Monographs in Prehistory, Ann Arbor, Michigan.

Valla, F. R., F. Le Mort, and H. Plisson
1991 Les fouilles en cours sur la Terrase d'Hayonim. In *The Natufian Culture in the Levant*, O. Bar-Yosef and F. R. Valla, eds., pp. 93–110. International Monographs in Prehistory, Ann Arbor, Michigan.

Valla, F. R., H. Plisson, and R. Buxom i Capdevila
1989 Notes préliminaires sur les fouilles en cours sur la Terrasse d'Hayonim. *Paléorient* 15(1): 245–257.

Valladas, H., N. Mercier, J.-L. Joron, and J.-L. Reyss
1998 GIF Laboratory dates for Middle Paleolithic Levant. In *Neanderthals and Modern Humans in Western Asia*, T. Akazawa, K. Aoki, and O. Bar-Yosef, eds., pp. 69–75. Plenum Press, New York.

Valladas, H., J. L. Reyss, J. L. Joron, G. Valladas, O. Bar-Yosef, and B. Vandermeersch
1988 Thermoluminsecence dating of Mousterian and "Proto-Cro-Magnon" remains from Israel and the origin of modern man. *Nature* 331: 614–616.

van Andel, T. H., W. Davies, B. Weninger, and O. Jöris
2003 Archaeological dates as proxies for the spatial and temporal human presence in Europe: A discourse on the method. In *Neandertals and Modern Humans in the European Landscape during the Last Glaciation*, T. H. van Andel and W. Davies, eds., pp. 21–29. McDonald Institute for Archaeological Research, University of Cambridge, Cambridge.

Vandermeersch, B.
1982 The first *Homo sapiens sapiens* in the Near East. In *The Transition from the Lower to the Middle Palaeolithic and the Origin of Modern Man*, A. Ronen, ed., pp. 297–300. British Archaeological Reports, International Series 151, Oxford.

Vandermeersch, B., and O. Bar-Yosef
1988 Evolution biologique et culturelle des populations du Levant au Paléolithique Moyen: Les donnees recentes de Kébara et Qafzeh. *Paléorient* 14:115–117.

Van Neer, W.
1986 Some notes on the fish remains from Wadi Kubbaniya (Upper Egypt, late Paleolithic). In *Fish and Archaeology: Studies in Osteometry, Taphonomy, Seasonality, and Fishing Methods*, D. C. Brinkhuizen and A. T. Clason, eds., pp. 103–113. British Archaeological Reports, International Series 294, Oxford.

Villa, P.
1991 Middle Pleistocene prehistory in southwestern Europe: The state of our knowledge and ignorance. *Journal of Anthropological Research* 47(2):193–217.

Villa, P., F. Bon, and J.-C. Castel
2001 Fuel, fire and fireplaces in the Paleolithic of western Europe. Review of *Économie des combustibles au Paléolithique, I*, Théry-Parisot, ed. *Review of Archaeology* 23(1):33–42.

Villa, P., and E. Mahieu
1991 Breakage patterns of human long bones. *Journal of Human Evolution* 21:27–48.

Villa, P., and M. Soressi
2000 Stone tools in carnivore sites: The case of Bois Roche. *Journal of Anthropological Research* 56: 187–215.

Wainwright, S. A., W. D. Briggs, J. D. Currey, and J. M. Gosline
1976 *Mechanical Design in Organisms*. Princeton University Press, Princeton, New Jersey.

Walker, W. F., Jr.
1973 The locomotor apparatus of Testudines. In *Biology of the Reptilia*, vol. 4, C. Gans, ed., pp. 1–59. Academic Press, London.

Walters, I.
1988 Fire and bones: Patterns of discard. In *Archaeology with Ethnography: An Australian Perspective*, B. Meehan and R. Jones, eds., pp. 215–221. Australian National University, Canberra, Australia.

Watanabe, H.
1968 Flake production in a transitional industry from the Amud Cave: A statistical approach to Paleolithic typology. In *La préhistoire: Problemes et tendances*, F. Bordes, ed., pp. 499–509. Éditions du CNRS, Paris.

Watson, P. J.
1995 Explaining the transition to agriculture. In *Last Hunters—First Farmers: New Perspectives on the Prehistoric Transition to Agriculture*, T. D. Price and A. B. Gebauer, eds., pp. 21–37. School of American Research Press, Santa Fe, New Mexico.

Weiner, S., and O. Bar-Yosef
1990 States of preservation of bones from prehistoric sites in the Near East: A survey. *Journal of Archaeological Science* 17:187–196.

Weiner, S., and P. Goldberg
1990 On-site Fourier transform infrared spectrometry at an archaeological excavation. *Spectroscopy* 5:46–50.

Weiner, S., P. Goldberg, and O. Bar-Yosef
1993 Bone preservation in Kebara Cave, Israel, using on-site Fourier transform infrared spectrometry. *Journal of Archaeological Science* 20: 613–627.
2002 Three-dimensional distribution of minerals in the sediments of Hayonim Cave, Israel: Diagenetic processes and archaeological implications. *Journal of Archaeological Science* 29(11): 1289–1308.

Weiner, S., and P. A. Price
1986 Disaggregation of bone into crystals. *Calcified Tissue Research* 39:365–375.

Weiner, S., S. Schiegl, P. Goldberg, and O. Bar-Yosef
1995 Mineral assemblages in Kebara and Hayonim caves, Israel: Excavation strategies, bone preservation, and wood ash remnants. *Israel Journal of Chemistry* 35:143–154.

Weiner, S., and W. Traub
1992 Bone structure: From angstroms to microns. *FASEB Journal* 6:879–885.

Weiner, S., and H. D. Wagner
1998 The material bone: Structure–mechanical function relations. *Annual Review of Material Science* 28:271–298.

Weiner, S., Q. Xu, P. Goldberg, J. Liu, and O. Bar-Yosef
1998 Evidence for the use of fire at Zhoukoudian, China. *Science* 281:251–253.

Weniger, G.-C.
1987 Magdalenian settlement pattern and subsistence in central Europe: The southwestern and central German cases. In *The Pleistocene Old World: Regional Perspectives*, O. Soffer, ed., pp. 201–215. Plenum Press, New York.

Whallon, R.
1989 Elements of cultural change in the later Palaeolithic. In *The Human Revolution: Behavioural and Biological Perspectives on the Origins of Modern Humans*, P. Mellars and C. Stringer, eds., pp. 433–454. Princeton University Press, Princeton, New Jersey.

White, R.
1982 Rethinking the Middle/Upper Paleolithic transition. *Current Anthropology* 23:169–192.
1993 Technological and social dimensions of "Aurignacian-age" body ornaments across Europe. In *Before Lascaux: The Complex Record of the Early Upper Paleolithic*, H. Knecht, A. Pike-Tay, and R. White, eds., pp. 277–300. CRC Press, Boca Raton, Florida.

White, T. E.
1953 A method of calculating the dietary percentage of various food animals utilized by aboriginal peoples. *American Antiquity* 19:396–398.

Wiessner, P.
1983 Style and social information in Kalahari San projectile points. *American Antiquity* 48:253–276.

Wilbur, H. M., and P. J. Morin
1988 Life history evolution in turtles. In *Biology of the Reptilia*, vol. 16, *Defense and Life History*, C. Gans and R. B. Huey, eds., pp. 387–439. Alan R. Liss, New York.

Wilbur, K. M.
1964 Shell formation and regeneration. In *Physiology of Mollusca*, K. M. Wilbur and C. M. Yonge, eds., pp. 243–282. Academic Press, New York.

Willemsen, R. E., and A. Hailey
1989 Review: Status and conservation of tortoises in Greece. *Herpetological Journal* 1:315–330.

Winterhalder, B.
1997 Gifts given, gifts taken: The behavioral ecology of nonmarket, intragroup exchange. *Journal of Anthropological Research* 5(2):121–168.
2001 Intragroup resource transfers: Comparative evidence, models, and implications for human evolution. In *Meat-Eating and Human Evolution*, C. Stanford and H. Bunn, eds., pp. 279–301. Oxford University Press, Oxford.

Winterhalder, B., W. Baillargeon, F. Cappelletto, I. R. Daniel, Jr., and C. Prescott
1988 The population ecology of hunter-gatherers and their prey. *Journal of Anthropological Archaeology* 7:289–328.

Winterhalder, B., and C. Goland
1993 On population, foraging efficiency, and plant domestication. *Current Anthropology* 34(5): 710–715.

Winterhalder, B., and F. Lu
1997 A forager-resource population ecology model and implications for indigenous conservation. *Conservation Biology* 11(6):1354–1364.

Wobst, H. M.
1974 Boundary conditions for Paleolithic social systems: A simulation approach. *American Antiquity* 39(2):147–178.
1977 Stylistic behavior and information exchange. In *Papers for the Director: Research Essays in Honor of James B. Griffin,* C. E. Cleland, ed., pp. 317–342. Anthropological Paper 61, Museum of Anthropology, University of Michigan, Ann Arbor.

Wolpoff, M.
1989 Multiregional evolution: The fossil alternative to Eden. In *The Human Revolution: Behavioural and Biological Perspectives on the Origins of Modern Humans,* P. Mellars and C. Stringer, eds., pp. 62–108. Princeton University Press, Princeton, New Jersey.

Wolverton, S.
2001 Environmental implications of zooarchaeological measures of resource depression. Ph.D. dissertation, Department of Anthropology, University of Missouri, Columbia.

Wreschner, E.
1976 The red hunters: Further thoughts on the evolution of speech. *Current Anthropology* 17(4): 717–719.

Wright, K. I.
1994 Ground-stone tools and hunter-gatherer subsistence in Southwest Asia: Implications for the transition to farming. *American Antiquity* 59(2):238–263.

Wrinn, P. J.
n.d. Reanalysis of the Pleistocene archaeofauna from Mugharet et'Aliya, Tangier, Morocco: Implications for the Aterian. In preparation.

Yellen, J. E.
1977 Cultural patterning in faunal remains: Evidence from the !Kung Bushmen. In *Experimental Archaeology,* D. Ingersoll, J. E. Yellen, and W. Macdonald, eds., pp. 271–331. Columbia University Press, New York.
1991a Small mammals: !Kung San utilization and the production of faunal assemblages. *Journal of Anthropological Archaeology* 10:1–26.
1991b Small mammals: Post-discard patterning of !Kung San faunal remains. *Journal of Anthropological Archaeology* 10:152–192.

Yellen, J. E., A. S. Brooks, E. Cornelissen, R. G. Klein, M. Mehlman, and K. Stewart
1995 A middle Stone Age worked bone industry from Katanda, upper Semliki River valley (Kivu), Zaire. *Science* 268:553–556.

Yesner, D. R.
1981 Archaeological applications of optimal foraging theory: Harvest strategies of Aleut hunter-gatherers. In *Hunter-Gatherer Foraging Strategies: Ethnographic and Archeological Analyses,* B. Winterhalder and E. A. Smith, eds., pp. 148–170. University of Chicago Press, Chicago.

Zeder, M. A., and B. Hesse
2000 The initial domestication of goats *(Capra hircus)* in the Zagros Mountains 10,000 years ago. *Science* 287:2254–2257.

Zilhão, J.
1990 The Portuguese Estremadura at 18,000 BP: The Solutrean. In *The World at 18,000 BP,* vol. 2: *Low Latitudes,* C. Gamble and O. Soffer, eds., pp. 109–125. Plenum Press, New York.

Zilhão, J., and F. d'Errico
1999 The chronology and taphonomy of the earliest Aurignacian and its implications for the understanding of Neandertal extinction. *Journal of World Prehistory* 13(1):1–68.

Ziv, V., and S. Weiner
1994 Bone crystal sizes: A comparison of transmission electron microscope and x-ray diffraction in width broadening techniques. *Connective Tissue Research* 30:165–175.

Zohary, D.
1969 The progenitors of wheat and barley in relation to domestication and agricultural dispersal in the Old World. In *The Domestication and Exploitation of Plants and Animals*, P. J. Ucko and G. W. Dimbleby, eds., pp. 47–66. Aldine, Chicago.

Index